Theory and Applications of Integral Transformations and Operational Calculus

Theory and Applications of Integral Transformations and Operational Calculus

Edited by
Jocelyn Cole

www.willfordpress.com

Published by Willford Press,
118-35 Queens Blvd., Suite 400,
Forest Hills, NY 11375, USA

ISBN: 978-1-64728-531-9

Cataloging-in-Publication Data

Theory and applications of integral transformations and operational calculus / edited by Jocelyn Cole.
p. cm.
Includes bibliographical references and index.
ISBN 978-1-64728-531-9
1. Integral transforms. 2. Calculus, Operational. 3. Transformations (Mathematics). I. Cole, Jocelyn.
QA432 .T44 2023
515.723--dc23

For information on all Willford Press publications
visit our website at www.willfordpress.com

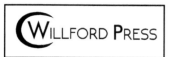

Contents

Preface

It is often said that books are a boon to mankind. They document every progress and pass on the knowledge from one generation to the other. They play a crucial role in our lives. Thus I was both excited and nervous while editing this book. I was pleased by the thought of being able to make a mark but I was also nervous to do it right because the future of students depends upon it. Hence, I took a few months to research further into the discipline, revise my knowledge and also explore some more aspects. Post this process, I begun with the editing of this book.

An integral transform maps a function from its original function space onto another function space by integration in which some of the features of the original function could be easier to manipulate and characterize than in the original function space. The function which is transformed can be traced back to the original function generally, through the inverse transform. An integral transform traces an equation from the original domain in the other domain; where solving and manipulating the equation is much easier as compared to the original domain. Operational calculus is a method which transforms differential equations into algebraic problems. The theory of integral transformations and related operational calculus has applications in a variety of fields including physical sciences, engineering, mathematical sciences, statistics and chemical sciences. This book explores all the important aspects of integral transformations and operational calculus. It is meant for students who are looking for an elaborate reference text on the theory and applications of integral transformations and operational calculus.

I thank my publisher with all my heart for considering me worthy of this unparalleled opportunity and for showing unwavering faith in my skills. I would also like to thank the editorial team who worked closely with me at every step and contributed immensely towards the successful completion of this book. Last but not the least, I wish to thank my friends and colleagues for their support.

Editor

1

A New Integral Transform: ARA Transform and its Properties and Applications

Rania Saadeh, Ahmad Qazza * and Aliaa Burqan

Department of Mathematics, Zarqa University, Zarqa 13132, Jordan; rsaadeh@zu.edu.jo (R.S.); aliaaburqan@zu.edu.jo (A.B.)

* Correspondence: aqazza@zu.edu.jo

Abstract: In this paper, we introduce a new type of integral transforms, called the ARA integral transform that is defined as: $\mathcal{G}_n[g(t)](s) = G(n,s) = s\int_0^\infty t^{n-1}e^{-st}g(t)dt,\ s > 0$. We prove some properties of ARA transform and give some examples. Also, some applications of the ARA transform are given.

Keywords: ARA transform; Fourier transform; Laplace transform; Sumudu transform; Elzaki transform; Mellin transform; Natural transform; Yang transform; Shehu transform; Novel transform; Ordinary differential equation; Integral equation

---

1. Introduction

The integral transforms play a vital role in finding solutions to initial value problems and initial boundary value problems. An integral transform T [1] has the form

$$T[g(t)](u) = \int_{t_1}^{t_2} g(t)K(t,u)dt, \tag{1}$$

where the input function of the transform is $g(t)$ and the output is $T[g(t)](u)$, and the function $K(t,u)$ is a kernel function. Moreover, the inverse transform related to the inverse kernel function is given by:

$$g(t) = \int_{u_1}^{u_2} T[g(t)](u)\, K^{-1}(u,t)du. \tag{2}$$

The integral transform was introduced by the French mathematician and physicist P.S. Laplace [2,3] in 1780. In 1822, J. Fourier [4] introduced the Fourier transform. Laplace and Fourier transforms form the foundation of operational analysis, a branch of mathematics that has very powerful applications, not only in applied mathematics but also in other branches of science like physics, engineering, astronomy, etc.

In recent years, mathematicians have been interested in developing and establishing new integral transforms. In 1993, Watugula [5] introduced the Sumudu transform. The natural transform was introduced by Khan and Khan [6] in 2008. In 2011, the Elzaki transform [7] was devised by Elzaki. Atangana and Kiliçman [8] in 2013, introduced the Novel transform. In 2015, Srivastava, Luo and Raina [9] introduced the M-transform. In 2016, many transforms were introduced, like the ZZ transform by Zafar [10], Ramadan Group (RG) transform [11], a polynomial transform by Barnes [12], also, a new integral transform was presented by Yang [13]. In the year 2017, other transforms were introduced, such as the Aboodh transform [14] and Rangaig transform [15], while the Shehu transform [16] was established in 2019, by Shehu and Weidong.

In this paper, we proclaim a new integral transform called the ARA integral transform.

This transform is a powerful and versatile generalization that unifies some variants of the classical Laplace transform, namely, the Sumudu transform, the Elzaki transform, the Natural transform, the Yang transform, and the Shehu transform.

In Section 2, we state our definition of the ARA transform and some related theorems. In Section 3, we provide the properties of the ARA transform, and in the last section, we give some applications.

2. Definitions and Theorems

Definition 1. *The ARA integral transform of order n of the continuous function $g(t)$ on the interval $(0,\infty)$ is defined as:*

$$\mathcal{G}_n[g(t)](s) = G(n,s) = s\int_0^\infty t^{n-1}e^{-st}g(t)dt,\ s>0 \tag{3}$$

Definition 2. *The inverse of the ARA transform is given by*

$$\begin{aligned} g(t) &= \mathcal{G}_{n+1}^{-1}[\mathcal{G}_{n+1}[g(t)]] \\ &= \frac{(-1)^n}{2\pi i}\int_{c-i\infty}^{c+i\infty} e^{st}\Bigg((-1)^n\Bigg(\frac{1}{s\Gamma(n-1)}\int_0^s (s-x)^{n-1}G(n \\ &\quad +1,x)dx + \sum_{k=0}^{n-1}\frac{s^k}{k!}\frac{\partial^k G(0)}{\partial s^k}\Bigg)\Bigg)ds, \end{aligned} \tag{4}$$

where

$$G(s) = \int_0^\infty e^{-st}\, g(t)\, dt,$$

is $(n-1)$ times differentiable.

In fact, from the definition of ARA transform of a function $g(t)$, we have

$$G(n+1,s) = \mathcal{G}_{n+1}[g(t)](s) = s\int_0^\infty t^n e^{-st}g(t)dt = (-1)^n\, s\,\frac{d^nG(s)}{ds^n},$$

thus

$$\frac{1}{s\,\Gamma(n-1)}\int_0^s (s-t)^{n-1}G(n+1,t)dt = (-1)^n\Bigg(G(s) - \sum_{k=0}^{n-1}\frac{s^k}{k!}\frac{\partial^k G(0)}{\partial s^k}\Bigg),$$

and so,

$$\frac{(-1)^n}{s\,\Gamma(n-1)}\int_0^s (s-t)^{n-1}G(n+1,t)dt + \sum_{k=0}^{n-1}\frac{s^k}{k!}\frac{\partial^k G(0)}{\partial s^k} = G(s).$$

It follows that:

$$\begin{aligned} &\frac{(-1)^n}{2\pi i}\int_{c-i\infty}^{c+i\infty} e^{st}\Bigg((-1)^n\Bigg(\frac{1}{s\Gamma(n-1)}\int_0^s (s-x)^{n-1}G(n+1,x)dx \\ &+\sum_{k=0}^{n-1}\frac{s^k}{k!}\frac{\partial^k G(0)}{\partial s^k}\Bigg)\Bigg)ds = \frac{(-1)^n}{2\pi i}\int_{c-i\infty}^{c+i\infty} e^{st}\big((-1)^n G(s)\big)ds. \end{aligned}$$

$$\mathcal{G}_{n+1}^{-1}[\mathcal{G}_{n+1}[g(t)]] = \frac{(-1)^{2n}}{2\pi i} \int_{c-i\infty}^{c+i\infty} e^{st} G(s) ds = g(t).$$

Theorem 1. *The sufficient condition for the existence of ARA transform. If the function $g(t)$ is piecewise continuous in every finite interval $0 \le t \le \alpha$ and satisfies:*

$$\left| t^{n-1} g(t) \right| \le K e^{\beta t}, \tag{5}$$

then ARA transform exists for all $s > \beta$.

Proof of Theorem 1. We have

$$s \int_0^{\infty} t^{n-1} e^{-st} g(t) dt = s \int_0^{\alpha} t^{n-1} e^{-st} g(t) dt + s \int_{\alpha}^{\infty} t^{n-1} e^{-st} g(t) dt,$$

since the function $g(t)$ is piecewise continuous then the first integral on the right side exists. Also, the second integral on the right side converges because:

$$\begin{aligned} \left| s \int_{\alpha}^{\infty} t^{n-1} e^{-st} g(t) dt \right| &\le s \int_{\alpha}^{\infty} e^{-st} \left| t^{n-1} g(t) \right| dt \le s \int_{\alpha}^{\infty} e^{-st} k e^{\beta t} dt \\ = sk \int_{\alpha}^{\infty} e^{\beta t - st} dt &= \lim_{b \to \infty} - sk \left. \frac{e^{-t(s-\beta)}}{s-\beta} \right|_{\alpha}^{b} = \frac{sk}{s-\beta} e^{-\alpha(s-\beta)}, \end{aligned}$$

and this improper integral is convergent for all $s > \beta$. Thus, $\mathcal{G}_{n+1}[g(t)](s)$ exists. □

3. Dualities between ARA Transform and Some Integral Transform:

Duality between ARA and Laplace Transforms [17]:

i:

$$\mathcal{G}_0[f(t)](s) = s \int_0^{\infty} t^{-1} e^{-st} f(t) dt = s\, \mathcal{L}\left[\frac{f(t)}{t}\right] = s \int_s^{\infty} F(u) du, \tag{6}$$

where $F(u) = \mathcal{L}[f(t)] = \int_0^{\infty} e^{-st} f(t) dt$

ii:

$$\begin{aligned} \mathcal{G}_1[f(t)](s) = s \int_0^{\infty} t^{1-1} e^{-st} f(t) dt = s \int_0^{\infty} e^{-st} f(t) dt = s\, \mathcal{L}[f(t)] \\ = s\, F(s). \end{aligned} \tag{7}$$

iii:

$$\mathcal{L}^{-1}[\mathcal{G}_n[f(t)](s)] = t^{n-2}(2H(t) - 1)((n-1) f(t) + t f'(t)) \tag{8}$$

where

$$H(t) = \int_{-\infty}^{t} \delta(s) ds,$$

is a Heaviside function (integral of Dirac delta).

There are some functions in which the Laplace transform does not exist.

Proof: Relations i and ii are obvious. Here, we prove relation iii.

$$\begin{aligned}
\mathcal{L}^{-1}[\mathcal{G}_n[f(t)](s)] &= \tfrac{1}{2\pi i}\int_{\alpha-i\infty}^{\alpha+i\infty} e^{st}\mathcal{G}_n[f(t)](s)ds \\
&= \tfrac{1}{2\pi i}\int_{\alpha-i\infty}^{\alpha+i\infty} e^{st}\left(s\int_0^{\infty} e^{-st}t^{n-1}f(t)dt\right)ds \\
&= \tfrac{1}{2\pi i}\int_{\alpha-i\infty}^{\alpha+i\infty} e^{st}s\,\mathcal{L}\left[t^{n-1}f(t)\right]ds = \mathcal{L}^{-1}\left[s\,\mathcal{L}\left[t^{n-1}f(t)\right]\right] \\
&= \mathcal{L}^{-1}[s]*\mathcal{L}^{-1}\left[\mathcal{L}\left[t^{n-1}f(t)\right]\right] = \delta'(t)*t^{n-1}f(t) \\
&= \int_0^t \delta'(t-\tau)\tau^{n-1}f(\tau)d\tau \\
&= t^{-2+n}(-1+2H(t))((-1+n)f(t)+tf'(t)).
\end{aligned}$$

and for relation iv, the Laplace transform for the function $\frac{e^{-t}}{t}$ does not exist, while:

$$\mathcal{G}_2\left[\frac{e^{-t}}{t}\right](s) = s\int_0^{\infty} t^{2-1}e^{-st}\frac{e^{-t}}{t}dt = \frac{s}{s+1}.$$

The duality between ARA and Laplace Carson transforms [18]

$$\mathcal{L}_*[g(t)] = \mathcal{G}_1[g(t)](s)$$

$$\mathcal{G}_n[g(t)](s) = \mathcal{L}_*\left[t^{n-1}g(t)\right]$$

where

$$\mathcal{L}_*[g(t)] = s\int_0^{\infty} e^{-st}g(t)dt.$$

Duality between ARA and Aboodh Transforms [19]

$$A[g(t)] = \frac{1}{s^2}\mathcal{G}_1[g(t)](s)$$

$$\mathcal{G}_n[g(t)](s) = s^2A\left[t^{n-1}\,g(t)\right]$$

where

$$A[g(t)] = \frac{1}{s}\int_0^{\infty} g(t)e^{-st}dt.$$

Duality between ARA and Mohand Transforms [20]:

$$M[g(t)] = s\mathcal{G}_1[g(t)](s)$$

$$\mathcal{G}_n[g(t)](s) = \frac{1}{s}M\left[t^{n-1}\,g(t)\right]$$

where

$$M[g(t)] = s^2\int_0^{\infty} g(t)e^{-st}dt.$$

□

4. Properties of ARA Transform

In this section, we establish some properties of the ARA transform, which enable us to calculate further transform of functions in applications.

Property 1. *(Linearity property) Let $u(t)$ and$v(t)$ be two functions in which ARA transform exists, then*

$$\mathcal{G}_n[\alpha u(t)+\beta v(t)](s)=\alpha\mathcal{G}_n[u(t)](s)+\beta\mathcal{G}_n[v(t)](s) \tag{9}$$

where α and β are nonzero constants.

Proof of Property 1:

$$\begin{aligned}\mathcal{G}_n[\alpha u(t)+\beta v(t)](s)&=s\int_0^\infty t^{n-1}e^{-st}(\alpha u(t)+\beta v(t))\,dt\\&=\alpha s\int_0^\infty t^{n-1}e^{-st}u(t)dt+\beta s\int_0^\infty t^{n-1}e^{-st}v(t)\,dt\\&=\alpha\mathcal{G}_n[u(t)](s)+\beta\mathcal{G}_n[v(t)](s).\end{aligned}$$

□

Property 2. *Change of scale property*

$$\mathcal{G}_n[g(at)](s)=\frac{1}{a^{n-1}}\mathcal{G}_n[g(t)]\left(\frac{s}{a}\right)=\frac{1}{a^{n-1}}G\left(n,\frac{s}{a}\right). \tag{10}$$

Proof of Property 2: Using the definition of ARA transform for $g(at)$, we get

$$\mathcal{G}_n[g(at)](s)=s\int_0^\infty t^{n-1}e^{-st}g(at)dt \tag{11}$$

and a substitution of $u=at$ in Equation (5) yields:

$$\begin{aligned}\mathcal{G}_n[g(at)](s)&=\tfrac{s}{a^n}\int_0^\infty u^{n-1}e^{-\frac{s}{a}u}g(u)du=\tfrac{1}{a^{n-1}}\tfrac{s}{a}\int_0^\infty u^{n-1}e^{-\frac{s}{a}u}g(u)du\\&=\tfrac{1}{a^{n-1}}G\left(n,\tfrac{s}{a}\right)\end{aligned}$$

□

Property 3. *Shifting in s– Domain*

$$\mathcal{G}_n\left[e^{-ct}g(t)\right](s)=\frac{s}{s+c}G(n,s+c). \tag{12}$$

Proof of Property 3:

$$\begin{aligned}\mathcal{G}_n\left[e^{-ct}g(t)\right](s)&=s\int_0^\infty t^{n-1}e^{-st}e^{-ct}g(t)dt=s\int_0^\infty t^{n-1}e^{-(s+c)t}g(t)dt\\&=\tfrac{s}{s+c}(s+c)\int_0^\infty t^{n-1}e^{-(s+c)t}g(t)dt\\&=\tfrac{s}{s+c}G(n,s+c).\end{aligned}$$

□

Property 4. *Shifting in n− Domain*

$$\mathcal{G}_n[t^m g(t)](s) = \mathcal{G}_{n+m}[g(t)] = G(n+m,s). \tag{13}$$

Proof of Property 4:

$$\mathcal{G}_n[t^m g(t)](s) = s\int_0^\infty t^{n-1}e^{-st}t^m g(t)dt = s\int_0^\infty t^{m+n-1}e^{-st}g(t)dt$$
$$= \mathcal{G}_{n+m}[g(t)](s) = G(n+m,s).$$

Also, $\mathcal{G}_n\left[\frac{g(t)}{t^m}\right] = \mathcal{G}_{n-m}[g(t)] = G(n-m,s)$. □

Property 5. *Shifting on t− domain*

$$\mathcal{G}_n[u_c(t)g(t-c)](s) = e^{-cs}\mathcal{G}_1\left[g(v)(v+c)^{n-1}\right] \tag{14}$$

Proof of Property 5:

$$\mathcal{G}_n[u_c(t)g(t-c)](s) = s\int_c^\infty t^{n-1}e^{-st}\ u_c(t)\ g(t-c)dt$$

$$= s\int_c^\infty t^{n-1}e^{-st}\ \ g(t-c)dt,$$

letting $t-c=v$ and substituting in the above equation we get

$$s\int_0^\infty (v+c)^{n-1}e^{-s(v+c)}\ \ g(v)dv = se^{-sc}\int_0^\infty e^{-sv}\ \ g(v)(v+c)^{n-1}dv$$

$$= e^{-sc}\mathcal{G}_1\left[g(v)(v+c)^{n-1}\right].$$

□

Property 6. *ARA transform for derivatives*

$$\mathcal{G}_n\left[f^{(m)}(t)\right](s) = (-1)^{n-1}s\frac{d^{n-1}}{ds^{n-1}}\left(\frac{\mathcal{G}_1\left[f^{(m)}(t)\right](s)}{s}\right) \tag{15}$$

Proof of Property 6:

$$\mathcal{G}_n\left[f^{(m)}(t)\right](s) = s\int_0^\infty t^{n-1}e^{-st}f^{(m)}(t)dt = s\int_0^\infty\left(t^{n-1}f^{(m)}(t)\right)e^{-st}dt$$
$$= \mathcal{G}_1\left[t^{n-1}f^{(m)}(t)\right](s)$$
$$= (-1)^{n-1}s\frac{d^{n-1}}{ds^{n-1}}\left(\frac{\mathcal{G}_1\left[f^{(m)}(t)\right](s)}{s}\right).$$

Moreover:

$$\mathcal{G}_n\left[f^{(n)}(t)\right](s) = (-1)^{n-1} s \frac{d^{n-1}}{ds^{n-1}}\left(s^{n-1}\mathcal{G}_1[f(t)](s) - \sum_{j=1}^{n} s^{n-j} f^{(j-1)}(0)\right) \tag{16}$$

Since

$$\begin{aligned}\mathcal{G}_n\left[f^{(n)}(t)\right](s) &= (-1)^{n-1} s \tfrac{d^{n-1}}{ds^{n-1}}\, \mathcal{L}\left[f^{(n)}(t)\right] \\ &= (-1)^{n-1} s \tfrac{d^{n-1}}{ds^{n-1}}\left(s^{n-1}\mathcal{G}_1[f(t)](s) - s^{n-1} f(0) - \cdots - f^{(n-1)}(0)\right) \\ &= (-1)^{n-1} s \tfrac{d^{n-1}}{ds^{n-1}}\left(s^{n-1}\mathcal{G}_1[f(t)](s) - \sum_{j=1}^{n} s^{n-j} f^{(j-1)}(0)\right).\end{aligned}$$

Using the properties of the convolution of Laplace transform, we get the following property. □

Property 7. *Convolution*

$$\mathcal{G}_n[f(t) * h(t)](s) = (-1)^{n-1} s \sum_{j=0}^{n-1} c_j^{n-1} F^{(j)}(s) \cdot H^{(n-1-j)}(s) \tag{17}$$

Proof of Property 7:

$$\begin{aligned}\mathcal{G}_n[f(t) * g(t)](s) &= s\int_0^\infty (f(t) * g(t)) t^{n-1} e^{-st} dt \\ = \mathcal{G}_1\left[t^{n-1} f(t) * g(t)\right] &= (-1)^{n-1} s \tfrac{d^{n-1}}{ds^{n-1}} \tfrac{\mathcal{G}_1[f(t)*g(t)]}{s} \\ &= (-1)^{n-1} s \sum_{j=0}^{n-1} c_j^{n-1} F^{(j)}(s) \cdot H^{(n-1-j)}(s),\end{aligned}$$

where c_n^k is the binomial coefficient. □

Now, we introduce some practical examples for finding ARA transform for some functions.
Example 4.1:

$$\begin{aligned}\mathcal{G}_n[1](s) = s\int_0^\infty t^{n-1} e^{-st} dt &= \Gamma(n)\left(\tfrac{1}{s}\right)^n s \int_0^\infty \tfrac{t^{n-1} e^{-st}}{\Gamma(n)\left(\frac{1}{s}\right)^n} dt = \Gamma(n)\left(\tfrac{1}{s}\right)^n s \\ &= \tfrac{\Gamma(n)}{s^{n-1}} = \tfrac{(n-1)!}{s^{n-1}}.\end{aligned}$$

Example 4.2:

$$\begin{aligned}\mathcal{G}_n[t](s) &= s\int_0^\infty t^{n-1} e^{-st} t\, dt = s\int_0^\infty t^n e^{-st}\, dt \\ &= \Gamma(n+1)\left(\tfrac{1}{s}\right)^{n+1} s \int_0^\infty \tfrac{t^n e^{-st}}{\Gamma(n+1)\left(\frac{1}{s}\right)^{n+1}}\, dt \\ &= \left(\tfrac{1}{s}\right)^{n+1} \Gamma(n+1) s = \tfrac{\Gamma(n+1)}{s^n}.\end{aligned}$$

Example 4.3:

$$\begin{aligned}\mathcal{G}_n[t^m](s) &= s\int_0^\infty t^{n-1} e^{-st} t^m\, dt = s\int_0^\infty t^{m+n-1} e^{-st}\, dt \\ &= \Gamma(m+n)\left(\tfrac{1}{s}\right)^{m+n} s \int_0^\infty \tfrac{t^{m+n-1} e^{-st}}{\Gamma(m+n)\left(\frac{1}{s}\right)^{m+n}}\, dt \\ &= \left(\tfrac{1}{s}\right)^{m+n} \Gamma(m+n) s = \tfrac{\Gamma(m+n)}{s^{m+n-1}}.\end{aligned}$$

Example 4.4:

$$\mathcal{G}_n\left[e^{at}\right](s) = s\int_0^\infty t^{n-1}e^{-t(s-a)}\,dt = \Gamma(n)\left(\tfrac{1}{s-a}\right)^n s\int_0^\infty \frac{t^{n-1}e^{-t(s-a)}}{\Gamma(n)\left(\frac{1}{s-a}\right)^n}\,dt$$
$$= \tfrac{s}{(s-a)^n}\Gamma(n), \text{ for all } s > a.$$

Example 4.5:

$$\begin{aligned}\mathcal{G}_n\left[t^m e^{at}\right](s) &= s\int_0^\infty t^{m+n-1}e^{-t(s-a)}\,dt\\ &= \Gamma(m+n)\left(\tfrac{1}{s-a}\right)^{m+n} s\int_0^\infty \frac{t^{m+n-1}e^{-t(s-a)}}{\Gamma(m+n)\left(\frac{1}{s-a}\right)^{m+n}}\,dt\\ &= \tfrac{s}{(s-a)^{m+n}}\Gamma(m+n).\end{aligned}$$

Example 4.6:

$$\begin{aligned}\mathcal{G}_n[\sin(at)](s) &= \mathcal{G}_n\left[\tfrac{e^{iat}-e^{-iat}}{2i}\right] = \tfrac{1}{2i}\left(\mathcal{G}_n\left[e^{iat}\right] - \mathcal{G}_n\left[-e^{-iat}\right]\right)\\ &= \tfrac{s}{2i}\Gamma(n)\left(\tfrac{1}{(s-ia)^n} - \tfrac{1}{(s+ia)^n}\right)\\ &= \tfrac{s}{2i}\Gamma(n)\left(\tfrac{2i}{(s^2+a^2)^{\frac{n}{2}}}\sin\left(n\tan^{-1}\left(\tfrac{a}{s}\right)\right)\right)\\ &= \left(1+\tfrac{a^2}{s^2}\right)^{-\frac{n}{2}} s^{1-n}\Gamma(n)\sin\left(n\tan^{-1}\left(\tfrac{a}{s}\right)\right).\end{aligned}$$

Example 4.7:

$$\mathcal{G}_n[\cos(at)](s) = \left(1+\frac{a^2}{s^2}\right)^{-\frac{n}{2}} s^{1-n}\Gamma(n)\cos\left(n\tan^{-1}\left(\frac{a}{s}\right)\right).$$

With similar arguments to example 4.6, we obtain the result.
Example 4.8:

$$\begin{aligned}\mathcal{G}_n[\sinh(at)] &= \mathcal{G}_n\left[\tfrac{e^{at}-e^{-at}}{2}\right] = \tfrac{1}{2}\left(\mathcal{G}_n\left[e^{at}\right] - \mathcal{G}_n\left[-e^{-at}\right]\right)\\ &= \tfrac{1}{2}\left(\tfrac{s}{(s-a)^n}\Gamma(n) - \tfrac{s}{(s+a)^n}\Gamma(n)\right)\\ &= \tfrac{s}{2}\Gamma(n)\left(\tfrac{1}{(s-a)^n} - \tfrac{1}{(s+a)^n}\right)\\ &= \tfrac{s}{2}\Gamma(n)\tfrac{1}{s^n}\left(\tfrac{1}{\left(1-\frac{a}{s}\right)^n} - \tfrac{1}{\left(1+\frac{a}{s}\right)^n}\right).\end{aligned}$$

Example 4.9:

$$\begin{aligned}\mathcal{G}_n[\cosh(at)](s) &= \mathcal{G}_n\left[\tfrac{e^{at}+e^{-at}}{2}\right] = \tfrac{1}{2}\left(\mathcal{G}_n\left[e^{at}\right] + \mathcal{G}_n\left[-e^{-at}\right]\right)\\ &= \tfrac{1}{2}\left(\tfrac{s}{(s-a)^n}\Gamma(n) + \tfrac{s}{(s+a)^n}\Gamma(n)\right)\\ &= \tfrac{s}{2}\Gamma(n)\left(\tfrac{1}{(s-a)^n} + \tfrac{1}{(s+a)^n}\right)\\ &= \tfrac{s}{2}\Gamma(n)\tfrac{1}{s^n}\left(\tfrac{1}{\left(1-\frac{a}{s}\right)^n} + \tfrac{1}{\left(1+\frac{a}{s}\right)^n}\right)\end{aligned}$$

Example 4.10:

$$\mathcal{G}_n\left[u_3(t)e^{t-3}\right](s) = e^{-3s}\mathcal{G}_1\left[g(v)(v+3)^{n-1}\right].$$

For $n = 1$

$$\mathcal{G}_1\left[u_3(t)e^{t-3}\right](s) = e^{-3s}\mathcal{G}_1\left[e^v(v+3)^{1-1}\right] = \frac{s}{s-1}\,e^{-3s}.$$

For $n = 2$

$$\mathcal{G}_2\left[u_3(t)e^{t-3}\right](s) = e^{-3s}\mathcal{G}_1\left[e^{v}(v+3)^{2-1}\right] = e^{-3s}\mathcal{G}_1\left[v\, e^{v} + 3\, e^{v}\right]$$
$$= s\, e^{-3s}\left(\tfrac{1}{(s-1)^2} + \tfrac{3}{s-1}\right).$$

We present a list of ARA transform of some special functions and General properties of the ARA transform in Table A1 (Appendix A).

5. Applications of the ARA Transform

In this section, we give some applications of ordinary differential equations, in which the efficiency and high accuracy of ARA transform are illustrated.

Example 5.1:

Consider the initial value problem:

$$y'(t) + y(t) = 0, \quad y(0) = 1. \tag{18}$$

Solution:

Applying ARA transform $\mathcal{G}_1$ on both side of Equation (18)

$$\mathcal{G}_1[y'(t)](s) + \mathcal{G}_1[y(t)](s) = 0$$

$$s\mathcal{G}_1[y(t)](s) - sy(0) + \mathcal{G}_1[y(t)](s) = 0$$

$$\mathcal{G}_1[y(t)](s) = \frac{s}{s+1}.$$

Taking the inverse ARA transform $\mathcal{G}_1^{-1}$ we get:

$$y(t) = e^{-t}.$$

Example 5.2:

Consider the initial value problem:

$$y'(t) - y(t) = e^{2t}, \quad y(0) = 1. \tag{19}$$

Solution:

Applying ARA transform $\mathcal{G}_1$ on both side of Equation (19)

$$\mathcal{G}_1[y'(t)](s) - \mathcal{G}_1[y(t)](s) = \mathcal{G}_1\left[e^{2t}\right](s)$$

$$s\mathcal{G}_1[y(t)](s) - sy(0) - \mathcal{G}_1[y(t)](s) = \frac{s}{s-2}$$

$$\mathcal{G}_1[y(t)](s) = \frac{1}{s-1}\left(\frac{s}{s-2} + s\right)$$

$$\mathcal{G}_1[y(t)](s) = \frac{s}{(s-2)(s-1)} + \frac{s}{s-1} = \frac{s}{s-2}.$$

Taking the inverse ARA transform $\mathcal{G}_1^{-1}$ we get:

$$y(t) = e^{2t}.$$

Example 5.3:

Consider the initial value problem:

$$y''(t) + y(t) = 0, \quad y(0) = 1, \; y'(0) = 1. \tag{20}$$

Solution:
First, we solve the initial value problem (20) applying $\mathcal{G}_1$:

$$\mathcal{G}_1[y''(t)](s) + \mathcal{G}_1[y(t)](s) = 0$$

$$s^2\mathcal{G}_1[y(t)](s) - s^2 y(0) - sy'(0) + \mathcal{G}_1[y(t)](s) = 0$$

$$s^2\mathcal{G}_1[y(t)](s) - s^2 - s + \mathcal{G}_1[y(t)](s) = 0$$

$$\left(s^2+1\right)\mathcal{G}_1[y(t)](s) = s^2 + s$$

$$\mathcal{G}_1[y(t)](s) = \frac{s^2}{s^2+1} + \frac{s}{s^2+1}.$$

Taking the inverse ARA transform $\mathcal{G}_1^{-1}$, we get

$$y(t) = \cos(t) + \sin(t).$$

Now, we solve the initial value problem (20) applying $\mathcal{G}_2$:

$$\mathcal{G}_2[y''(t)] + \mathcal{G}_2[y(t)] = 0$$

$$sy(0) - s^2\mathcal{G}_1'[y(t)](s) - s\mathcal{G}_1[y(t)](s) + \frac{\mathcal{G}_1[y(t)](s)}{s} - \mathcal{G}_1'[y(t)](s) = 0$$

$$s^2 - s^3\mathcal{G}_1'[y(t)](s) - s^2\mathcal{G}_1[y(t)](s) + \mathcal{G}_1[y(t)](s) - s\mathcal{G}_1'[y(t)](s) = 0$$

$$-\left(s^3+s\right)\mathcal{G}_1'[y(t)](s) - \left(s^2-1\right)\mathcal{G}_1[y(t)](s) = -s^2$$

$$\mathcal{G}_1'[y(t)](s) + \frac{s^2-1}{s^3+s}\mathcal{G}_1[y(t)](s) = \frac{s^2}{s^3+s}$$

$$\frac{d}{ds}\left(\mathcal{G}_1[y(t)](s)\frac{s^2+1}{s}\right) = \frac{s^2}{s^3+s}\,\frac{s^2+1}{s} = 1$$

$$\mathcal{G}_1[y(t)](s) = \frac{s(s+c)}{s^2+1}.$$

Taking the inverse ARA transform $\mathcal{G}_1^{-1}$ and using the second initial condition $y'(0) = 1$, we get:

$$y(t) = \cos t + \sin t.$$

Example 5.4:

$$y'' - y' + 6y = 8e^{2t},\ y(0) = 0,\ y'(0) = -3. \tag{21}$$

Solution:
Applying ARA transform $\mathcal{G}_2$ on both side of Equation (21)

$$\mathcal{G}_2[y''(t)] - \mathcal{G}_2[y'(t)] + 6\mathcal{G}_2[y(t)] = 8\mathcal{G}_2\left[e^{2t}\right]$$

$$sy(0) - s^2\mathcal{G}_1'[y(t)](s) - s\mathcal{G}_1[y(t)](s) + s\mathcal{G}_1'[y(t)](s) + 6\left(\frac{\mathcal{G}_1[y(t)](s)}{s} - \mathcal{G}_1'[y(t)](s)\right) = \frac{8s}{(s-2)^2}$$

$$-s^3\mathcal{G}_1'[y(t)](s) - s^2\mathcal{G}_1[y(t)](s) + s^2\mathcal{G}_1'[y(t)](s) + 6\mathcal{G}_1[y(t)](s) - 6s\mathcal{G}_1'[y(t)](s) = \frac{8s^2}{(s-2)^2}$$

$$-\left(s^3 - s^2 + 6s\right)\mathcal{G}_1'[y(t)](s) - \left(s^2-6\right)\mathcal{G}_1[y(t)](s) = \frac{8s^2}{(s-2)^2}$$

$$\mathcal{G}_1'[y(t)](s) + \frac{s^2-6}{s^3-s^2+6s}\mathcal{G}_1[y(t)](s) = \frac{-8s^2}{(s-2)^2(s^3-s^2+6s)}$$

$$\frac{d}{ds}\left(\mathcal{G}_1[y(t)](s)\frac{s^2-s+6}{s}\right) = \frac{-8}{(s-2)^2}$$

$$\begin{aligned}\mathcal{G}_1[y(t)](s) &= \left(8(s-2)^{-1}+c\right)\left(\tfrac{s}{s^2-s+6}\right)\\ &= \tfrac{8}{(s-2)(s^2-s+6)} + \tfrac{cs}{s^2-s+6}\\ &= \tfrac{1}{s-2} - \tfrac{s+1}{s^2-s+6} + \tfrac{cs}{s^2-s+6}.\end{aligned}$$

Taking the inverse ARA transform $\mathcal{G}_1^{-1}$ and using second initial condition $y'(0) = -3$, we get:

$$y(t) = e^{2t} - e^{\frac{1}{2}t}\cos\left(\frac{\sqrt{23}}{2}t\right) - \frac{9}{\sqrt{23}}e^{\frac{1}{2}t}\sin\left(\frac{\sqrt{23}}{2}t\right).$$

Example 5.5:

$$y'' + 2y' + 5y = e^{-t}\sin t,\ y(0) = 0,\ y'(0) = 1. \tag{22}$$

Solution:
Applying ARA transform $\mathcal{G}_2$ on both side of the differential Equation (22):

$$\mathcal{G}_2[y''(t)] + 2\mathcal{G}_2[y'(t)] + 5\mathcal{G}_2[y(t)] = \mathcal{G}_2\left[e^{-t}\sin t\right]$$

$$\begin{aligned}sy(0) - s^2\mathcal{G}_1'[y(t)](s) - s\mathcal{G}_1[y(t)](s) + 2\left(-s\mathcal{G}_1'[y(t)](s)\right)\\ +5\left(\tfrac{\mathcal{G}_1[y(t)](s)}{s} - \mathcal{G}_1'[y(t)](s)\right)\\ = \left(1+\tfrac{1}{(-1-s)^2}\right)^{-2/2} s(1\\ +s)^{-2}\Gamma(2)\sin\left(2\tan^{-1}\left(\tfrac{1}{1+s}\right)\right),\end{aligned}$$

$$\begin{aligned}-s^3\mathcal{G}_1'[y(t)](s) - s^2\mathcal{G}_1[y(t)](s) - 2s^2\mathcal{G}_1'[y(t)](s) + 5\mathcal{G}_1[y(t)](s)\\ -5s\mathcal{G}_1'[y(t)](s)\\ = s\left(\tfrac{(s+1)^2+1}{(s+1)^2}\right)^{-1}\tfrac{s}{(1+s)^2}2\sin\left(\tan^{-1}\left(\tfrac{1}{1+s}\right)\right)\cos\left(\tan^{-1}\left(\tfrac{1}{1+s}\right)\right)\\ = \tfrac{2s^2}{1+(s+1)^2}\,\tfrac{s+1}{1+(s+1)^2} = \tfrac{2s^2(s+1)}{\left(1+(s+1)^2\right)^2}\end{aligned}$$

$$-\left(s^3+2s^2+5s\right)\mathcal{G}_1'[y(t)](s) - \left(s^2-5\right)\mathcal{G}_1[y(t)](s) = \frac{2s^2(s+1)}{\left(1+(s+1)^2\right)^2}$$

$$\begin{aligned}\mathcal{G}_1'[y(t)](s) + \tfrac{s^2-5}{s^3+2s^2+5s}\mathcal{G}_1[y(t)](s)\\ = \tfrac{-2s^2(s+1)}{(s^3+2s^2+5s)\left(1+(s+1)^2\right)^2}\end{aligned}$$

$$\begin{aligned}\tfrac{d}{ds}\left(\mathcal{G}_1[y(t)](s)\tfrac{s^2+2s+5}{s}\right)\\ = \tfrac{-2s^2(s+1)}{(s^3+2s^2+5s)\left(1+(s+1)^2\right)^2}\,\tfrac{s^2+2s+5}{s}\end{aligned}$$

$$\mathcal{G}_1[y(t)](s) = \frac{1}{1+(s+1)^2}\,\frac{s}{s^2+2s+5} = s\frac{1}{1+(s+1)^2}\,\frac{1}{4+(s+1)^2}.$$

Taking the inverse ARA transform $\mathcal{G}_1^{-1}$ and using the second initial condition $y'(0) = 1$ we get:

$$y(t) = \frac{1}{3}e^{-t}\sin t + \frac{2}{3}e^{-t}\sin 2t.$$

Example 5.6:
Consider the initial boundary value problem

$$u_t = u_{xx} \tag{23}$$

$$u(0,t) = u(1,t) = 0$$

$$u(x,0) = \sin(2\pi x).$$

Solution:
Applying ARA transform $\mathcal{G}_1$ on both side of the Equation (23)

$$\mathcal{G}_1[u_t] = \mathcal{G}_1[u_{xx}]$$

$$s\mathcal{G}_1[u(x,t)](s) - su(x,0) = \frac{d^2}{dx^2}(\mathcal{G}_1[u(x,t)](s))$$

$$\mathcal{G}_1''[u(x,t)](s) - s\mathcal{G}_1[u(x,t)](s) = -s\sin(2\pi x). \tag{24}$$

The general solution of Equation (24) can be written as

$$\mathcal{G}_1[u(x,t)](s) = \mathcal{G}_{1c}[u(x,t)](s) + \mathcal{G}_{1p}[u(x,t)](s),$$

where $\mathcal{G}_{1c}[u(x,t)](s) = c_1 e^{\sqrt{s}x} + c_2 e^{-\sqrt{s}x}$, and the solution of the nonhomogeneous part is given by

$$\mathcal{G}_{1p}[u(x,t)](s) = \alpha_1 \sin(2\pi x) + \alpha_2 \cos(2\pi x),$$

after simple calculations, we get $\alpha_2 = 0$.
Hence:

$$\mathcal{G}_1[u(x,t)](s) = c_1 e^{\sqrt{s}x} + c_2 e^{-\sqrt{s}x} + \frac{s}{4\pi^2 + s}\sin(2\pi x).$$

Using boundary conditions, we get $c_1 = c_2 = 0$:

$$\mathcal{G}_1[u(x,t)](s) = \frac{s}{4\pi^2 + s}\sin(2\pi x).$$

Taking the inverse ARA transform $\mathcal{G}_1^{-1}$:

$$u(x,t) = e^{-4\pi^2 t}\sin(2\pi x).$$

Example 5.7
Consider the initial boundary value problem:

$$u_{tt} = u_{xx} + \sin(\pi x), \tag{25}$$

$$u(0,t) = u(1,t) = 0$$

$$u(x,0) = u_t(x,0) = 0.$$

Solution:
Applying ARA transform $\mathcal{G}_1$ on both sides of the Equation (25)

$$\mathcal{G}_1[u_{tt}] = \mathcal{G}_1[u_{xx}] + \mathcal{G}_1[\sin(\pi x)]$$

$$s^2\mathcal{G}_1[u(x,t)](s) - s^2 u(x,0) - su_t(x,0)$$
$$= \tfrac{d^2}{dx^2}(\mathcal{G}_1[u(x,t)](s)) + \sin(\pi x)$$

$$\frac{d^2}{dx^2}(\mathcal{G}_1[u(x,t)](s)) - s^2\mathcal{G}_1[u(x,t)](s) = -\sin(\pi x). \tag{26}$$

The general solution of Equation (26) can be written as

$$\mathcal{G}_1[u(x,t)](s) = \mathcal{G}_{1c}[u(x,t)](s) + \mathcal{G}_{1p}[u(x,t)](s),$$

where $\mathcal{G}_{1c}[u(x,t)](s) = c_1e^{sx} + c_2e^{-sx}$, and the solution of the nonhomogeneous part is given by:

$$\mathcal{G}_{1p}[u(x,t)](s) = A\sin(\pi x) + B\cos(\pi x).$$

To find A and B, we substituting $\mathcal{G}_{1p}$ in Equation (26) to get:

$$\mathcal{G}_{1p}[u(x,t)](s) = \frac{1}{\pi^2+s^2}\sin(\pi x)$$

$$\mathcal{G}_1[u(x,t)](s) = c_1e^{sx} + c_2e^{-sx} + \frac{1}{\pi^2+s^2}\sin(\pi x).$$

Using boundary conditions, we get:

$$\mathcal{G}_1[u(x,t)](s) = \frac{1}{\pi^2+s^2}\sin(\pi x).$$

Taking the inverse ARA transform $\mathcal{G}_1^{-1}$ we get the solution:

$$u(x,t) = \frac{\sin(\pi x)}{\pi^2}(1-\cos(\pi t)).$$

6. Conclusions

In this paper, we introduced a new integral operator transform called the ARA transform. We presented its existence and inverse transform. We presented some properties and their application in the solving of ordinary and partial differential equations that arise in some branches of science like physics, engineering, etc.

Author Contributions: Data curation, R.S., A.Q. and A.B.; Formal analysis, R.S., A.Q. and A.B.; Investigation, R.S., A.Q. and A.B.; Methodology, R.S. and A.Q.; Project administration, A.Q.; Resources, R.S.; Writing—original draft, A.Q.; Writing—review & editing, R.S. and A.B. All authors have read and agreed to the published version of the manuscript.

Acknowledgments: The authors express their gratitude to the dear unknown referees and the editor for their helpful suggestions, which improved the final version of this paper.

Appendix A

Table A1. Here we present a list of ARA transform of some special functions and General properties of the ARA transform.

$g(t)$	$G(n,s)$
1	$s^{1-n}\Gamma(n)$
t	$s^{-n}\Gamma(1+n)$
$\sqrt{t}$	$s^{\frac{1}{2}-n}\Gamma\left(\frac{1}{2}+n\right)$
$\sqrt{\frac{\pi}{t}}$	$\sqrt{\pi}s^{\frac{3}{2}-n}\Gamma\left(-\frac{1}{2}+n\right)$

Table A1. *Cont.*

$g(t)$	$G(n,s)$
t^2	$s^{-1-n}\Gamma(2+n)$
t^m	$s^{1-m-n}\Gamma(m+n)$
$t^{m-1/2}$	$s^{\frac{3}{2}-m-n}\Gamma\left(-\frac{1}{2}+m+n\right)$
$e^{\alpha t}$	$s(s-\alpha)^{-n}\Gamma(n)$
$e^{-\alpha t}$	$s(s+\alpha)^{-n}\Gamma(n)$
$te^{\alpha t}$	$s(s-\alpha)^{-1-n}\Gamma(1+n)$
$t^m e^{\alpha t}$	$s(s-\alpha)^{-m-n}\Gamma(m+n)$
$\sin(at)$	$\frac{s}{2i}\Gamma(n)\left(\frac{1}{(s-ia)^n}-\frac{1}{(s+ia)^n}\right)=\left(1+\frac{a^2}{s^2}\right)^{-\frac{n}{2}}s^{1-n}\Gamma(n)\sin\left(n\tan^{-1}\left(\frac{a}{s}\right)\right)$
$\cos(at)$	$\frac{s}{2i}\Gamma(n)\left(\frac{1}{(s-ia)^n}+\frac{1}{(s+ia)^n}\right)=\left(1+\frac{a^2}{s^2}\right)^{-\frac{n}{2}}s^{1-n}\Gamma(n)\cos\left(n\tan^{-1}\left(\frac{a}{s}\right)\right)$
$t\sin(at)$	$\frac{\left(1+\frac{a^2}{s^2}\right)^{\frac{1}{2}(-1-n)}s^{-n}\Gamma(n+2)\sin\left((1+n)\tan^{-1}\left(\frac{a}{s}\right)\right)}{1+n}$
$t\cos(at)$	$\left(1+\frac{a^2}{s^2}\right)^{\frac{1}{2}(-1-n)}s^{-n}\Gamma(n+1)\cos\left((1+n)\tan^{-1}\left(\frac{a}{s}\right)\right)$
$\sin(at)-a\,t\cos(at)$	$\left(1+\frac{a^2}{s^2}\right)^{\frac{1}{2}(-1-n)}s^{-n}\Gamma(n)\left(-a\,n\cos\left((1+n)\tan^{-1}\left(\frac{a}{s}\right)\right)+\sqrt{1+\frac{a^2}{s^2}}s\,\sin\left(n\tan^{-1}\left(\frac{a}{s}\right)\right)\right)$
$\sin(at)+a\,t\cos(at)$	$\left(1+\frac{a^2}{s^2}\right)^{\frac{1}{2}(-1-n)}s^{-n}\Gamma(n)\left(a\,n\cos\left((1+n)\tan^{-1}\left(\frac{a}{s}\right)\right)+\sqrt{1+\frac{a^2}{s^2}}s\,\sin\left(n\tan^{-1}\left(\frac{a}{s}\right)\right)\right)$
$\cos(at)-a\,t\sin(at)$	$\left(1+\frac{a^2}{s^2}\right)^{\frac{1}{2}(-1-n)}s^{-n}\Gamma(n)\left(s\cos\left((1+n)\tan^{-1}\left(\frac{a}{s}\right)\right)-a(-1+n)\,\sin\left((1+n)\tan^{-1}\left(\frac{a}{s}\right)\right)\right)$
$\cos(at)+a\,t\sin(at)$	$\left(1+\frac{a^2}{s^2}\right)^{\frac{1}{2}(-1-n)}s^{-n}\Gamma(n)\left(s\,\cos\left((1+n)\tan^{-1}\left(\frac{a}{s}\right)\right)+a(1+n)\,\sin\left((1+n)\tan^{-1}\left(\frac{a}{s}\right)\right)\right)$
$\sin(at+b)$	$\left(1+\frac{a^2}{s^2}\right)^{-n/2}s^{1-n}\,\Gamma(n)\,\sin\left(b+n\,\tan^{-1}\left(\frac{a}{s}\right)\right)$
$\cos(at+b)$	$\left(1+\frac{a^2}{s^2}\right)^{-n/2}s^{1-n}\,\Gamma(n)\,\cos\left(b+n\,\tan^{-1}\left(\frac{a}{s}\right)\right)$
$e^{at}\sin(bt)$	$\left(1+\frac{b^2}{(a-s)^2}\right)^{-n/2}s(-a+s)^{-n}\Gamma(n)\,\sin\left(n\tan^{-1}\left(\frac{b}{-a+s}\right)\right)$
$e^{at}\cos(bt)$	$\left(1+\frac{b^2}{(a-s)^2}\right)^{-n/2}s(-a+s)^{-n}\Gamma(n)\,\cos\left(n\tan^{-1}\left(\frac{b}{a-s}\right)\right)$
$\sinh(at)$	$\frac{1}{2}s\left(-\frac{a^2}{s}+s\right)^{-n}\Gamma(n)\left(-\left(1-\frac{a}{s}\right)^n+\left(\frac{a+s}{s}\right)^n\right)$
$\cosh(at)$	$\frac{1}{2}s\left(-a^2+s^2\right)^{-n}\Gamma(n)\left((s-\lvert a\rvert)^n+(s+\lvert a\rvert)^n\right)$
$e^{at}\sinh(at)$	$\frac{1}{2}\left(1-\frac{b^2}{(a-s)^2}\right)^{-n}s(-a+s)^{-n}\Gamma(n)\left(-\left(1+\frac{b}{a-s}\right)^n+\left(1+\frac{b}{-a+s}\right)^n\right)$
$e^{at}\cosh(at)$	$\frac{1}{2}s\left(-a^2+s^2\right)^{-n}\Gamma(n)\left(\left((1+\frac{\sqrt{b^2}}{a-s}\right)^{-n}+\left(1+\frac{\sqrt{b^2}}{-a+s}\right)^{-n}\right)$
$\mathcal{G}_2[y(t)](s)$	$-s\frac{d}{ds}\left(\frac{\mathcal{G}_1[y(t)](s)}{s}\right)$
$\mathcal{G}_3[y(t)](s)$	$s\frac{d^2}{ds^2}\left(\frac{\mathcal{G}_1[y(t)](s)}{s}\right)$
$\mathcal{G}_n[y(t)](s)$	$s(-1)^{n-1}\frac{d^{n-1}}{ds^{n-1}}\left(\frac{\mathcal{G}_1[y(t)](s)}{s}\right)$
$\mathcal{G}_1[t\,y(t)](s)$	$-s\frac{d}{ds}\left(\frac{\mathcal{G}_1[y(t)](s)}{s}\right)$
$\mathcal{G}_1[y'(t)](s)$	$s\mathcal{G}_1[y](s)-sy(0)$
$\mathcal{G}_1[y''](s)$	$s^2\mathcal{G}_1[y(t)](s)-s^2y(0)-sy'(0)$
$\mathcal{G}_2[y(t)](s)$	$\frac{\mathcal{G}_1[y(t)](s)}{s}-\mathcal{G}_1'[y(t)](s)$
$\mathcal{G}_2[y'(t)](s)$	$-s\frac{d}{ds}(\mathcal{G}_1[y(t)](s))$
$\mathcal{G}_2[y''(t)](s)$	$sy(0)-s^2\frac{d}{ds}(\mathcal{G}_1[y(t)](s))-s\mathcal{G}_1[y(t)](s)$
$\mathcal{G}_n[\alpha u(t)+\beta v(t)](s)$	$\alpha\mathcal{G}_n[u(t)](s)+\beta\mathcal{G}_n[v(t)](s)$
$\mathcal{G}_n[g(at)](s)$	$\frac{1}{a^{n-1}}G\left(n,\frac{s}{a}\right)$
$\mathcal{G}_n\left[e^{-ct}g(t)\right](s)$	$\frac{s}{s+c}G(n,s+c)$
$\mathcal{G}_n[t^m g(t)](s)$	$G(n+m,s)$
$\mathcal{G}_n\left[\frac{g(t)}{t^m}\right]$	$G(n-m,s)$
$\mathcal{G}_n\left[f^{(m)}(t)\right](s)$	$(-1)^{n-1}s\frac{d^{n-1}}{ds^{n-1}}\left(\frac{\mathcal{G}_1\left[f^{(m)}(t)\right](s)}{s}\right)$
$\mathcal{G}_n\left[f^{(n)}(t)\right](s)$	$(-1)^{n-1}s\frac{d^{n-1}}{ds^{n-1}}\left(s^{n-1}\mathcal{G}_1[f(t)](s)-\sum_{j=1}^{n}s^{n-j}f^{(j-1)}(0)\right)$
$\mathcal{G}_n[f(t)*h(t)](s)$	$(-1)^{n-1}s\sum_{j=0}^{n-1}c_j^{n-1}F^{(j)}(s)\cdot H^{(n-1-j)}(s)$
$\mathcal{G}_n[u_c(t)g(t-c)](s)$	$e^{-cs}\mathcal{G}_1\left[g(u)(u+c)^{n-1}\right]$

References

1. Debnath, L.; Bhatta, D. *Integral Transforms and Their Applications*, 2nd ed.; C.R.C. Press: London, UK, 2007.
2. Widder, D.V. *The Laplace Transform*; Princeton University Press: London, UK, 1946.
3. Spiegel, M.R. *Theory and Problems of Laplace Transforms*; Schaums Outline Series; McGraw-Hill: New York, NY, USA, 1965.

4. Bochner, S.; Chandrasekharan, K. *Fourier Transforms*; Princeton University Press: London, UK, 1949.
5. Watugula, G.K. Sumudu transform: A new integral transform to solve differential equations and control engineering problems. *Int. J. Math. Edu. Sci. Technol.* **1993**, *24*, 35–43. [CrossRef]
6. Khan, Z.H.; Khan, W.A. Natural transform-properties and applications. *NUST J. Eng. Sci.* **2008**, *1*, 127–133.
7. Elzaki, T.M. The new integral transform "Elzaki transform". *Glob. J. Pure Appl. Math.* **2011**, *7*, 57–64.
8. Atangana, A.; Kiliçman, A. A novel integral operator transform and its application to some FODE and FPDE with some kind of singularities. *Math. Probl. Eng.* **2013**. [CrossRef]
9. Srivastava, H.; Luo, M.; Raina, R. A new integral transform and its applications. *Acta Math. Sci.* **2015**, *35B*, 1386–1400. [CrossRef]
10. Zafar, Z. ZZ transform method. *Int. J. Adv. Eng. Glob. Technol.* **2016**, *4*, 1605–1611.
11. Ramadan, M.; Raslan, K.; El-Danaf, T.; Hadhoud, A. A new general integral transform: Some properties and remarks. *J. Math. Comput. Sci.* **2016**, *6*, 103–109.
12. Barnes, B. Polynomial integral transform for solving differential Equations. *Eur. J. Pure Appl. Math.* **2016**, *9*, 140–151.
13. Yang, X.J. A new integral transform with an application in heat transf. *Therm. Sci.* **2016**, *20*, 677–681. [CrossRef]
14. Aboodh, K.; Abdullahi, I.; Nuruddeen, R. On the Aboodh transform connections with some famous integral transforms. *Int. J. Eng. Inform. Syst.* **2017**, *1*, 143–151.
15. Rangaig, N.; Minor, N.; Penonal, G.; Filipinas, J.; Convicto, V. On Another Type of Transform Called Rangaig Transform. *Int. J. Partial. Diff. Equ. Appl.* **2017**, *5*, 42–48. [CrossRef]
16. Maitama, S.; Zhao, W. New integral transform: Shehu transform a generalization of Sumudu and Laplace transform for solving differential equations. *Int. J. Anal. Appl.* **2019**, *17*, 167–190.
17. Aggarwal, S.; Chaudhary, R. A comparative study of Mohand and Laplace transforms. *J. Emerg. Technol. Innov. Res.* **2019**, *6*, 230–240.
18. Aggarwal, S.; Gupta, A.R.; Sharma, S.D.; Chauhan, R.; Sharma, N. Mahgoub transform (Laplace-Carson transform) of error function. *Int. J. Latest Technol. Eng. Manag. Appl. Sci.* **2019**, *8*, 92–98.
19. Aggarwal, S.; Singh, G.P. Aboodh transform of error function. *Univers. Rev.* **2019**, *10*, 137–150.
20. Aggarwal, S.; Gupta, A.R.; Kumar, D. Mohand transform of error function. *Int. J. Res. Advent Technol.* **2019**, *7*, 224–231. [CrossRef]

2

Exploiting the Symmetry of Integral Transforms for Featuring Anuran Calls

Amalia Luque [1,*], Jesús Gómez-Bellido [1], Alejandro Carrasco [2] and Julio Barbancho [3]

[1] Ingeniería del Diseño, Escuela Politécnica Superior, Universidad de Sevilla, 41004 Sevilla, Spain; jesgombel@outlook.es
[2] Tecnología Electrónica, Escuela Ingeniería Informática, Universidad de Sevilla, 41004 Sevilla, Spain; acarrasco@us.es
[3] Tecnología Electrónica, Escuela Politécnica Superior, Universidad de Sevilla, 41004 Sevilla, Spain; jbarbancho@us.es
* Correspondence: amalialuque@us.es

Abstract: The application of machine learning techniques to sound signals requires the previous characterization of said signals. In many cases, their description is made using cepstral coefficients that represent the sound spectra. In this paper, the performance in obtaining cepstral coefficients by two integral transforms, Discrete Fourier Transform (DFT) and Discrete Cosine Transform (DCT), are compared in the context of processing anuran calls. Due to the symmetry of sound spectra, it is shown that DCT clearly outperforms DFT, and decreases the error representing the spectrum by more than 30%. Additionally, it is demonstrated that DCT-based cepstral coefficients are less correlated than their DFT-based counterparts, which leads to a significant advantage for DCT-based cepstral coefficients if these features are later used in classification algorithms. Since the DCT superiority is based on the symmetry of sound spectra and not on any intrinsic advantage of the algorithm, the conclusions of this research can definitely be extrapolated to include any sound signal.

Keywords: spectrum symmetry; DCT; MFCC; audio features; anuran calls

---

1. Introduction

Automatic processing of sound signals is a very active topic in many fields of science and engineering which find applications in multiple areas, such as speech recognition [1], speaker identification [2,3], emotion recognition [4], music classification [5], outlier detection [6], classification of animal species [7–9], detection of biomedical disease [10], and design of medical devices [11]. Sound processing is also applied in urban and industrial contexts, such as environmental noise control [12], mining [13], and transportation [14,15].

These applications typically include, among their first steps, the characterization of the sound: a process which is commonly known as feature extraction [16]. A recent survey of techniques employed in sound feature extraction can be found in [17], of which Spectrum-Temporal Parameters (STPs) [18], Linear Prediction Coding (LPC) coefficients [19], Linear Frequency Cepstral Coefficients (LFCC) [20], Pseudo Wigner-Ville Transform (PWVT) [21], and entropy coefficients [22] are of note.

Nevertheless, the Mel-Frequency Cepstral Coefficients (MFCC) [23] are probably the most widely employed set of features in sound characterization and the majority of the sound processing applications mentioned above are based on their use. Additionally, these features have also been successfully employed in other fields, such as analysis of electrocardiogram (ECG) signals [24], gait analysis [25,26], and disturbance interpretation in power grids [27].

On the other hand, the processing and classification of anuran calls have attracted the attention of the scientific community for biological studies and as indicators of climate change. This taxonomic group is regarded as an outstanding gauge of biodiversity. Nevertheless, frog populations have suffered a significant decrease in the last years due to habitat loss, climate change and invasive species [28]. So, the continual monitoring of frog populations is becoming increasingly important to develop adequate conservation policies [29].

It should be mentioned that the system of sound production in ectotherms is strongly affected by the ambient temperature. Therefore, the temperature can significantly influence the patterns of calling songs by modifying the beginning, duration, and intensity of calling episodes and, thus, the anuran reproductive activity. The presence or absence of certain anuran calls in a certain territory, and their evolution over time, can therefore be used as an indicator of climate change.

In our previous work, several classifiers for anuran calls are proposed that use non-sequential procedures [30] or temporally-aware algorithms [31], or that consider score series [32], mainly using a set of MPEG-7 features [33]. MPEG-7 is an ISO/IEC standard developed by MPEG (Moving Picture Experts Group). In [34], the comparison of MPEG-7 and MFCC are undertaken both in terms of classification performance and computational cost. Finally, the optimal values of MFCC options for the classification of anuran calls are derived in [35].

State of the art classification of sound relies on Convolutional Neural Networks (CNN) that take input from some form of the spectrogram [36] or even the raw waveform [37]. Moreover, CNN deep learning approaches have also been used in the identification of anuran sound [38]. In spite of that, studying and optimizing the process of extracting MFCC features is of great interest at least for three reasons. First, because sound processing goes beyond the classification task, including procedures such as compression, segmentation, semantic description, sound database retrieval, etc. Secondly, because the spectrograms that feed the state-of-the-art deep CNN classifiers can be constructed using MFCC [39]. And finally due to the fact that CNN classifiers based on spectrograms or raw waveforms require intensive computing resources which makes them unsuitable for implementation in low-cost low-power-consumption distributed nodes, as is the usual case in environmental monitoring networks [35].

As presented in greater detail later, the MFCC features are a representation of the sounds in the cepstral domain. They are derived after a first integral transform (from time to frequency domain), which obtains the sound spectrum, and then a second integral transform is carried out (from frequency to cepstral domain). In this paper, we will show that, by exploiting the symmetry of the sound spectra, it is possible to obtain a more accurate representation of the anuran calls and the derived features will therefore more precisely reflect the sound.

The main contribution of the paper is to offer a better understanding of the reason (symmetry) that justify and quantify why Discrete Cosine Transform (DCT) has been extensively used to compute MFCC. In more detail, the paper will show that DCT-based sound features yielded to a significantly lower error representing spectra, which is a very convenient result for several applications such as sound compression. Additionally, through the paper it will be demonstrated that symmetry-based features (DCT) are less correlated, which is an advantage to be exploited in later classification algorithms.

2. Materials and Methods

2.1. Extracting MFCC

The process of extracting the MFCC features from the n samples of a certain sound requires 7 steps in 3 different domains, which are depicted in Figure 1, and can be summarized as follows:

1. Pre-emphasis (time domain): The sound's high frequencies are increased to compensate for the fact that the Signal-to-Noise Ratio (SNR) is usually lower at these frequencies.

2. Framing (time domain): The n samples of the full-length sound segment are split into frames of short duration (N samples, $N \ll n$). These frames are commonly obtained using non-rectangular overlapping windows (for instance, Hamming windows [40]). The subsequent steps are executed on the N samples of each frame.
3. Log-energy spectral density (spectral domain): Using the Discrete Fourier Transform (DFT) or its faster version, the Fast Fourier Transform (FFT), the N samples of each frame are converted into the N samples of an energy spectral density, which are usually represented in a log-scale.
4. Mel bank filtering (spectral domain): The N samples of each frame's spectrum are grouped into M banks of frequencies, using M triangular filters centred according to the mel scale [41] and the mel Filter Bank Energy (mel-FBE) is obtained.
5. Integral transform (cepstral domain): The M samples of the mel-FBE (in the spectral domain) are converted into M samples in the cepstral domain using an integral transform. In this article, it will be shown that the exploitation of the symmetry of the DFT integral transform obtained in step 3 yields a cepstral integral transform with a better performance.
6. Reduction of cepstral coefficients (cepstral domain): The M samples of the cepstrum are reduced to C coefficients by discarding the least significant coefficients.
7. Liftering (cepstral domain): The C coefficients of the cepstrum are finally liftered to compensate for the fact that high quefrency coefficients are usually much smaller than their low quefrency counterparts.

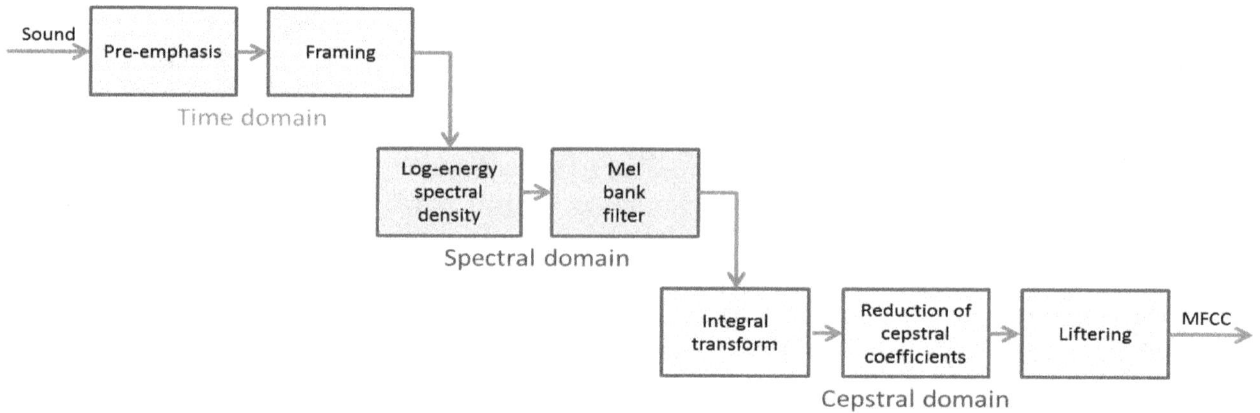

Figure 1. The process of extracting the Mel-Frequency Cepstral Coefficients (MFCC) features from a certain sound.

In this process, integral transforms are used twice: in step 3 to move from the time domain into the spectral domain; and in step 5 to move forward into the cepstral domain. In this paper, the symmetric properties of the DFT integral transform in step 3 will be exploited for the selection of the most appropriate integral transform required in step 5.

2.2. *Integral Transforms of Non-Symmetric Functions*

As detailed in the previous subsection, a sound spectrum is featured in order to obtain the MFCC of a sound, specifically by characterizing the logarithm of its energy spectral density. In short, this would be a particular case of the characterization of a function $f(x)$ by means of a reduced set of values where, in this case, $f(x)$ is the spectrum of a sound. To address this problem, which is none other than that of the compression of information, several techniques have been proposed, from among which the frequency representation of the function stands out. In effect, the idea underlying this type of technique is to consider the original signal, expand it in Fourier series, and then approximate the function by means of a few terms of its expansion. Thus, instead of having to supply the values of the function corresponding to each value of x, only the amplitude values (and eventually also the phase) of a reduced number of harmonics are provided.

Let us consider an arbitrary example function $f(x)$, such as that shown in Figure 2, of which we know only one fragment in the interval $[x_0, x_0 + P]$ (dashed line). Now let us consider that this function is sampled, and the values only at specific points for $x = x_n$, separated at intervals Δx, are known. By denoting N as the total number of points (samples) in a period, we know that $\Delta x = P/N$. The sampled function will be called $\hat{f}(x_n) = f_n$ where the hat (^) above f represents a sampled function.

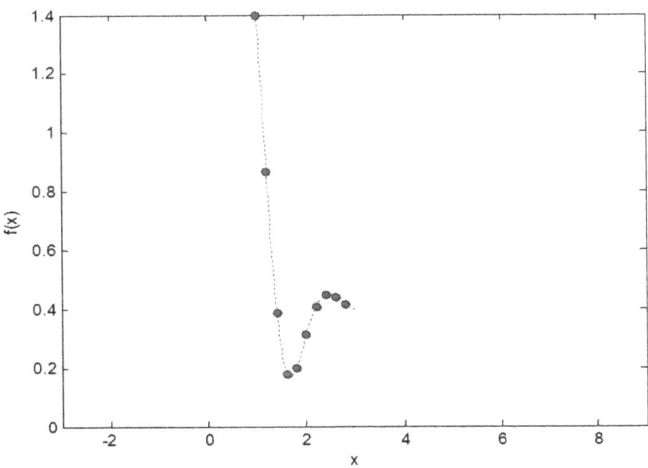

Figure 2. Known fragment of an example function $f(x)$ (dashed line) and its corresponding sampled function $\hat{f}(x_n)$ (dots).

The usual way to obtain the spectrum of that function is to define a periodic function $f_p(x)$ of period P that coincides with the previous function in the known interval (see Figure 3), and to proceed to compute the spectrum of that new function. The spectral representation of the function $f_p(x)$ is composed of the complex coefficients of the Fourier series expansion given by [42].

$$c_k = \frac{1}{P} \int_{x_0}^{x_0+P} f(x) e^{-j\frac{2\pi k x}{P}} dx. \tag{1}$$

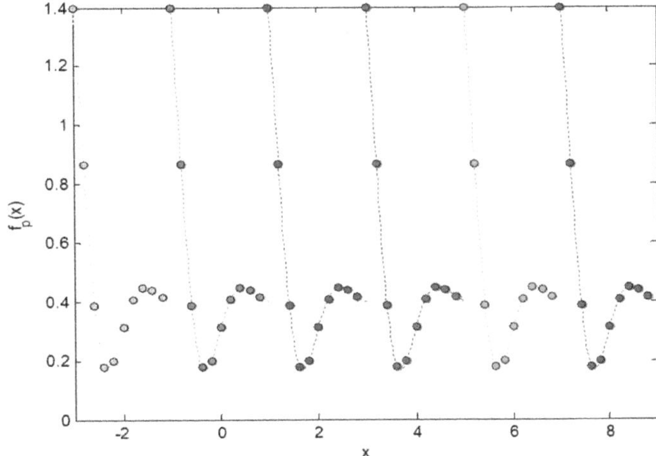

Figure 3. Periodic function $f_p(x)$ obtained by repetition of the known fragment of $f(x)$.

On the other hand, the sampled function, $\hat{f}(x_n) = f_n$, will have a spectral representation $\hat{c}_k$ that corresponds to c_k, when the sampling of the variable x is taken into account. Now let us call $I(x)$ the integrand of Equation (1), i.e.,

$$I(x) = f(x)e^{-j\frac{2\pi kx}{P}}, \tag{2}$$

and hence the spectral representation of the non-sampled function $f_P(x)$ is featured by the coefficients

$$c_k = \frac{1}{P}\int_{x_0}^{x_0+P} I(x)dx. \tag{3}$$

in order to obtain the values $\hat{c}_k$ that take into account the sampling of the variable x, the continuous calculation of the area that supposes the integral of the previous expression is substituted with the sum of the rectangles corresponding to the discrete values (sum of Riemann). In Figure 4, the calculation of the real part of $\hat{c}_1$ is depicted for the example function $f_P(x)$.

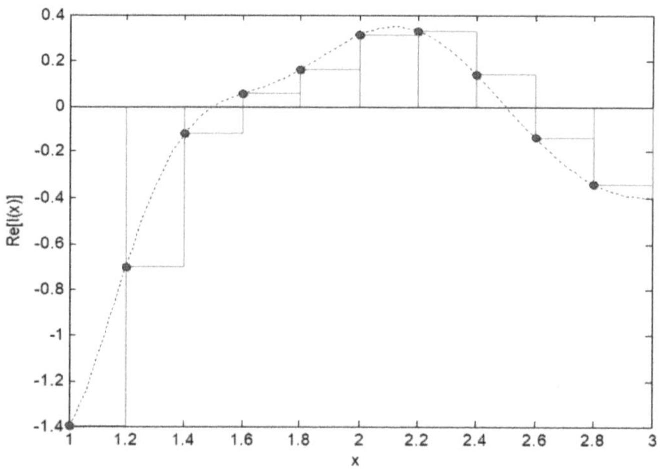

Figure 4. Integration of sampled functions (sum of Riemann).

Therefore,

$$\hat{c}_k \equiv [c_k]_{x=x_n} = \left[\frac{1}{P}\int_{x_0}^{x_0+P} I(x)dx\right]_{x=x_n}. \tag{4}$$

From this equation it can be derived (see supplementary material) that

$$\hat{c}_k = \frac{1}{N}e^{-j\frac{2\pi kx_0}{N\Delta x}}\sum_{n=0}^{N-1} f_n\, e^{-j\frac{2\pi kn}{N}}. \tag{5}$$

It can be observed that the spectral representation $\hat{c}_k$ depends on the point x_0 selected as the origin of coordinates, due to the factor $e^{-j\frac{2\pi kx_0}{N\Delta x}}$. This factor does not affect the amplitude spectrum (since its modulus is 1), but it does affect the phase spectrum corresponding to the known time-shift property of the Fourier Transform. For practical purposes, the origin of coordinates is usually considered to be the starting point of the sequence, that is, at $x_0 = 0$, and hence the spectral representation finally becomes

$$\hat{c}_k = \frac{1}{N}\sum_{n=0}^{N-1} f_n\, e^{-j\frac{2\pi kn}{N}}. \tag{6}$$

This expression coincides with the usual definition of the Discrete Fourier Transform (DFT) [43]. In other words: The Discrete Fourier Transform of a known fragment of a function presupposes the periodic repetition of that fragment.

2.3. Integral Transforms of Symmetric Functions

Let us now again consider the function $f(x)$ of which we know only sampled values of a fragment f_n in the interval $[x_0, x_0 + P]$, as shown in Figure 2. An alternative way of representing its spectrum to that of periodically repeating the values f_n as in Figure 3, lies in defining a sequence of values g_n of length $2P$ that coincides with f_n in the interval $[x_0, x_0 + P]$, which is its symmetric in the interval $[x_0 - P, x_0]$, as depicted in Figure 5.

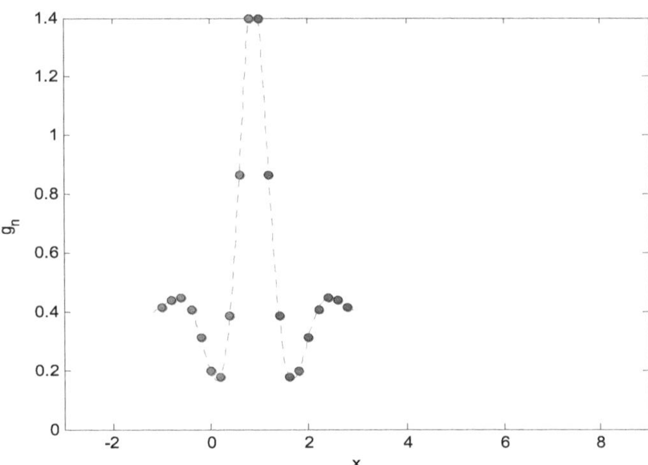

Figure 5. Known fragment of a symmetric example function $g(x)$ (dashed line) and its corresponding sampled function $\hat{g}(x_n)$ (dots). These functions are obtained by considering the original fragment of the example function $f(x)$ (blue) and its symmetric (green).

It can be observed that

$$\begin{array}{c} g_n = f_n \; \forall n \in [0, N-1] \\ g_n = f_{-n-1} \; \forall n \in [-N, -1] \end{array} . \quad (7)$$

Subsequently, a sequence of periodic values h_n of period $P' = 2P$ is defined that coincides with g_n in the interval $[x_0 - P, x_0 + P]$, as shown in Figure 6.

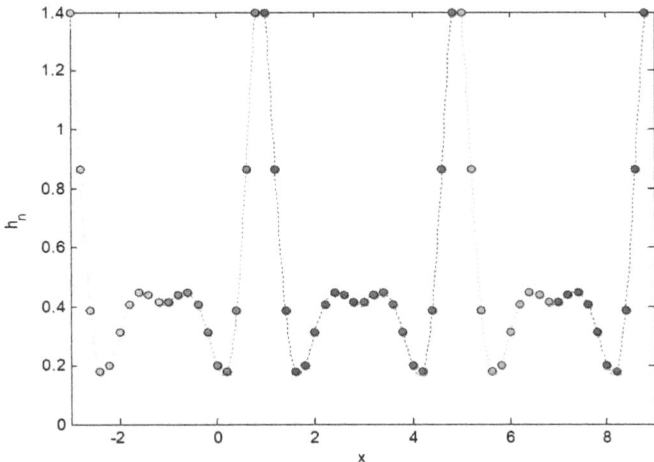

Figure 6. Periodic function h_n obtained by repetition of the known fragment of g_n.

In order to obtain the spectrum of the sequence of values h_n it can be written that

$$\hat{c}_k = \frac{1}{P\prime} \sum_{x_n = x_0 - P}^{x_n = x_0 + P - \Delta x} h_n \, e^{-j\frac{2\pi k x_n}{P\prime}} \Delta x. \quad (8)$$

From this equation it can be derived (see supplementary material) that

$$\hat{c}_k = \frac{1}{2N} e^{-j\frac{\pi k x_0}{N\Delta x}} \left[e^{j\frac{\pi k}{N}} \sum_{n=0}^{N-1} f_n \, e^{j\frac{\pi k n}{N}} + \sum_{n=0}^{N-1} f_n \, e^{-j\frac{\pi k n}{N}} \right]. \tag{9}$$

As can be observed, due to the factor $e^{-j\frac{2\pi k x_0}{N\Delta x}}$, the spectral representation $\hat{c}_k$ depends on the point x_0 where the origin of coordinates is defined. This factor does not affect the amplitude spectrum (since its modulus is 1), but it does affect the phase spectrum, which corresponds to the known time-shifting property of the Fourier transform. For practical purposes, the origin of coordinates is usually considered to be located the midpoint of the symmetric sequence g_n, that is, $x_0 = \Delta x/2$, as shown in Figure 7.

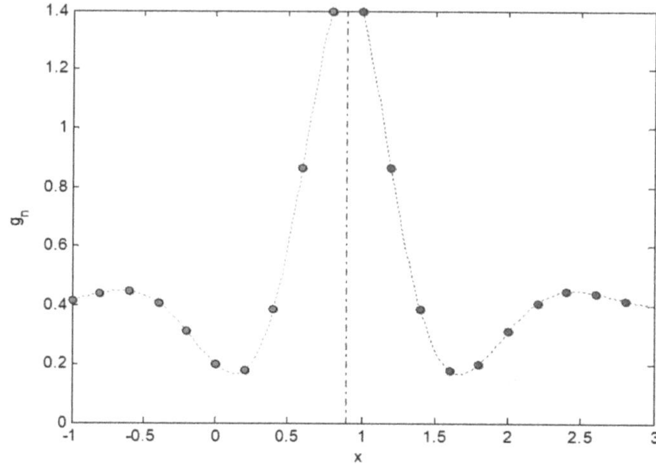

Figure 7. Defining the origin of coordinates.

Finally, the spectral representation becomes (see supplementary material)

$$\hat{c}_k = \frac{1}{N} \sum_{n=0}^{N-1} f_n \cos\left[\frac{\pi k}{N} \left(n + \frac{1}{2} \right) \right]. \tag{10}$$

This expression coincides with the usual definition of the Discrete Cosine Transform (DCT) [44]. In other words, the Discrete Cosine Transform of a known fragment of a function presupposes the periodic repetition of that fragment and its symmetric.

2.4. Representing Anuran Call Spectra

With this digression, we can now address the question posed at the beginning of Section 2.2 concerning the best way to characterize the spectrum of a sound by using the sum of its harmonics. Note that it is necessary to compute the spectrum (step 5) of a spectrum (step 4), that is, the trans-spectrum or the cepstrum, as previously discussed. The decision regarding whether this trans-spectrum (cepstrum) should be derived using either the Fourier transform, or the cosine transform, is based on the form of the fragment f_n (in this case the spectral values of the sound). That is, it should be considered whether the best approximation to the spectrum is either a periodic repetition of f_n or, in contrast, a periodic repetition of f_n and its symmetric.

Although this is a general question, we have addressed it in the context of a specific application by featuring anuran calls for their further classification. The dataset employed contains 1 hour and 13 minutes of sounds which have been recorded at five different locations (four in Spain, and one in Portugal) [32] and they were subsequently sampled at 44.1 kHz. The recordings include 4 types of anuran calls and, since they have been taken in their natural habitat, are affected by highly significant surrounding environmental noise (such as that of wind, water, rain, traffic, and voices).

In this paper, the duration of the frames (step 2) was set to 10 ms, such that each frame has $N = 441$ data points and a total of $W = 434,313$ frames are considered. The log-energy spectral density (step 3) is obtained using a standard FFT algorithm, which obtains a spectrum with $N = 441$ values. The mel-scaling (step 4) employs a set of $M = 23$ filters, and hence the mel-FBE spectrum is characterised by this number of values ($M = 23$). In step 5, two different approaches for obtaining the cepstrum are used and compared: DFT and DCT. The results are then analysed for a different number of cepstral coefficients ($1 \le C \le M$).

In order to carry out a more systematic study of the spectrum approximation error, let us call $E_i(n)$ the original mel-FBE spectrum of the i-th frame (the result of step 4), where n is the filter index (equivalent to the frequency in mel scale). Let us also call $H_i(m)$ the spectrum of $E_i(n)$, that is, the cepstrum as obtained in step 5, where m is the cepstral index (equivalent to the quefrency in mel scale). It can be written that $H_i(m) = \mathcal{F}[E_i(n)]$, where $\mathcal{F}$ represents either the DFT or the DCT Fourier expansions.

After reducing the number of cepstral coefficients to a value of $C \le M$, the resulting approximate cepstrum (step 6) will be called $\widetilde{H}_i(m)$, where the tilde ($\widetilde{\ }$) above the H represents an approximation. Using these C values in the corresponding Fourier expansion leads to an approximation of the mel-FBE, that is, $\widetilde{E}_i(n) = \mathcal{F}^{-1}\left[\widetilde{H}_i(m)\right]$. The approximation error for the i-th frame is therefore $\varepsilon_i(n) = E_i(n) - \widetilde{E}_i(n)$, that is, a different error for each value of n, the filter index (or frequency in mel-scale). An error measure for the overall spectrum of the i-th frame can be obtained using the Root Mean Square Error ($RMSE_i$) defined as:

$$RMSE_i \equiv \sqrt{\frac{1}{M}\sum_{n=0}^{M-1}[\varepsilon_i(n)]^2} = \sqrt{\frac{1}{M}\sum_{n=0}^{M-1}\left[E_i(n) - \widetilde{E}_i(n)\right]^2}. \tag{11}$$

In this paper, an arbitrary selected single frame is first considered, mainly for illustration purposes. Its time-domain representation is depicted in Figure 8A while its spectrum is plotted in Figure 8B. Some other examples can be found in [32].

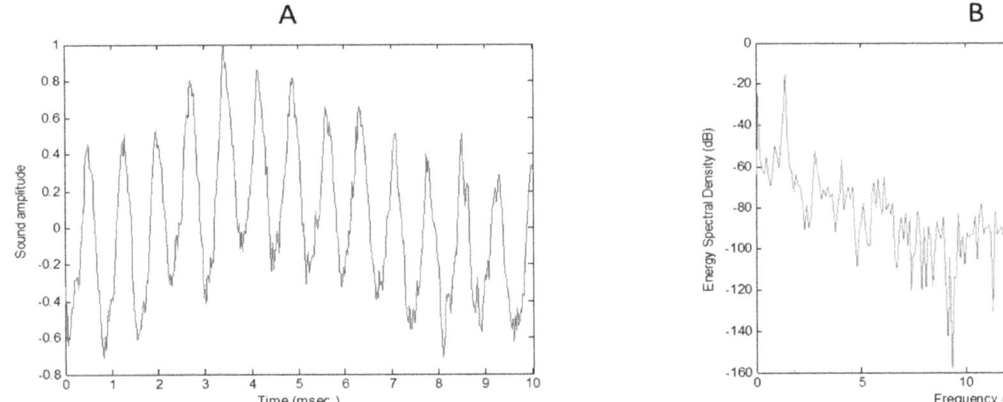

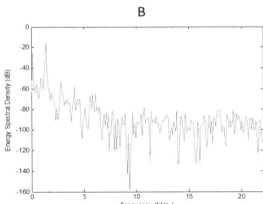

Figure 8. Sound amplitude for an arbitrarily selected frame of an anuran call (**A**); and its log-scale Energy Spectral Density (**B**).

Additionally, in order to compare the performance of the 2 competing algorithms obtaining the cepstrum, an overall metric for the whole dataset is considered and defined as the mean RMSE for every frame, that is,

$$RMSE \equiv \frac{1}{W}\sum_{i=1}^{W} RMSE_i = \frac{1}{W}\sum_{i=1}^{W}\sqrt{\frac{1}{M}\sum_{n=0}^{M-1}[\varepsilon_i(n)]^2}. \tag{12}$$

3. Results

Let us first consider a single frame, arbitrarily selected from the whole sound dataset. Although these results are limited to that specific sound frame, very similar results are obtained if a different frame is selected. Moreover, at the end of this section, the overall sound dataset is considered.

For the case of the single frame, the mel-FBE spectrum obtained in step 4 is depicted in Figure 9. This is the $f(x)$ function whose spectrum (cepstrum in this case) must be computed in step 5.

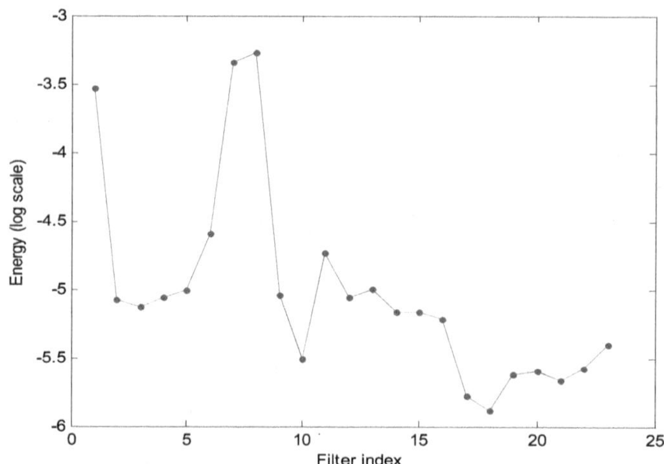

Figure 9. Mel Filter Bank Energy (mel-FBE) spectrum for an arbitrarily selected frame of an anuran call.

For this frame, let us consider whether it is better to use either a DFT or a DCT. The decision depends on whether the function $f(x)$ can be considered as a fragment of a periodic repetition of: (A) the fragment, as shown in Figure 10A, or (B) the function and its symmetric, as shown in Figure 10B. In the first case, the DFT should be more appropriate, while in the second case the DCT would obtain better results.

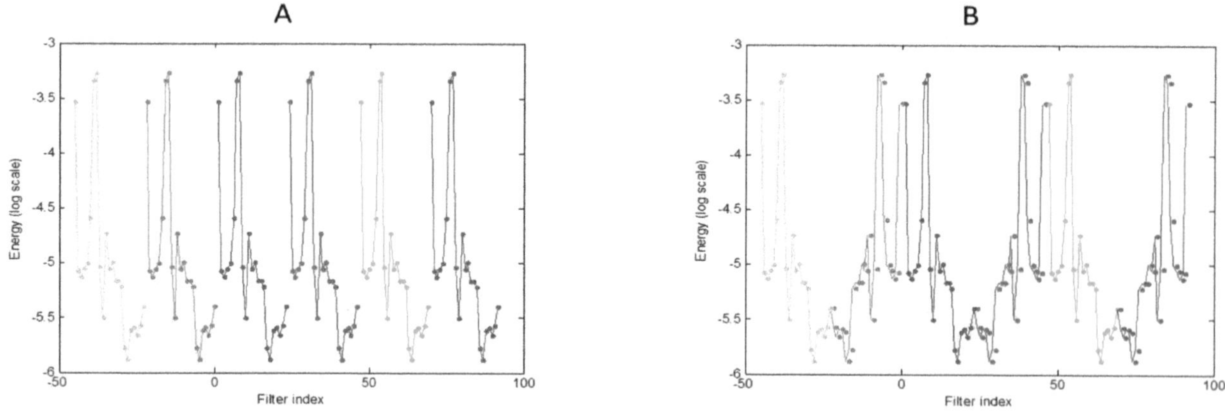

Figure 10. Periodic repetition of the mel-FBE spectrum (**A**); and the mel-FBE spectrum and its symmetric (**B**).

However, the mel-FBE is nothing but a rescaled and compressed way of presenting a spectrum. On the other hand, it is a well-known fact that the spectrum of a real signal is symmetric with respect to the vertical axis [43]. And finally, it is also known that the spectrum of a sampled signal is periodic [45]. For this reason, the repetition of the fragment of Figure 9 corresponds to Figure 10B and, therefore, using the DCT to compute its trans-spectrum (or cepstrum) should obtain better results. This hypothesis is verified in the following paragraphs for the selected frame, and, later in this section, it is verified for the whole dataset.

The number of coefficients obtained by applying either DCT or DFT is $M = 23$, that is, they have the same number of values that define the mel-FBE. The resulting cepstrum for the selected frame is shown in Figure 11.

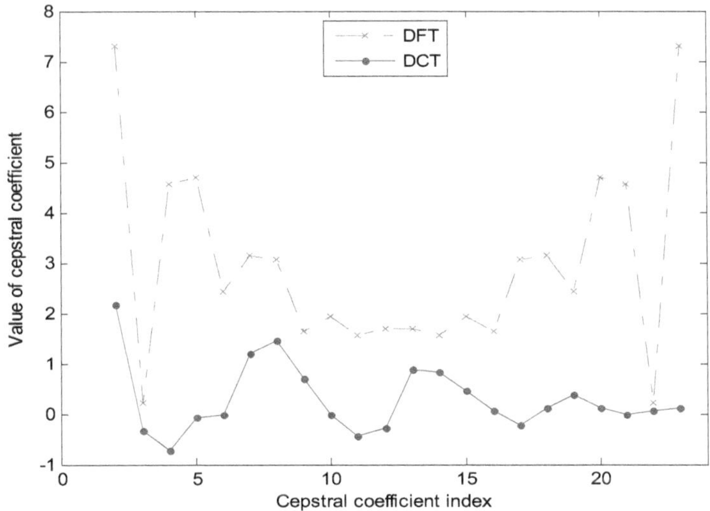

Figure 11. Cepstral representation of the mel-FBE spectrum (cepstrum).

The ability to compress information of the Fourier transforms (either in the DFT or DCT version) lies in the fact that it is not necessary to consider the full set of the M coefficients of the Fourier expansion to obtain a good approximation of the original function. In Figure 12, the original mel-FBE spectrum is depicted for the example frame, and those spectra recovered using $C \leq M$ cepstral coefficients obtained using DCT.

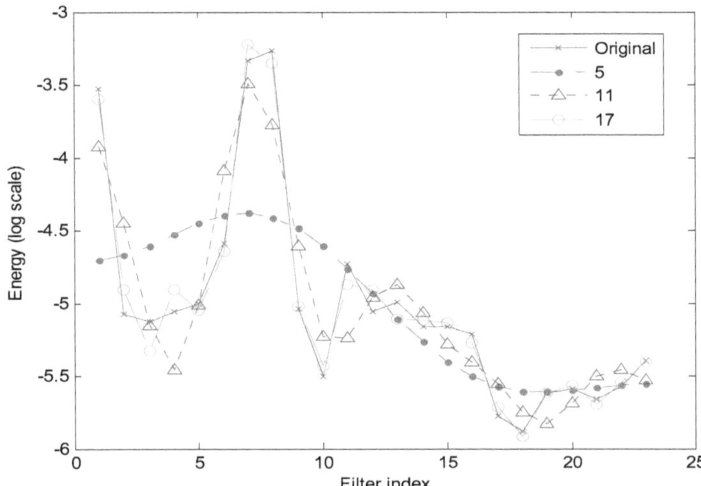

Figure 12. Mel-FBE spectrum for an arbitrarily selected frame of an anuran call. Original spectrum and recovered spectra using a different number of Discrete Cosine Transform (DCT) cepstral coefficients.

Additionally, as expected, the DCT achieves approximations to the original spectrum that are, in general, significantly better than those obtained for the DFT with the same number of coefficients. In Figure 13, the original mel-FBE spectrum is depicted for the example frame, and those spectra recovered using $C = 11$ cepstral coefficients obtained using DFT and DCT.

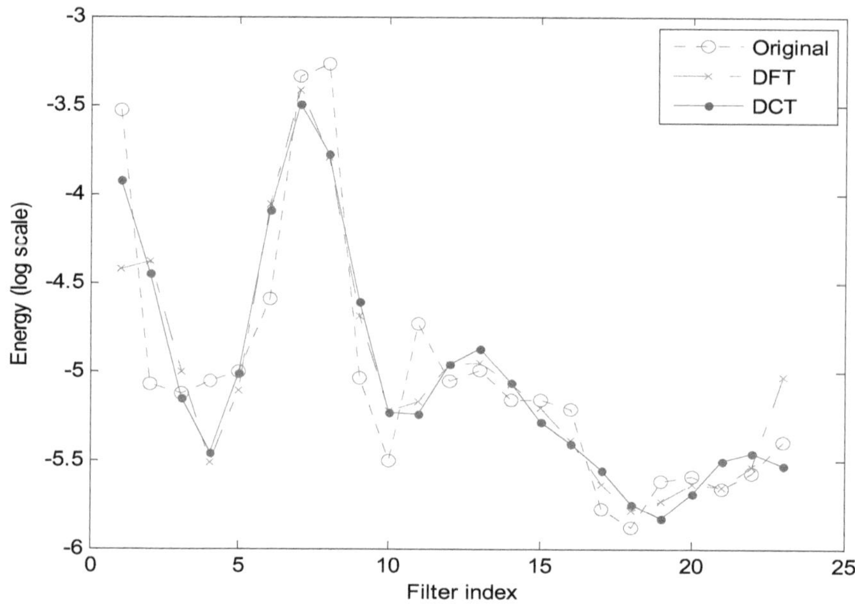

Figure 13. Mel-FBE spectrum for an arbitrarily selected frame of an anuran call. Original spectrum and recovered spectrum using $C = 11$ coefficients obtained using Discrete Fourier Transform (DFT) and DCT.

In order to quantify the error of recovering the selected mel-FBE spectrum using $C \leq M$ cepstral coefficients, the Root Mean Square Error (RMSE) is computed in accordance with Equation (11). The value of RMSE as a function of the number C of cepstral coefficients used for the recovery of the spectrum is depicted in Figure 14, both for DFT and DCT.

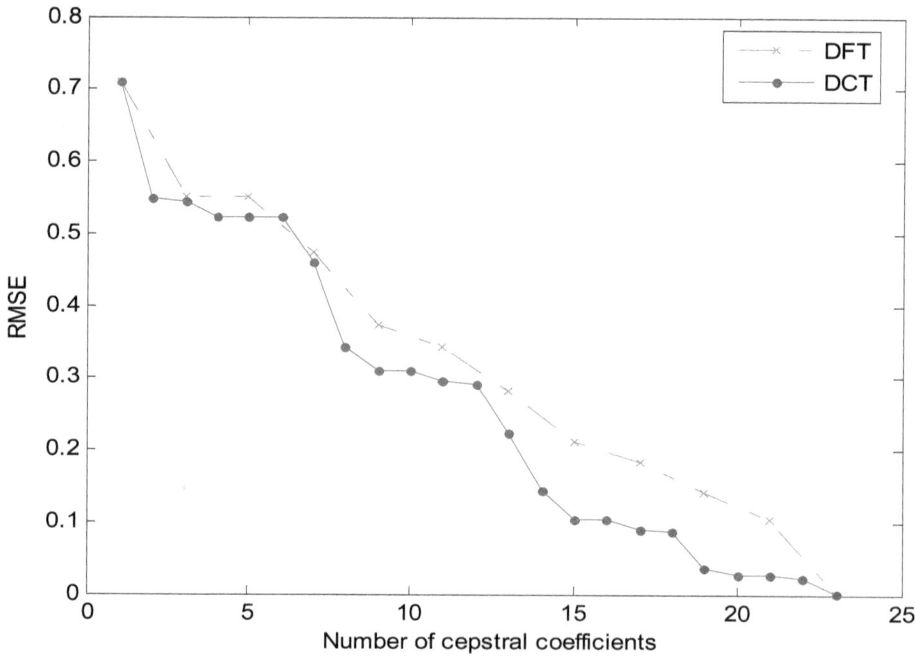

Figure 14. Root Mean Square Error recovering the original mel-FBE spectrum when a different number of C cepstral coefficients are used. The cepstral coefficients are obtained applying either DFT or DCT.

This analysis can be extended to include the computation of the RMSE for the whole dataset in accordance with Equation (12). The value of RMSE as a function of the number C of cepstral coefficients used for the recovery of the spectrum is depicted in Figure 15 for DFT and DCT separately.

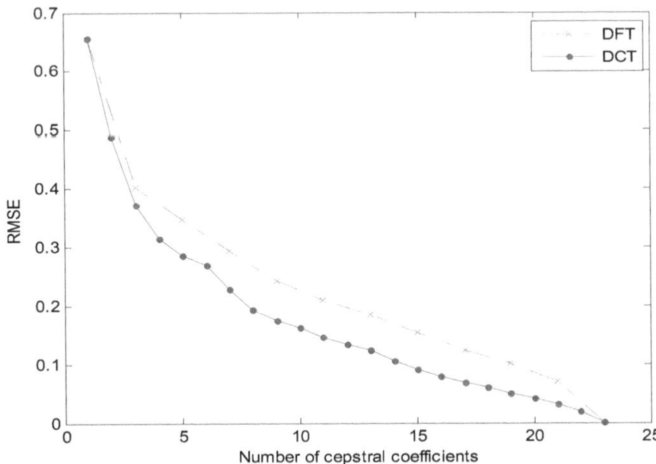

Figure 15. Root Mean Square Error for the whole dataset when either DFT or DCT is employed.

4. Discussion

Let us first consider the $RMSE_i$ for a single frame as depicted in Figure 14. Let us now regard the case where, for instance, the number of values required to describe the mel-FBE spectrum ($M = 23$) is halved, and hence the number of cepstral coefficients used for the recovering an approximation of the spectrum is $C = 11$ (in accordance with Equations (6) and (10)).

In this case, it can be observed that $RMSE_i$ is 0.34 for DFT, and 0.30 for DCT. On the other hand, as depicted in Figure 9, the values of the mel-FBE spectrum lie within the range $[-6, -3]$, with a mean value of -5.02. This means that the relative error of the spectrum representation is only 6.84% for DFT (5.36% for DCT) when the number of values employed for that representation are halved.

Let us now focus on the RMSE when the DFT is used (green line), either for a single frame (Figure 14) or for the whole dataset (Figure 15). In both cases, it can be observed that RMSE has values only for an odd number of cepstral coefficients. This fact can be explained by recalling that, according to Equation (6), every DFT cepstral coefficient $\hat{c}_k$ is a complex number for $1 \leq k \leq M - 1$ and a real number for $k = 0$. On the other hand, according to Equation (10), the DCT cepstral coefficients $\hat{c}_k$ are real numbers for every value of k. Additionally, it has to be considered that DFT cepstrum is symmetric (green line in Figure 11). Therefore, for $k > 0$, it can be written that $\hat{c}_k = \hat{c}_{M-k+1}$ and, therefore, only one of these 2 terms have to be kept for recovery purposes. These circumstances jointly explain the odd number of DFT cepstral coefficients.

To clarify this idea, let us consider an example where $M = 23$ and $C = 5$. The DCT cepstrum is then described using $\hat{c}_0$, $\hat{c}_1$, $\hat{c}_2$, $\hat{c}_3$ and $\hat{c}_4$, that is, 5 real numbers which can be employed to approximately recover the mel-FBE spectrum. On the other hand, the DFT cepstrum is described using $\hat{c}_0$, which is a real number, and $\hat{c}_1$ and $\hat{c}_2$, which are complex numbers, that is, although 3 terms are used, a total of 5 values (coefficients) are required. However, to approximately recover the mel-FBE spectrum, the terms $\hat{c}_0$, $\hat{c}_1$, $\hat{c}_2$, $\hat{c}_{23}$ and $\hat{c}_{22}$ can be used since $\hat{c}_1 = \hat{c}_{23}$ and $\hat{c}_2 = \hat{c}_{22}$.

As regards the results obtained for the whole dataset (Figure 15), it can be seen that DCT is better at describing the mel-FBE spectra than is its DFT counterpart. This improvement (decrease of the RMSE), can be measured by defining $\Delta RMSE \equiv RMSE_{DFT} - RMSE_{DCT}$ (Figure 16A) or its relative value $\Delta RMSE(\%) \equiv 100 \cdot \Delta RMSE / RMSE_{DFT}$ (Figure 16B). For example, for $C = 11$, the RMSE is reduced from 0.209 (DFT) to 0.146, which involves an improvement of approximately 30%. For the degenerated cases where $C = 1$ and $C = M$, there is no improvement. In the first case, only $\hat{c}_0$ is used which, according to Equations (6) and (10), is the mean value of the mel-FBE spectrum, that is, the DFT and DCT recovering methods have the same error. On the other hand, if $C = M$ then no reduction on the number of coefficients is achieved, and both equations exactly recover the original spectrum (no error).

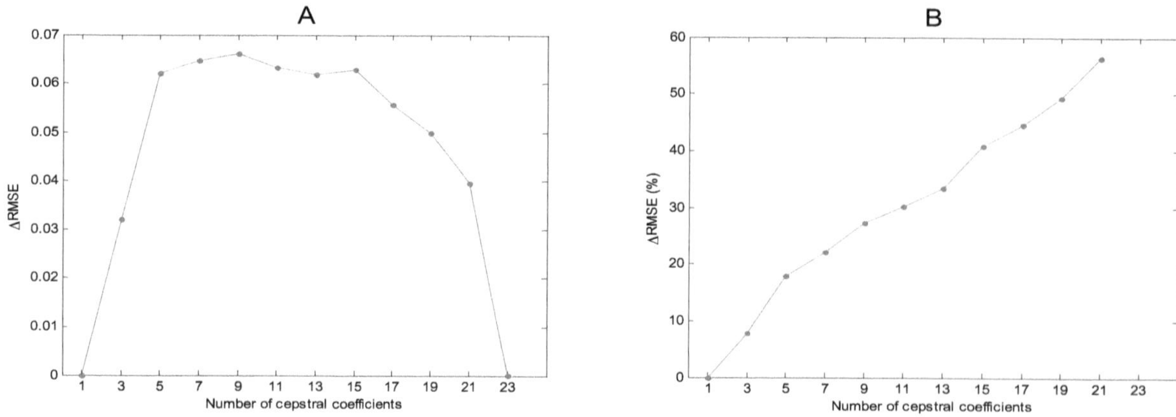

Figure 16. Improvement of DCT over DFT describing mel-FBE spectra. (**A**): $\Delta RMSE$. (**B**): $\Delta RMSE(\%)$.

The above results concern the mean improvement of DCT over DFT for every frame in the dataset. In a more in-depth analysis, let us also compute its probability density function (pdf). The results are depicted in Figure 17. In panel A, the pdf is shown for several values of the number of cepstral coefficients (C). In panel B, the value of the pdf is colour-coded as a function of the improvement ($\Delta Error$) and of the number of cepstral coefficients (C). It can be observed that only a negligible number of the frames present a significant negative improvement, thereby demonstrating that DCT is superior to DFT.

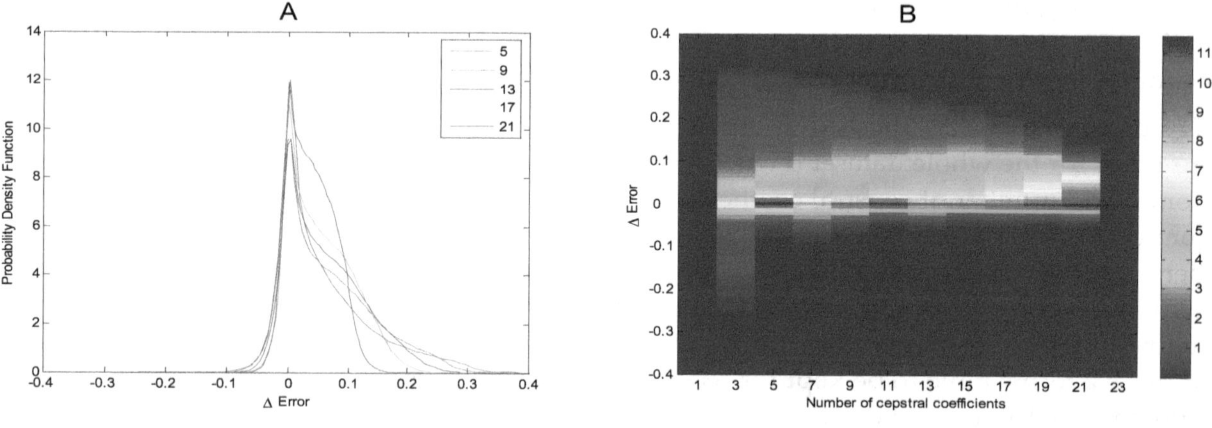

Figure 17. Improvement of DCT over DFT in describing mel-FBE spectra. (**A**): Probability density function for several values of the number of cepstral coefficients. (**B**): Probability density function for each value of the number of cepstral coefficients.

The higher performance of DCT over DFT is due to the fact that the mel-FBE spectra are a special type of function derived from symmetric sound spectra. Consequently, if DCT and DFT were compared in the task of recovering arbitrary functions, they would each present equal performance. To demonstrate this claim, one million M-value arbitrary functions are randomly generated ($M = 23$), and DFT and DCT are then employed to recover the original function with a reduced set of C coefficients to measure the errors of that recovery. Finally, the improvement of DCT over DFT is computed. The results are depicted in Figure 18 where it can be observed that positive and negative improvements are symmetrically distributed around a zero-mean improvement. Therefore, it can be concluded that DCT and DFT have similar performance in describing arbitrary functions.

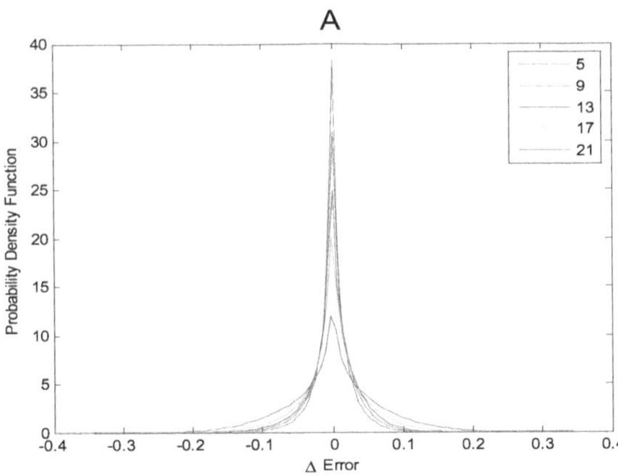

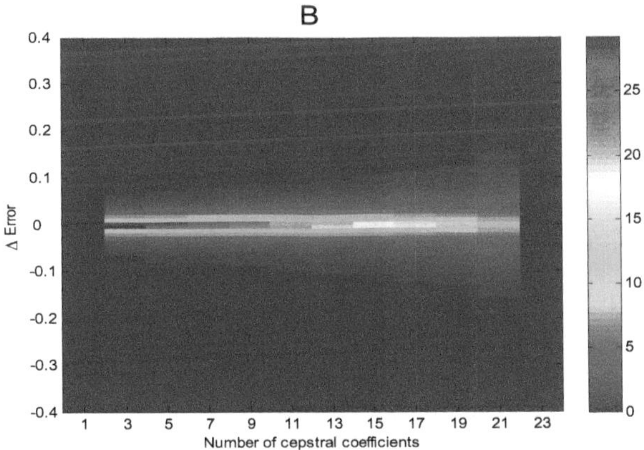

Figure 18. Improvement of DCT over DFT in describing arbitrary function. (**A**): Probability density function for several values of the number of cepstral coefficients. (**B**): Probability density function for each value of the number of cepstral coefficients.

From the above results, it is clear that DCT offers superior performance featuring mel-FBE spectra and, therefore offers superior performance featuring sounds. When the purpose of these features is to be used as input to some kind of classifier, then DCT offers an additional advantage. It is a well-established result that classifiers obtain better results if their input features are low-correlated. The reason is clear: a classification algorithm that includes a new feature that is highly correlated with previous features adds almost no new information and, therefore, almost no classification improvement should be expected. Let us therefore examine the correlation between coefficients obtained by DFT and those by DCT.

Let us call μ_u the mean value of the u-th coefficient $\hat{c}_{ui}$ describing the i-th frame, obtained by

$$\mu_u = \frac{1}{W}\sum_{i=1}^{W}\hat{c}_{ui}, \tag{13}$$

where W is the total number of frames in the dataset. The variance σ_u^2 of the u-th coefficient can be obtained by

$$\sigma_u^2 = \frac{1}{W-1}\sum_{i=1}^{W}(\hat{c}_{ui} - \mu_u)^2. \tag{14}$$

The correlation ρ_{uv} between the u-th and the v-th coefficient for the whole dataset is therefore given by

$$\rho_{uv} = \frac{1}{W-1}\sum_{i=1}^{W}\frac{\hat{c}_{ui} - \mu_u}{\sigma_u}\cdot\frac{\hat{c}_{vi} - \mu_v}{\sigma_v}. \tag{15}$$

In Figure 19, the absolute values of the correlation are shown, whereby the values for the case $M = 23$ are colour-coded. The correlations corresponding to the DFT are shown in panel A and those corresponding to DCT in panel B. In the DFT case, each $\hat{c}_{ui}$ factor is a complex number, and hence the total number of values is 46, whereby the first 23 coefficients represent the real parts and the last 23 the imaginary parts. By simply considering the colours in that figure, it is clear that DCT coefficients are less correlated.

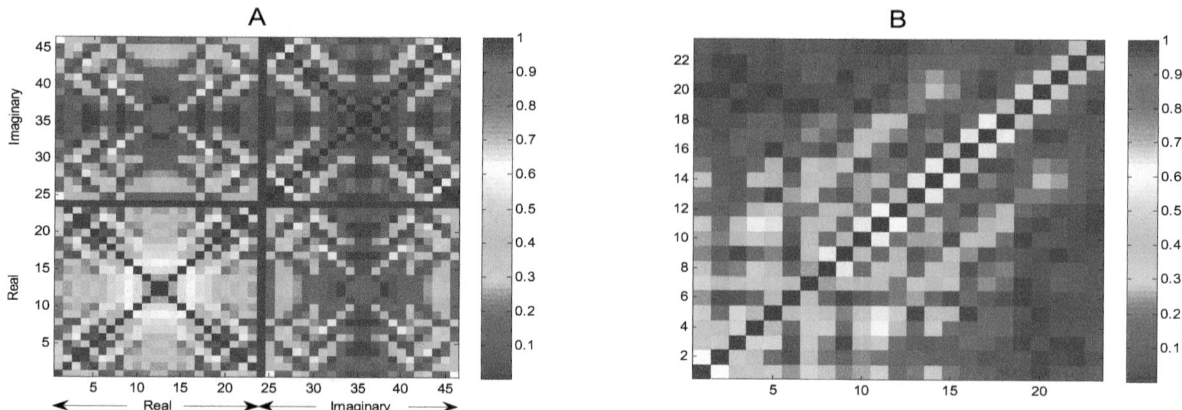

Figure 19. Correlation between cepstral coefficients describing mel-FBE spectra for DFT (panel **A**) and DCT (panel **B**).

An alternative way to present this result is by using a histogram of the values of the correlation coefficients, as depicted in Figure 20. Those corresponding to DCT are more frequent for the low values of correlation, that is, DCT-obtained features are less correlated than those obtained using DFT. Hence, classifiers of a more efficient nature should be expected from using DCT.

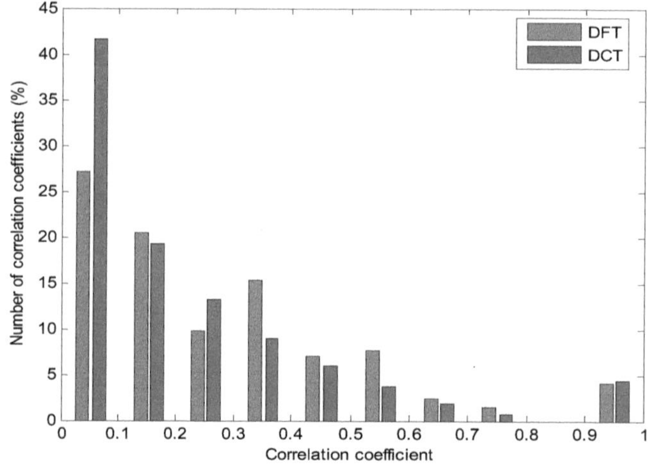

Figure 20. Histogram of the correlation among cepstral coefficients describing mel-FBE spectra for DFT and DCT.

When the MFCC features are used as input of a later classification algorithm, the lower correlation of DCT-obtained features should yield to a better classification performance. The results obtained classifying anuran calls [35] do confirm a slight advantage for the DCT as it is reflected in Table 1. This table has been produced taking the best result (geometric mean of sensitivity and specificity) obtained through a set of ten classification procedures: minimum distance, maximum likelihood, decision trees, k-nearest neighbors, support vector machine, logistic regression, neural networks, discriminant function, Bayesian classifiers and hidden Markov models.

Table 1. Classification performance metrics for DCT and DFT.

Cepstral Transform	ACC	PRC	F_1
DFT	94.27%	74.46%	77.67%
DCT	94.85%	76.76%	78.93%

Let us finally consider the computing efforts required for these two algorithms which mainly depend on the number of samples defining the mel-FBE spectra. Fast versions of DFT and DCT algorithms have been tested on a conventional desktop personal computer. The results are depicted in Figure 21. It can be seen that DCT is about one order of magnitude slower than DFT. Although this fact is certainly a drawback of DCT it has a limited impact on conventional MFCC extraction process because the number of values describing the mel-FBE spectra is usually very low (about 20). Additional studies on processing times for anuran sounds classification can be found in [34].

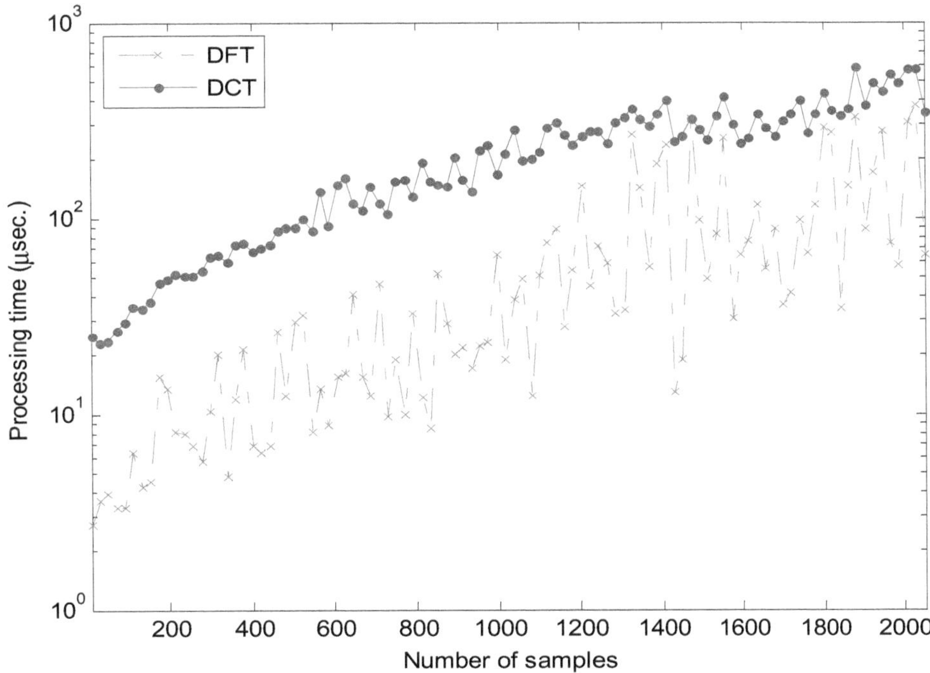

Figure 21. Processing time required to compute the DFT and DCT vs. the number of samples describing mel-FBE spectra.

5. Conclusions

In this article, it has been shown that DCT outperforms DFT in the task of representing sound spectra. It has also been shown that this improvement is due to the symmetry of the spectrum and not to any intrinsic advantage of DCT.

In representing the mel-FBE spectra required to obtain the MFCC features of anuran calls, DCT errors are approximately 30% lower than DFT errors. This type of spectra is therefore much better represented using DCT.

Additionally, it has been shown than MFCC features obtained using DCT are remarkably less correlated than those obtained using DFT. This result will make DCT-based MFCC features more powerful in later classification algorithms.

Although only one specific dataset has been analysed herein, the advantage of DCT can easily be extrapolated to include any sound since this advantage is based on the symmetry of the spectrum of the sound

Author Contributions: Conceptualization, A.L.; investigation, A.L., J.G.-B., A.C. and J.B.; writing—original draft, A.L., J.G.-B., A.C. and J.B.

Acknowledgments: The authors would like to thank Rafael Ignacio Marquez Martinez de Orense (Museo Nacional de Ciencias Naturales) and Juan Francisco Beltrán Gala (Faculty of Biology, University of Seville) for their collaboration and support.

References

1. Haridas, A.V.; Marimuthu, R.; Sivakumar, V.G. A critical review and analysis on techniques of speech recognition: The road ahead. *Int. J. Knowl.-Based Intell. Eng. Syst.* **2018**, *22*, 39–57. [CrossRef]
2. Gómez-García, J.A.; Moro-Velázquez, L.; Godino-Llorente, J.I. On the design of automatic voice condition analysis systems. Part II: Review of speaker recognition techniques and study on the effects of different variability factors. *Biomed. Signal Process. Control* **2019**, *48*, 128–143. [CrossRef]
3. Vo, T.; Nguyen, T.; Le, C. Race Recognition Using Deep Convolutional Neural Networks. *Symmetry* **2018**, *10*, 564. [CrossRef]
4. Dahake, P.P.; Shaw, K.; Malathi, P. Speaker dependent speech emotion recognition using MFCC and Support Vector Machine. In Proceedings of the 2016 International Conference on Automatic Control and Dynamic Optimization Techniques (ICACDOT), Pune, India, 9–10 September 2016; pp. 1080–1084.
5. Chakraborty, S.S.; Parekh, R. Improved Musical Instrument Classification Using Cepstral Coefficients and Neural Networks. In *Methodologies and Application Issues of Contemporary Computing Framework*; Springer: Singapore, 2018; pp. 123–138.
6. Panteli, M.; Benetos, E.; Dixon, S. A computational study on outliers in world music. *PLoS ONE* **2017**, *12*, e0189399. [CrossRef] [PubMed]
7. Noda, J.J.; Sánchez-Rodríguez, D.; Travieso-González, C.M. A Methodology Based on Bioacoustic Information for Automatic Identification of Reptiles and Anurans. In *Reptiles and Amphibians*; IntechOpen: London, UK, 2018.
8. Desai, N.P.; Lehman, C.; Munson, B.; Wilson, M. Supervised and unsupervised machine learning approaches to classifying chimpanzee vocalizations. *J. Acoust. Soc. Am.* **2018**, *143*, 1786. [CrossRef]
9. Malfante, M.; Mars, J.I.; Dalla Mura, M.; Gervaise, C. Automatic fish sounds classification. *J. Acoust. Soc. Am.* **2018**, *143*, 2834–2846. [CrossRef]
10. Wang, Y.; Sun, B.; Yang, X.; Meng, Q. Heart sound identification based on MFCC and short-term energy. In Proceedings of the 2017 Chinese Automation Congress (CAC), Jinan, China, 20–22 October 2017; pp. 7411–7415.
11. Usman, M.; Zubair, M.; Shiblee, M.; Rodrigues, P.; Jaffar, S. Probabilistic Modeling of Speech in Spectral Domain using Maximum Likelihood Estimation. *Symmetry* **2018**, *10*, 750. [CrossRef]
12. Cao, J.; Cao, M.; Wang, J.; Yin, C.; Wang, D.; Vidal, P.P. Urban noise recognition with convolutional neural network. *Multimed. Tools Appl.* **2018**. [CrossRef]
13. Xu, J.; Wang, Z.; Tan, C.; Lu, D.; Wu, B.; Su, Z.; Tang, Y. Cutting Pattern Identification for Coal Mining Shearer through Sound Signals Based on a Convolutional Neural Network. *Symmetry* **2018**, *10*, 736. [CrossRef]
14. Lee, J.; Choi, H.; Park, D.; Chung, Y.; Kim, H.Y.; Yoon, S. Fault detection and diagnosis of railway point machines by sound analysis. *Sensors* **2016**, *16*, 549. [CrossRef]
15. Choi, Y.; Atif, O.; Lee, J.; Park, D.; Chung, Y. Noise-Robust Sound-Event Classification System with Texture Analysis. *Symmetry* **2018**, *10*, 402. [CrossRef]
16. Guyon, I.; Elisseeff, A. An introduction to feature extraction. In *Feature Extraction*; Springer: Berlin/Heidelberg, Germany, 2006; pp. 1–25.
17. Alías, F.; Socoró, J.; Sevillano, X. A review of physical and perceptual feature extraction techniques for speech, music and environmental sounds. *Appl. Sci.* **2016**, *6*, 143. [CrossRef]
18. Zhang, H.; McLoughlin, I.; Song, Y. Robust sound event recognition using convolutional neural networks. In Proceedings of the 2015 IEEE International Conference on Acoustics, Speech and Signal Processing (ICASSP), Brisbane, Australia, 19–24 April 2015; pp. 559–563.
19. Dave, N. Feature extraction methods LPC, PLP and MFCC in speech recognition. *Int. J. Adv. Res. Eng. Technol.* **2013**, *1*, 1–4.
20. Paul, D.; Pal, M.; Saha, G. Spectral features for synthetic speech detection. *IEEE J. Sel. Top. Signal Process.* **2017**, *11*, 605–617. [CrossRef]
21. Taebi, A.; Mansy, H.A. Analysis of seismocardiographic signals using polynomial chirplet transform and smoothed pseudo Wigner-Ville distribution. In Proceedings of the 2017 IEEE Signal Processing in Medicine and Biology Symposium (SPMB), Philadelphia, PA, USA, 2 December 2017; pp. 1–6.

22. Dayou, J.; Han, N.C.; Mun, H.C.; Ahmad, A.H.; Muniandy, S.V.; Dalimin, M.N. Classification and identification of frog sound based on entropy approach. In Proceedings of the 2011 International Conference on Life Science and Technology, Mumbai, India, 7–9 January 2011; Volume 3, pp. 184–187.
23. Zheng, F.; Zhang, G.; Song, Z. Comparison of different implementations of MFCC. *J. Comput. Sci. Technol.* **2001**, *16*, 582–589. [CrossRef]
24. Hussain, H.; Ting, C.M.; Numan, F.; Ibrahim, M.N.; Izan, N.F.; Mohammad, M.M.; Sh-Hussain, H. Analysis of ECG biosignal recognition for client identifiction. In Proceedings of the 2017 IEEE International Conference on Signal and Image Processing Applications (ICSIPA), Kuching, Malaysia, 12–14 September 2017; pp. 15–20.
25. Nickel, C.; Brandt, H.; Busch, C. Classification of Acceleration Data for Biometric Gait Recognition on Mobile Devices. *Biosig* **2011**, *11*, 57–66.
26. Muheidat, F.; Tyrer, W.H.; Popescu, M. Walk Identification using a smart carpet and Mel-Frequency Cepstral Coefficient (MFCC) features. In Proceedings of the 2018 40th Annual International Conference of the IEEE Engineering in Medicine and Biology Society (EMBC), Honolulu, HI, USA, 18–21 July 2018; pp. 4249–4252.
27. Negi, S.S.; Kishor, N.; Negi, R.; Uhlen, K. Event signal characterization for disturbance interpretation in power grid. In Proceedings of the 2018 First International Colloquium on Smart Grid Metrology (SmaGriMet), Split, Croatia, 24–27 April 2018; pp. 1–5.
28. Xie, J.; Towsey, M.; Zhang, J.; Roe, P. Frog call classification: A survey. *Artif. Int. Rev.* **2018**, *49*, 375–391. [CrossRef]
29. Colonna, J.G.; Nakamura, E.F.; Rosso, O.A. Feature evaluation for unsupervised bioacoustic signal segmentation of anuran calls. *Expert Syst. Appl.* **2018**, *106*, 107–120. [CrossRef]
30. Luque, A.; Romero-Lemos, J.; Carrasco, A.; Barbancho, J. Non-sequential automatic classification of anuran sounds for the estimation of climate-change indicators. *Expert Syst. Appl.* **2018**, *95*, 248–260. [CrossRef]
31. Luque, A.; Romero-Lemos, J.; Carrasco, A.; Gonzalez-Abril, L. Temporally-aware algorithms for the classification of anuran sounds. *PeerJ* **2018**, *6*, e4732. [CrossRef]
32. Luque, A.; Romero-Lemos, J.; Carrasco, A.; Barbancho, J. Improving Classification Algorithms by Considering Score Series in Wireless Acoustic Sensor Networks. *Sensors* **2018**, *18*, 2465. [CrossRef] [PubMed]
33. Romero, J.; Luque, A.; Carrasco, A. Anuran sound classification using MPEG-7 frame descriptors. In Proceedings of the XVII Conferencia de la Asociación Española para la Inteligencia Artificial (CAEPIA), Salamanca, Spain, 14–16 September 2016; pp. 801–810.
34. Luque, A.; Gómez-Bellido, J.; Carrasco, A.; Personal, E.; Leon, C. Evaluation of the processing times in anuran sound classification. *Wireless Communications and Mobile Computing* **2017**. [CrossRef]
35. Luque, A.; Gómez-Bellido, J.; Carrasco, A.; Barbancho, J. Optimal Representation of Anuran Call Spectrum in Environmental Monitoring Systems Using Wireless Sensor Networks. *Sensors* **2018**, *18*, 1803. [CrossRef] [PubMed]
36. Hershey, S.; Chaudhuri, S.; Ellis, D.P.; Gemmeke, J.F.; Jansen, A.; Moore, R.C.; Plakal, M.; Platt, D.; Saurous, R.A.; Seybold, B.; et al. CNN architectures for large-scale audio classification. In Proceedings of the 2017 IEEE International Conference on Acoustics, Speech and Signal Processing (ICASSP), New Orleans, LA, USA, 5–9 March 2017; pp. 131–135.
37. Dai, W.; Dai, C.; Qu, S.; Li, J.; Das, S. Very deep convolutional neural networks for raw waveforms. In Proceedings of the 2017 IEEE International Conference on Acoustics, Speech and Signal Processing (ICASSP), New Orleans, LA, USA, 5–9 March 2017; pp. 421–425.
38. Strout, J.; Rogan, B.; Seyednezhad, S.M.; Smart, K.; Bush, M.; Ribeiro, E. Anuran call classification with deep learning. In Proceedings of the 2017 IEEE International Conference on Acoustics, Speech and Signal Processing (ICASSP), New Orleans, LA, USA, 5–9 March 2017; pp. 2662–2665.
39. Colonna, J.; Peet, T.; Ferreira, C.A.; Jorge, A.M.; Gomes, E.F.; Gama, J. Automatic classification of anuran sounds using convolutional neural networks. In Proceedings of the Ninth International Conference on Computer Science & Software Engineering, Porto, Portugal, 20–22 July 2016; pp. 73–78.
40. Podder, P.; Khan, T.Z.; Khan, M.H.; Rahman, M.M. Comparative performance analysis of hamming, hanning and blackman window. *Int. J. Comput. Appl.* **2014**, *96*, 1–7. [CrossRef]
41. O'shaughnessy, D. *Speech Communication: Human and Machine*, 2nd ed.; Wiley-IEEE Press: Hoboken, NJ, USA, 1999; ISBN 978-0-7803-3449-6.
42. Bhatia, R. *Fourier Series*; American Mathematical Society: Providence, RI, USA, 2005.

43. Broughton, S.A.; Bryan, K. *Discrete Fourier Analysis and Wavelets: Applications to Signal and Image Processing*; John Wiley & Sons: Hoboken, NJ, USA, 2018.
44. Rao, K.R.; Yip, P. *Discrete Cosine Transform: Algorithms, Advantages, Applications*; Academic Press: Cambridge, MA, USA, 2014.
45. Tan, L.; Jiang, J. *Digital Signal Processing: Fundamentals and Applications*; Academic Press: Cambridge, MA, USA, 2018.

3

Sufficiency Criterion for a Subfamily of Meromorphic Multivalent Functions of Reciprocal Order with Respect to Symmetric Points

Shahid Mahmood [1,*], Gautam Srivastava [2,3], Hari Mohan Srivastava [4,5], Eman S. A. Abujarad [6], Muhammad Arif [7] and Fazal Ghani [7]

1 Department of Mechanical Engineering, Sarhad University of Science & I. T Landi Akhun Ahmad, Hayatabad Link. Ring Road, Peshawar 25000, Pakistan
2 Department of Mathematics and Computer Science, Brandon University, 270 18th Street, Brandon, MB R7A 6A9, Canada; srivastavag@brandonu.ca
3 Research Center for Interneural Computing, China Medical University, Taichung 40402, Taiwan
4 Department of Mathematics and Statistics, University of Victoria, Victoria, BC V8W 3R4, Canada; harimsri@math.uvic.ca
5 Department of Medical Research, China Medical University Hospital, China Medical University, Taichung 40402, Taiwan
6 Department of Mathematics, Aligarh Muslim University, Aligarh 202002, India; emanjarad2@gmail.com
7 Department of Mathematics, Abdul Wali Khan University Mardan, Mardan 23200, Pakistan; marifmaths@awkum.edu.pk (M.A.); fazalghanimaths@gmail.com (F.G.)
* Correspondence: shahidmahmood757@gmail.com

Abstract: In the present research paper, our aim is to introduce a new subfamily of meromorphic p-valent (multivalent) functions. Moreover, we investigate sufficiency criterion for such defined family.

Keywords: meromorphic multivalent starlike functions; subordination

---

1. Introduction

Let the notation Ω_p be the family of meromorphic p-valent functions f that are holomorphic (analytic) in the region of punctured disk $\mathbb{E} = \{z \in \mathbb{C} : 0 < |z| < 1\}$ and obeying the following normalization

$$f(z) = \frac{1}{z^p} + \sum_{j=1}^{\infty} a_{j+p}\, z^{j+p} \ \ (z \in \mathbb{E}). \tag{1}$$

In particular $\Omega_1 = \Omega$, the familiar set of meromorphic functions. Further, the symbol $\mathcal{MS}^*$ represents the set of meromorphic starlike functions which is a subfamily of Ω and is given by

$$\mathcal{MS}^* = \left\{ f : f(z) \in \Omega \text{ and } \Re\left(\frac{zf'(z)}{f(z)}\right) < 0 \ \ (z \in \mathbb{E}) \right\}.$$

Two points p and p' are said to be symmetrical with respect to o if o' is the midpoint of the line segment pp'. This idea was further nourished in [1,2] by introducing the family $\mathcal{MS}_s^*$ which is defined in set builder form as;

$$\mathcal{MS}_s^* = \left\{ f : f(z) \in \Omega \text{ and } \Re\left(\frac{-2zf'(z)}{f(-z) - f(z)}\right) < 0 \ \ (z \in \mathbb{E}) \right\}.$$

Now, for $-1 \leq t < s \leq 1$ with $s \neq 0 \neq t$, $0 < \xi < 1$, λ is real with $|\lambda| < \frac{\pi}{2}$ and $p \in \mathbb{N}$, we introduce a subfamily of Ω_p consisting of all meromorphic p-valent functions of reciprocal order ξ, denoted by $\mathcal{NS}_p^{\lambda}(s,t,\xi)$, and is defined by

$$\mathcal{NS}_p^{\lambda}(s,t,\xi) = \left\{ f : f(z) \in \Omega_p \text{ and } \Re\left(e^{-i\lambda} \frac{ps^p t^p}{s^p - t^p} \frac{f(sz) - f(tz)}{zf'(z)} \right) > \xi \cos\lambda \ (z \in \mathbb{E}) \right\}.$$

We note that for $p = s = 1$ and $t = -1$, the class $\mathcal{NS}_p^{\lambda}(s,t,\xi)$ reduces to the class $\mathcal{NS}_1^{\lambda}(1,-1,\xi) = \mathcal{NS}_*^{\lambda}(\xi)$ and is represented by

$$\mathcal{NS}_*^{\lambda}(\xi) = \left\{ f : f(z) \in \Omega \text{ and } \Re\left(e^{-i\lambda} \frac{f(-z) - f(z)}{2zf'(z)} \right) > \xi \cos\lambda \ (z \in \mathbb{E}) \right\}.$$

For detail of the related topics, see the work of Al-Amiri and Mocanu [3], Rosihan and Ravichandran [4], Aouf and Hossen [5], Arif [6], Goyal and Prajapat [7], Joshi and Srivastava [8], Liu and Srivastava [9], Raina and Srivastava [10], Sun et al. [11], Shi et al. [12] and Owa et al. [13], see also [14–16].

For simplicity and ignoring the repetition, we state here the constraints on each parameter as $0 < \xi < 1$, $-1 \leq t < s \leq 1$ with $s \neq 0 \neq t$, λ is real with $|\lambda| < \frac{\pi}{2}$ and $p \in \mathbb{N}$.

We need to mention the following lemmas which will use in the main results.

Lemma 1. *"Let $H \subset \mathbb{C}$ and let $\Phi : \mathbb{C}^2 \times \mathbb{E}^* \to \mathbb{C}$ be a mapping satisfying $\Phi(ia, b : z) \notin H$ for $a, b \in \mathbb{R}$ such that $b \leq -n\frac{1+a^2}{2}$. If $p(z) = 1 + c_n z^n + \cdots$ is regular in $\mathbb{E}^*$ and $\Phi(p(z), zp'(z) : z) \in H \ \forall\, z \in \mathbb{E}^*$, then $\Re(p(z)) > 0$."*

Lemma 2. *"Let $p(z) = 1 + c_1 z + \cdots$ be regular in $\mathbb{E}^*$ and η be regular and starlike univalent in $\mathbb{E}^*$ with $\eta(0) = 0$. If $zp'(z) \prec \eta(z)$, then*

$$p(z) \prec 1 + \int_0^z \frac{\eta(t)}{t} dt.$$

This result is the best possible."

2. Sufficiency Criterion for the Family $\mathcal{NS}_p^{\lambda}(s,t,\xi)$

In this section, we investigate the sufficiency criterion for any meromorphic p-valent functions belonging to the introduced family $\mathcal{NS}_p^{\lambda}(s,t,\xi)$:

Now, we obtain the necessary and sufficient condition for the p-valent function f to be in the family $\mathcal{NS}_p^{\lambda}(s,t,\xi)$ as follows:

Theorem 1. *Let the function $f(z)$ be the member of the family Ω_p. Then*

$$f(z) \in \mathcal{NS}_p^{\lambda}(s,t,\xi) \Leftrightarrow \left| \frac{e^{i\lambda}}{\mathcal{G}(z)} - \frac{1}{2\xi\cos\lambda} \right| < \frac{1}{2\xi\cos\lambda}, \tag{2}$$

where

$$\mathcal{G}(z) = \frac{p\, s^p t^p}{(s^p - t^p)} \frac{f(sz) - f(tz)}{zf'(z)}. \tag{3}$$

Proof. Suppose that inequality (2) holds. Then, we have

$$\begin{aligned}
\left|\frac{2\xi\cos\lambda - e^{-i\lambda}\mathcal{G}(z)}{2\xi\cos\lambda e^{-i\lambda}\mathcal{G}(z)}\right| &< \frac{1}{2\xi\cos\lambda} \\
&\Leftrightarrow \left|\frac{2\xi\cos\lambda - e^{-i\lambda}\mathcal{G}(z)}{2\xi\cos\lambda e^{-i\lambda}\mathcal{G}(z)}\right|^2 < \frac{1}{4\xi^2\cos^2\lambda} \\
&\Leftrightarrow \left(2\xi\cos\lambda - e^{-i\lambda}\mathcal{G}(z)\right)\left(\overline{2\xi\cos\lambda - e^{-i\lambda}\mathcal{G}(z)}\right) < \left(e^{i\lambda}\overline{\mathcal{G}(z)}\right)e^{-i\lambda}\mathcal{G}(z) \\
&\Leftrightarrow 4\xi^2\cos^2\lambda - 2\xi\cos\lambda\left(e^{i\lambda}\overline{\mathcal{G}(z)} + e^{-i\lambda}\mathcal{G}(z)\right) < 0 \\
&\Leftrightarrow 2\xi\cos\lambda - 2\Re\left(e^{-i\lambda}\mathcal{G}(z)\right) < 0 \\
&\Leftrightarrow \Re\left(e^{-i\lambda}\mathcal{G}(z)\right) > \xi\cos\lambda,
\end{aligned}$$

and hence the result follows. ☐

Next, we investigate the sufficient condition for the p-valent function f to be in the family $\mathcal{NS}_p^\lambda(s,t,\xi)$ in the following theorem:

Theorem 2. *If $f(z)$ belongs to the family Ω_p of meromorphic p-valent functions and obeying*

$$\sum_{n=p+1}^{\infty}\left|\left(\frac{s^n - t^n}{s^p - t^p}s^p t^p - \frac{n\beta\cos\lambda}{p}e^{i\lambda}\right)\right||a_n| < \frac{1}{2}\left(1 - \left|1 - 2\beta\cos\lambda e^{i\lambda}\right|\right), \tag{4}$$

then $f(z) \in \mathcal{NS}_p^\lambda(s,t,\xi)$.

Proof. To prove the required result we only need to show that

$$\left|\frac{2e^{i\lambda}\xi\cos\lambda z f'(z)/p - \frac{s^p t^p}{(t^p - s^p)}(f(tz) - f(sz))}{\frac{s^p t^p}{(t^p - s^p)}(f(tz) - f(sz))}\right| < 1. \tag{5}$$

Now consider the left hand side of (5), we get

$$\begin{aligned}
LHS &= \left|\frac{2e^{i\lambda}\xi\cos\lambda z f'(z)/p - \frac{s^p t^p}{(t^p - s^p)}(f(tz) - f(sz))}{\frac{s^p t^p}{(t^p - s^p)}(f(tz) - f(sz))}\right| \\
&= \left|\frac{(2e^{i\lambda}\xi\cos\lambda - 1) + \sum_{n=p+1}^{\infty}\left(\frac{s^n - t^n}{s^p - t^p}s^p t^p - \frac{2n\xi\cos\lambda}{p}e^{i\lambda}\right)a_n z^{n+p}}{1 + \sum_{n=p+1}^{\infty}\left(\frac{s^n - t^n}{s^p - t^p}\right)s^p t^p a_n z^{n+p}}\right| \\
&\leq \frac{|2e^{i\lambda}\xi\cos\lambda - 1| + \sum_{n=p+1}^{\infty}\left|\left(\frac{s^n - t^n}{s^p - t^p}s^p t^p - 2\beta\cos\lambda e^{i\lambda}\frac{n}{p}\right)\right||a_n||z^{n+p}|}{1 - \sum_{n=p+1}^{\infty}\left|\left(\frac{s^n - t^n}{s^p - t^p}\right)s^p t^p\right||a_n||z^{n+p}|} \\
&\leq \frac{|2e^{i\lambda}\xi\cos\lambda - 1| + \sum_{n=p+1}^{\infty}\left|\left(\frac{s^n - t^n}{s^p - t^p}s^p t^p - 2\beta\cos\lambda e^{i\lambda}\frac{n}{p}\right)\right||a_n|}{1 - \sum_{n=p+1}^{\infty}\left|\left(\frac{s^n - t^n}{s^p - t^p}\right)s^p t^p\right||a_n|}.
\end{aligned}$$

By virtue of inequality (4), we at once get the desired result. ☐

Also, we obtain another sufficient condition for the p-valent function f to be in the family $\mathcal{NS}_p^{\lambda}(s,t,\xi)$ by using Lemma 1, in the following theorem:

Theorem 3. *If $f(z) \in \Omega_p$ satisfies*

$$\Re\left\{e^{-i\lambda}\left(\alpha z\frac{\mathcal{G}'(z)}{\mathcal{G}(z)}+1\right)\mathcal{G}(z)\right\} > \beta\cos\lambda - \frac{n}{2}\left((1-\beta)\,\alpha\cos\lambda\right),$$

then $f(z) \in \mathcal{NS}_p^{\lambda}(s,t,\xi)$, where $\mathcal{G}(z)$ is defined in Equation (3).

Proof. Let we choose the function $q(z)$ by

$$q(z) = \frac{e^{-i\lambda}\mathcal{G}(z) - \beta\cos\lambda + i\sin\lambda}{(1-\beta)\cos\lambda}, \tag{6}$$

then Equation (6) shows that $q(z)$ is holomorphic in $\mathbb{E}$ and also normalized by $q(0) = 1$.

From Equation (6), we can easily obtain that

$$e^{-i\lambda}\mathcal{G}(z)\left(1+\alpha z\frac{\mathcal{G}'(z)}{\mathcal{G}(z)}\right) = \Phi\left(q(z), zq'(z), z\right),$$

where

$$\Phi\left(q(z), zq'(z), z\right) = \left[(1-\beta)\,\alpha zq'(z) + (1-\beta)\,q(z) + \beta\right]\cos\lambda - i\sin\lambda.$$

Now for all $a, b \in \mathbb{R}$ satisfying $2y \leq -n\left(1+a^2\right)$, we have

$$\begin{aligned}\Re\left\{\Phi(ia,b,z)\right\} &\leq \beta\cos\lambda - \frac{n}{2}\left(1+a^2\right)(1-\beta)\,\alpha\cos\lambda \\ &\leq \beta\cos\lambda - \frac{n}{2}(1-\beta)\,\alpha\cos\lambda.\end{aligned}$$

Now, let us define a set as

$$H = \left\{\zeta : \Re(\zeta) > \beta\cos\lambda - \frac{n}{2}\left((1-\beta)\,\alpha\cos\lambda\right)\right\},$$

then, we see that $\Phi(ia,b,z) \notin H$ and $\Phi\left(q(z), zq'(z), z\right) \in H$. Therefore, by using Lemma 1, we obtain that $\Re\left(q(z)\right) > 0$.

□

Further, in the next theorem, we obtain the sufficient condition for the p-valent function f to be in the family $\mathcal{NS}_p^{\lambda}(s,t,\xi)$ by using Lemma 2.

Theorem 4. *If $f(z)$ is a member of the family Ω_p of meromorphic p-valent functions and satisfies*

$$\left|\frac{e^{i\lambda}}{\mathcal{G}(z)}\left(\frac{z\mathcal{G}'(z)}{\mathcal{G}(z)}\right)\right| < \frac{1}{\beta\cos\lambda} - 1, \tag{7}$$

then $f(z) \in \mathcal{NS}_p^{\lambda}(s,t,\xi)$, where $\mathcal{G}(z)$ is given by Equation (3).

Proof. In order to prove the required result, we need to define the following function

$$q(z)\cos\lambda = e^{-i\lambda}\mathcal{G}(z) + i\sin\lambda,$$

then, Equation (6) shows that th function $q(z)$ is holomorphic in $\mathbb{E}$ and also normalized by $q(0) = 1$.

Now, by routine computations, we get

$$\frac{zq'(z)}{q(z) - i\tan\lambda} = \frac{z\mathcal{G}'(z)}{\mathcal{G}(z)}.$$

Now, let us consider $z\left(\frac{1}{q(z)\cos\lambda - i\sin\lambda}\right)'$ and then by using inequality (7), we have

$$\left|z\left(\frac{1}{q(z)\cos\lambda - i\sin\lambda}\right)'\right| = \left|\frac{e^{i\lambda}}{\mathcal{G}(z)}\left(\frac{z\mathcal{G}'(z)}{\mathcal{G}(z)}\right)\right| < \frac{1}{\beta\cos\lambda} - 1,$$

therefore

$$z\left(\frac{1}{q(z)\cos\lambda - i\sin\lambda}\right)' \prec \frac{(1-\beta\cos\lambda)z}{\beta\cos\lambda}.$$

Using Lemma 2, we have

$$\frac{1}{(q(z) - i\tan\lambda)\cos\lambda} \prec 1 + \frac{(1-\beta\cos\lambda)}{\beta\cos\lambda}z,$$

equivalently

$$(q(z) - i\tan\lambda)\cos\lambda \prec \frac{\beta\cos\lambda}{\beta\cos\lambda + (1-\beta\cos\lambda)z} = H(z)\,(say). \tag{8}$$

After simplifications, we get

$$1 + \Re\left(\frac{zH''(z)}{H'(z)}\right) = 2\beta\cos\lambda - 1 > 0, \;\; for\, \frac{1}{2} < \beta < 1.$$

The region $H(\mathbb{E})$ shows that it is symmetric about the real axis and also $H(z)$ is convex. Hence

$$\Re\,(\mathcal{G}(z)) \geq H(1) > 0,$$

or

$$\Re\,(q(z)\cos\lambda - i\sin\lambda) > \beta\cos\lambda,$$

or

$$\Re\left(e^{-i\lambda}\mathcal{G}(z)\right) > \beta\cos\lambda, \;\; for\, \frac{1}{2} < \beta < 1.$$

□

Finally, we investigate the sufficient condition for the p-valent function f to be in the family $\mathcal{NS}_p^{\lambda}(s,t,\xi)$ in the following theorem:

Theorem 5. *If $f(z) \in \Omega_p$ satisfies*

$$\left|\left(\frac{2\beta\cos\lambda e^{i\lambda}}{\mathcal{G}(z)} - 1\right)'\right| \leq \eta\,|z|^{\gamma}, \, for\;\; 0 < \eta \leq \gamma + 1, \tag{9}$$

then $f(z) \in \mathcal{NS}_p^{\lambda}(s,t,\xi)$, where $\mathcal{G}(z)$ is defined in Equation (3).

Proof. Let us put

$$G(z) = z\left(\frac{2\beta\cos\lambda e^{i\lambda}}{\mathcal{G}(z)} - 1\right).$$

Then $G(0) = 0$ and $G(z)$ is analytic in $\mathbb{E}$. Using inequality (9), we can write

$$\left|\left(\frac{G(z)}{z}\right)'\right| = \left|\left(\frac{2\beta\cos\lambda e^{i\lambda}}{\mathcal{G}(z)} - 1\right)'\right| \leq \eta\,|z|^{\gamma}.$$

Now,

$$\left|\left(\frac{G(z)}{z}\right)\right| = \left|\int_0^z \left(\frac{G(t)}{t}\right)' dt\right| \leq \int_0^{|z|} \left|\left(\frac{G(t)}{t}\right)'\right| dt \leq \int_0^{|z|} \eta\,|t|^{\gamma}\,dt = \frac{\eta\,|z|^{\gamma+1}}{\gamma+1} < 1,$$

and this implies that

$$\left|\frac{2\beta\cos\lambda e^{i\lambda}}{\mathcal{G}(z)} - 1\right| < 1.$$

Now by using Theorem 1, we get the result which we needed. □

3. Conclusions

In our results, a new subfamily of meromorphic p-valent (multivalent) functions were introduced. Further, various sufficient conditions for meromorphic p-valent functions belonging to these subfamilies were obtained and investigated.

Author Contributions: Conceptualization, H.M.S. and M.A.; Formal analysis, H.M.S. and S.M.; Funding acquisition, S.M. and G.S.; Investigation, E.S.A.A. and S.M.; Methodology, M.A. and F.G.; Supervision, H.M.S. and M.A.; Validation, M.A. and S.M.; Visualization, G.S. and E.S.A.A.; Writing original draft, M.A., S.M. and F.G.; Writing review and editing, M.A., F.G. and S.M.

Acknowledgments: The authors would like to thank the reviewers of this paper for their valuable comments on the earlier version of the paper. They would also like to acknowledge Salim ur Rehman, the Vice Chancellor, Sarhad University of Science & I.T, for providing excellent research environment and his financial support.

References

1. Srivastava, H.M.; Yang, D.-G.; Xu, N.-E. Some subclasses of meromorphically multivalent functions associated with a linear operator. *Appl. Math. Comput.* **2008**, *195*, 11–23. [CrossRef]
2. Wang, Z.-G.; Jiang, Y.-P.; Srivastava, H.M. Some subclasses of meromorphically multivalent functions associated with the generalized hypergeometric function. *Comput. Math. Appl.* **2009**, *57*, 571–586. [CrossRef]
3. Al-Amiri, H.; Mocanu, P.T. Some simple criteria of starlikeness and convexity for meromorphic functions. *Mathematica (Cluj)* **1995**, *37*, 11–21.
4. Ali, R.M.; Ravichandran, V. Classes of meromorphic α-convex functions. *Taiwanese J. Math.* **2010**, *14*, 1479–1490. [CrossRef]
5. Aouf, M.K.; Hossen, H.M. New criteria for meromorphic p-valent starlike functions. *Tsukuba J. Math.* **1993**, *17*, 481–486. [CrossRef]
6. Arif, M. On certain sufficiency criteria for p-valent meromorphic spiralike functions. In *Abstract and Applied Analysis*; Hindawi: London, UK, 2012.
7. Goyal, S.P.; Prajapat, J.K. A new class of meromorphic multivalent functions involving certain linear operator. *Tamsui Oxf. J. Math. Sci.* **2009**, *25*, 167–176.
8. Joshi, S.B.; Srivastava, H.M. A certain family of meromorphically multivalent functions. *Comput. Math. Appl.* **1999**, *38*, 201–211. [CrossRef]
9. Liu, J.-L.; Srivastava, H.M. A linear operator and associated families of meromorphically multivalent functions. *J. Math. Anal. Appl.* **2001**, *259*, 566–581. [CrossRef]
10. Raina, R.K.; Srivastava, H.M. A new class of mermorphically multivalent functions with applications of generalized hypergeometric functions. *Math. Comput. Model.* **2006**, *43*, 350–356. [CrossRef]

11. Sun, Y.; Kuang, W.-P.; Wang, Z.-G. On meromorphic starlike functions of reciprocal order α. *Bull. Malays. Math. Sci. Soc.* **2012**, *35*, 469–477.
12. Shi, L.; Wang, Z.-G.; Yi, J.-P. A new class of meromorphic functions associated with spirallike functions. *J. Appl. Math.* **2012**, *2012*, 1–12. [CrossRef]
13. Owa, S.; Darwish, H.E.; Aouf, M.A. Meromorphically multivalent functions with positive and fixed second coefficients. *Math. Japon.* **1997**, *46*, 231–236.
14. Arif, M.; Ahmad, B. New subfamily of meromorphic starlike functions in circular domain involving q-differential operator. *Math. Slovaca* **2018**, *68*, 1049–1056. [CrossRef]
15. Arif, M.; Raza, M.; Ahmad, B. A new subclass of meromorphic multivalent close-to-convex functions. *Filomat* **2016**, *30*, 2389–2395. [CrossRef]
16. Arif, M.; Sokół J.; Ayaz, M. Sufficient condition for functions to be in a class of meromorphic multivalent Sakaguchi type spiral-like functions. *Acta Math. Sci.* **2014**, *34*, 1–4. [CrossRef]

4

Generalized Mittag-Leffler Input Stability of the Fractional Differential Equations

Ndolane Sene [1] and Gautam Srivastava [2,3,*]

[1] Laboratoire Lmdan, Département de Mathématiques de la Décision, Université Cheikh Anta Diop de Dakar, Faculté des Sciences Economiques et Gestion, BP 5683 Dakar Fann, Senegal; ndolanesene@yahoo.fr
[2] Department of Mathematics and Computer Science, Brandon University, Brandon, MB R7A 6A9, Canada
[3] Research Center for Interneural Computing, China Medical University, Taichung 40402, Taiwan
* Correspondence: srivastavag@brandonu.ca

Abstract: The behavior of the analytical solutions of the fractional differential equation described by the fractional order derivative operators is the main subject in many stability problems. In this paper, we present a new stability notion of the fractional differential equations with exogenous input. Motivated by the success of the applications of the Mittag-Leffler functions in many areas of science and engineering, we present our work here. Applications of Mittag-Leffler functions in certain areas of physical and applied sciences are also very common. During the last two decades, this class of functions has come into prominence after about nine decades of its discovery by a Swedish Mathematician Mittag-Leffler, due to the vast potential of its applications in solving the problems of physical, biological, engineering, and earth sciences, to name just a few. Moreover, we propose the generalized Mittag-Leffler input stability conditions. The left generalized fractional differential equation has been used to help create this new notion. We investigate in depth here the Lyapunov characterizations of the generalized Mittag-Leffler input stability of the fractional differential equation with input.

Keywords: fractional differential equations with input; Mittag-Leffler stability; left generalized fractional derivative; ρ-Laplace transforms

---

1. Introduction

The behavior of the analytical solutions of the fractional differential equation described by the fractional order derivative operators is the main subject in stability problems [1]. There exist many stability notions introduced in fractional calculus. Some examples are asymptotic stability, global asymptotic uniform stability, synchronization problems, stabilization problems, Mittag-Leffler stability and fractional input stability. In this paper, we extend the Mittag-Leffler input stability in the context of the fractional differential equations described by the left generalized fractional derivative. We note here that the left generalized fractional derivative is the generalization of the Liouville-Caputo fractional derivative and the Riemann-Liouville fractional derivative [2]. There exist many works related to stability problems. In [3], Souahi et al. propose some new Lyapunov characterizations of fractional differential equations described by the conformable fractional derivative. In [4], Sene proposes a new stability notion and introduce the Lyapunov characterization of the conditional asymptotic stability. In [5,6], Sene proposes some applications of the fractional input stability to the electrical circuits described the Liouville-Caputo fractional derivative and the Riemann-Liouville fractional derivative. In [7], Li et al. introduce the Mittag-Leffler stability of the fractional differential equations described by the Liouville-Caputo fractional derivative [8]. In [9], Song et al. analyze the stability of the fractional differential equations with time variable impulses. In [10], Tuan et al. propose a novel methodology for studying the stability of the fractional differential equations using the Lyapunov direct method. In [11], Makhlouf studies the stability with respect to part of the variables of nonlinear Caputo fractional differential equations. In [12], Alidousti et al. propose a new stability analysis of the fractional differential equation described by the Liouville-Caputo fractional derivative. Many other works related to the stability analysis exist in literature, we direct our readers to the References section for more related literature.

The generalized Mittag-Leffler input stability is a new stability notion. This new stability notion studies the behavior of the analytical solution of the fractional differential equations with exogenous input described by the left generalized fractional derivative [13]. We know from previous work in stability problems, it is not trivial to get analytical solutions. The issue is to propose a method to analyze the stability of the fractional differential equations with exogenous input. Classically, the most popular method is the Lyapunov direct method as given in [14–18]. We propose the Lyapunov characterization of the generalized Mittag-Leffler input stability here in this work. As we will be able to show, the generalized Mittag-Leffler input stability generates three properties:

- the converging-input converging-state
- the bounded-input bounded-state
- the uniform global asymptotic stability of the trivial solution of the unforced fractional differential equation (fractional differential equation without exogenous input).

We note here that the fractional differential equation with exogenous input is said to be generalized Mittag-Leffler input stable when the Euclidean norm of its solution is bounded, by a generalized Mittag-Leffler function, plus a quantity which is proportional to the exogenous input bounded when the input is bounded and converging when the input converges in time. The fractional input stability and its consequences are a good compromise in stability problems of the fractional differential equations described by the fractional order derivative operators.

We organize the rest of the paper as follows. In Section 2, we recall the definition of the fractional derivative operators with or without singular kernels. In Section 3, we propose our motivations for studying the generalized Mittag-Leffler input stability. In Section 4, we give the Lyapunov characterizations for the generalized Mittag-Leffler input stability of the fractional differential equations with exogenous inputs. In Section 5, we provide numerical examples for illustrating the main results of this paper. Finally, we finish with some concluding remarks in Section 6.

2. Background on Fractional Derivatives

Let us first recall the fractional derivative operators and the comparison functions [19]. We will use them throughout this paper. There exist many fractional derivative operators in fractional calculus. There exist two types of fractional derivative operators. The first is fractional derivatives with singular kernels and the second is fractional derivatives without singular kernels. With regards to fractional derivatives with singular kernels, we cite the Riemann-Liouville fractional derivative [2], the Liouville-Caputo fractional derivative [2], the Hilfer fractional derivative [20], the Hadamard fractional derivative [2], and Erdélyi-Kober fractional derivative [21]. We note here that all previous fractional derivatives are associated to their fractional integrals [2,20]. As fractional derivatives without singular kernels we cite the Atangana-Baleanu-Liouville-Caputo derivative [22], the Caputo-Fabrizio fractional derivative [23], and the Prabhakar fractional derivative [24]. We note here that all previous fractional derivatives are associated to their fractional integrals [21–24]. Recently, the generalization of the Riemann-Liouville and the Liouville-Caputo fractional derivative were introduced in the literature by Udita [25]. Namely, the generalized fractional derivative and the Liouville-Caputo generalized fractional derivative. Let us now observe the comparison functions used in this paper.

Definition 1. *The class $\mathcal{PD}$ function denotes the set of all continuous functions $\alpha : \mathbb{R}_{\geq 0} \to \mathbb{R}_{\geq 0}$ satisfying $\alpha(0) = 0$, and $\alpha(s) > 0$ for all $s > 0$. A class $\mathcal{K}$ function is an increasing $\mathcal{PD}$ function. The class $\mathcal{K}_\infty$ represents the set of all unbounded $\mathcal{K}$ functions [17].*

Definition 2. *A continuous function $\beta : \mathbb{R}_{\geq 0} \to \mathbb{R}_{\geq 0}$ is said to be of class $\mathcal{L}$ if β is non-increasing and tends to zero as its arguments tend to infinity [17].*

Definition 3. *Let the function $f : [0, +\infty[\longrightarrow \mathbb{R}$, the Liouville-Caputo derivative of the function f of order α is expressed in the form*

$$D_{\alpha}^{c} f(t) = \frac{1}{\Gamma(1-\alpha)} \int_{0}^{t} \frac{f'(s)}{(t-s)^{\alpha}} ds, \tag{1}$$

for all $t > 0$, where the order $\alpha \in (0,1)$ and $\Gamma(.)$ is the gamma function [2,26–29].

Definition 4. *Let the function $f : [0, +\infty[\longrightarrow \mathbb{R}$, the Riemann-Liouville derivative of the function f of order α is expressed in the form*

$$D_{\alpha}^{RL} f(t) = \frac{1}{\Gamma(1-\alpha)} \frac{d}{dt} \int_{0}^{t} \frac{f(s)}{(t-s)^{\alpha}} ds, \tag{2}$$

for all $t > 0$, where the order $\alpha \in (0,1)$ and $\Gamma(.)$ is the gamma function [2,26–30].

Definition 5. *Let the function $f : [0, +\infty[\longrightarrow \mathbb{R}$, the Liouville-Caputo generalized derivative of the function f of order α is expressed in the form*

$$\left(D_{c}^{\alpha,\rho} f\right)(t) = \frac{1}{\Gamma(1-\alpha)} \int_{0}^{t} \left(\frac{t^{\rho} - s^{\rho}}{\rho}\right)^{-\alpha} f'(s) \frac{ds}{s^{1-\rho}}, \tag{3}$$

for all $t > 0$, where the order $\alpha \in (0,1)$ and $\Gamma(.)$ is the gamma function [2,26,28,29,31].

Definition 6. *Let the function $f : [0, +\infty[\longrightarrow \mathbb{R}$, the left generalized derivative of the function f of order α is expressed in the form*

$$\left(D^{\alpha,\rho} f\right)(t) = \frac{1}{\Gamma(1-\alpha)} \left(\frac{d}{dt}\right) \int_{0}^{t} \left(\frac{t^{\rho} - s^{\rho}}{\rho}\right)^{-\alpha} f(s) \frac{ds}{s^{1-\rho}}, \tag{4}$$

for all $t > 0$, where the order $\alpha \in (0,1)$ and $\Gamma(.)$ is the gamma function [2,26,28,29,31].

Definition 7. *Let us take the function $f : [0, +\infty[\longrightarrow \mathbb{R}$, the Caputo-Fabrizio fractional derivative of the function f of order α is expressed in the form*

$$D_{\alpha}^{CF} f(t) = \frac{M(\alpha)}{1-\alpha} \int_{0}^{t} f'(s) \exp\left(-\frac{\alpha}{1-\alpha}(t-s)\right) ds, \tag{5}$$

for all $t > 0$, where the order $\alpha \in (0,1)$ and $\Gamma(.)$ is the gamma function [22].

Definition 8. *Let the function $f : [0, +\infty[\longrightarrow \mathbb{R}$, the Caputo-Fabrizio fractional derivative of the function f of order α is expressed in the form*

$$D_{\alpha}^{ABC} f(t) = \frac{AB(\alpha)}{1-\alpha} \int_{0}^{t} f'(s) E_{\alpha}\left(-\frac{\alpha}{1-\alpha}(t-s)^{\alpha}\right) ds, \tag{6}$$

for all $t > 0$, where the order $\alpha \in (0,1)$ and $\Gamma(.)$ is the gamma function [22,30].

Definition 9. *Let us consider the function $f : [0, +\infty[\longrightarrow \mathbb{R}$, the Erdélyi-Kober fractional integral of the function f of order $\alpha > 0$, $\eta > 0$ and $\gamma \in \mathbb{R}$ is expressed in the form*

$$I_{\eta}^{\gamma,\alpha} f(t) = \frac{t^{-\eta(\gamma+\alpha)}}{\Gamma(\alpha)} \int_{0}^{t} \tau^{\eta\gamma} \left(t^{\eta} - \tau^{\eta}\right)^{\alpha-1} f(\tau) d(\tau^{\eta}), \tag{7}$$

for all $t > 0$, and $\Gamma(.)$ is the gamma function [21].

Definition 10. *Let us consider the function $f : [0, +\infty[\longrightarrow \mathbb{R}$, the Erdélyi-Kober fractional derivative of the function f of order $\alpha > 0$, $\eta > 0$ and $\gamma \in \mathbb{R}$ is expressed in the form*

$$D_{\eta}^{\gamma,\alpha} f(t) = \prod_{j=1}^{n} \left(\gamma + j + \frac{1}{\eta} \frac{d}{dt} \right) \left(I_{\eta}^{\gamma+\alpha,n-\mu} f(t) \right), \tag{8}$$

for all $t > 0$, and where $n - 1 < \alpha \leq n$ [21].

Some special cases can be recovered with the above definitions. In Definition 8, when $\rho = 1$, we recover the Liouville-Caputo fractional derivative. In Definition 9, when $\rho = 1$, we recover the Riemann-Liouville fractional derivative. In Definition 10, when $\gamma = -\alpha$ and $\eta = 1$, we obtain the relation existing between Erdélyi-Kobar fractional derivative and Riemann-Liouville fractional derivative expressed in the form $D_{1}^{-\alpha,\alpha} f(t) = t^{\alpha} D^{\alpha,1} f(t)$.

The Laplace transform will be used for solving a class of the fractional differential equations. The ρ-Laplace transform was recently introduced by Fahd et al. in order to solve differential equations in the frame of conformable derivatives to extend the possibility of working in a large class of functions [2]. We encourage readers to refer to [2] for more detailed information about ρ-Laplace transforms and their applications.

The ρ-Laplace transform of the generalized fractional derivative in the Liouville-Caputo sense is expressed in the following form

$$\mathcal{L}_{\rho} \left\{ \left(D_{c}^{\alpha,\rho} f \right) (t) \right\} = s^{\alpha} \mathcal{L}_{\rho} \left\{ f(t) \right\} - s^{\alpha-1} f(0), \tag{9}$$

The ρ-Laplace transform of the function f is given in the form

$$\mathcal{L}_{\rho} \left\{ f(t) \right\} (s) = \int_{0}^{\infty} e^{-s\frac{t^{\rho}}{\rho}} f(t) \frac{dt}{t^{1-\rho}}. \tag{10}$$

Definition 11. *The Mittag–Leffler function with two parameters is defined as the following series*

$$E_{\alpha,\beta}(z) = \sum_{k=0}^{\infty} \frac{z^{k}}{\Gamma(\alpha k + \beta)} \tag{11}$$

where $\alpha > 0$, $\beta \in \mathbb{R}$ and $z \in \mathbb{C}$. The classical exponential function is obtained with $\alpha = \beta = 1$. Here we see that when α and β are strictly positive, the series is convergent [14].

3. New Stability Notion of the Fractional Differential Equations

In this section, we introduce a new stability notion for the fractional differential equation with exogenous input described by the left generalized fractional derivative. Historically, the fractional input stability and the Mittag-Leffler input stability of the fractional differential equation represented by the Liouville-Caputo fractional derivative were stated in previous works [5,18]. Moreover, the idea of a discrete version of fractional derivatives is studied in the seminal work [32]. The Lyapunov characterizations of these new stability notions have been provided in [15,18]. In this section, we extend the Mittag-Leffler input stability involving the left generalized fractional derivative. We provide some modifications in the structure of the definitions, however the idea is not modified. The new stability notion addressed in this paper is called the generalized Mittag-Leffler input stability. In the literature there exist many stability notions related to the fractional differential equations without exogenous inputs such as the asymptotic stability [7,14], the practical stability [12,33], the Mittag-Leffler stability [7] and many others notions. Let us consider the fractional differential equations with exogenous inputs. In fractional calculus, we have not seen a lot of work related to the stability of the fractional differential equations with inputs. The stabilization problems [3] of the fractional differential equations with exogenous inputs is one of the most popular notion existing in the known literature. The challenge consists of finding possible values of the input under which the trivial solution of the obtained fractional differential equation is asymptotically stable. In this paper, we adopt a new method. Let us consider the fractional differential equation with exogenous input described by the left generalized fractional derivative

$$D^{\alpha,\rho}x = Ax + Bu, \tag{12}$$

where $x \in \mathbb{R}^n$ is a state variable, the matrix $A \in \mathbb{R}^{n\times n}$ satisfies the property $|\arg(\lambda(A))| > \frac{\alpha\pi}{2}$, the matrix $B \in \mathbb{R}^{n\times n}$ and $u \in \mathbb{R}^n$ represents the exogenous input. The initial boundary condition is defined by $\left(I^{1-\alpha,\rho}x\right)(0) = x_0$. Firstly, we give the analytical solution of the fractional differential equation with exogenous input described by the left generalized fractional derivative defined by Equation (12). Applying the ρ-Laplace transform to both sides of Equation (12), we obtain

$$\begin{aligned} \mathcal{L}_\rho\left(D^{\alpha,\rho}x(t)\right) - \left(I^{1-\alpha,\rho}x\right)(0) &= A\mathcal{L}_\rho\left(x(t)\right) + \mathcal{L}_\rho\left(Bu\right) \\ s^\alpha \bar{x}(s) - x_0 &= A\bar{x}(s) + B\bar{u}(s) \\ \bar{x}(s) - x_0\left(s^\alpha I_n - A\right)^{-1} &= \left(s^\alpha I_n - A\right)^{-1} B\bar{u}(s), \end{aligned} \tag{13}$$

where $\bar{x}$ denotes the Laplace transform of the function x and $\bar{u}$ denotes the Laplace transform of the function u. Applying the inverse of the ρ-Laplace transform to both sides of Equation (13), we obtain

$$\begin{aligned} x(t) &= x_0\left(\frac{t^\rho - t_0^\rho}{\rho}\right)^{\alpha-1} E_{\alpha,\alpha}\left(A\left(\frac{t^\rho - t_0^\rho}{\rho}\right)^\alpha\right) \\ &+ \int_{t_0}^t \left(\frac{t^\rho - s^\rho}{\rho}\right)^{\alpha-1} E_{\alpha,\alpha}\left(A\left(\frac{t^\rho - t_0^\rho}{\rho}\right)^\alpha\right) Bu(s)\frac{ds}{s^{1-\rho}}. \end{aligned} \tag{14}$$

Applying the Euclidean norm to both sides of Equation (14), we obtain the following relationship

$$\begin{aligned} \|x(t)\| &\leq \|x_0\|\left\|\left(\frac{t^\rho - t_0^\rho}{\rho}\right)^{\alpha-1} E_{\alpha,\alpha}\left(A\left(\frac{t^\rho - t_0^\rho}{\rho}\right)^\alpha\right)\right\| \\ &+ \|B\|\,\|u\|\int_{t_0}^t \left\|\left(\frac{t^\rho - s^\rho}{\rho}\right)^{\alpha-1} E_{\alpha,\alpha}\left(A\left(\frac{t^\rho - t_0^\rho}{\rho}\right)^\alpha\right)\frac{ds}{s^{1-\rho}}\right\|. \end{aligned} \tag{15}$$

From assumption $|\arg(\lambda(A))| > \frac{\alpha\pi}{2}$, there exist a positive number $M > 0$ [4,18,34] such that, we have

$$\int_{t_0}^t \left\|\left(\frac{t^\rho - s^\rho}{\rho}\right)^{\alpha-1} E_{\alpha,\alpha}\left(A\left(\frac{t^\rho - t_0^\rho}{\rho}\right)^\alpha\right)\frac{ds}{s^{1-\rho}}\right\| \leq M. \tag{16}$$

This inequality is a classic condition in stability analysis of fractional derivatives shown in [34]. Finally, the solution of the fractional differential Equation (12) described by the left generalized fractional derivative with exogenous input satisfies the following relationship

$$\|x(t)\| \leq \|x_0\|\left\|\left(\frac{t^\rho - t_0^\rho}{\rho}\right)^{\alpha-1} E_{\alpha,\alpha}\left(A\left(\frac{t^\rho - t_0^\rho}{\rho}\right)^\alpha\right)\right\| + \|B\|\,\|u\|\,M. \tag{17}$$

We first notice, when the exogenous input of the fractional differential Equation (12) described by the left generalized fractional derivative is null $\|u\| = 0$. The solution obtained in Equation (17) becomes

$$\|x(t)\| \leq \|x_0\| \left\| \left(\frac{t^\rho - t_0^\rho}{\rho} \right)^{\alpha-1} E_{\alpha,\alpha}\left(A \left(\frac{t^\rho - t_0^\rho}{\rho} \right)^{\alpha} \right) \right\|. \tag{18}$$

It corresponds to the classical Mittag-Leffler stability of the trivial solution of the fractional differential equation without input $D^{\alpha,\rho}x = Ax$ described by the left generalized fractional derivative.

Secondly, let us consider the exogenous input converging to zero when t tends to infinity. We know when the identity $|\arg(\lambda(A))| > \frac{\alpha\pi}{2}$ is held, we have

$$E_{\alpha,\alpha}\left(A \left(\frac{t^\rho - t_0^\rho}{\rho} \right)^{\alpha} \right) \longrightarrow 0. \tag{19}$$

From which we obtain $\|x(t)\| \longrightarrow 0$. Summarizing, we have the following

$$\|u\| \longrightarrow 0 \Longrightarrow \|x(t)\| \longrightarrow 0. \tag{20}$$

In other words, a converging input generates a converging state. This property is called the CICS property, derived in [15,18].

Finally, let us consider the exogenous input bounded ($\|u\| \leq \eta$). The solution of the fractional differential Equation (12) described by the left generalized fractional derivative satisfies the following relationship

$$\|x(t)\| \leq \|x_0\| \left\| \left(\frac{t^\rho - t_0^\rho}{\rho} \right)^{\alpha-1} E_{\alpha,\alpha}\left(A \left(\frac{t^\rho - t_0^\rho}{\rho} \right)^{\alpha} \right) \right\| + \|B\| \eta M. \tag{21}$$

Furthermore, we consider the function $\left(\frac{t^\rho - t_0^\rho}{\rho} \right)^{\alpha-1} E_{\alpha,\alpha}\left(A \left(\frac{t^\rho - t_0^\rho}{\rho} \right)^{\alpha} \right) \in \mathcal{L}$, thus there exist $\sigma > 0$ such that we have the following relationship

$$\left(\frac{t^\rho - t_0^\rho}{\rho} \right)^{\alpha-1} E_{\alpha,\alpha}\left(A \left(\frac{t^\rho - t_0^\rho}{\rho} \right)^{\alpha} \right) \leq \sigma. \tag{22}$$

Thus Equation (21) can be expressed in the following form

$$\|x(t)\| \leq \|x_0\| \sigma + \|B\| \eta M. \tag{23}$$

Thus, the solution of the fractional differential given in Equation (12) described by the left generalized fractional derivative is bounded as well. A bounded input for Equation (12), we obtain a bounded state for Equation (12). This property is called the BIBS property, created in [15,18]. The objective in this paper is to introduce a new stability notion taking into account a few things; namely the converging input, the converging state, the bounded input bounded state and the generalized Mittag-Leffler stability of the trivial solution of the unforced fractional differential equation. This stability notion we refer to as the generalized Mittag-Leffler input stability. In other words, the fractional differential equation described by the Left generalized fractional derivative is said generalized Mittag-Leffler stable, when its solution is bounded by a class $\mathcal{KL}$ function (contain a Mittag-Leffler function) plus a class $\mathcal{K}_\infty$ function proportional to the input of the given fractional differential equation. A similar derivation leading to Equation (23) has also recently been applied to the study of fixed-time stability in [35].

In this section, we introduce new stability notion in the context of the fractional differential equations described by the left generalized fractional derivative operator. The fractional differential equation under consideration is expressed in the following form

$$D^{\alpha,\rho}x = f(t,x,u) \tag{24}$$

where the function $f:\mathbb{R}^+\times\mathbb{R}\times\mathbb{R}^m\to\mathbb{R}^n$ is a continuous locally Lipschitz function, the function $x\in\mathbb{R}^n$ is a state variable, and furthermore the condition $f(t,0,0)=0$ is held. Given an initial condition $x_0\in\mathbb{R}^n$, the solution of the fractional differential Equation (24) starting at x_0 at time t_0 is represented by $x(.)=x(.,x_0,u)$.

Definition 12. *The solution $x=0$ of the fractional differential equation described by the left generalized fractional derivative defined by Equation (24) is said to be generalized Mittag-Leffler stable if, for any initial condition $\|x_0\|$ and initial time t_0, its solution satisfies the following condition*

$$\|x(t,\|x_0\|)\| \le \left[m(\|x_0\|)\left(\frac{t^\rho-t_0^\rho}{\rho}\right)^{\alpha-1}E_{\alpha,\alpha}\left(\eta\left(\frac{t^\rho-t_0^\rho}{\rho}\right)^{\alpha}\right)\right]^a, \tag{25}$$

where $a>0$, $\eta<0$ and the function m is locally Lipschitz on a domain contained in $\mathbb{R}^n$ and satisfies $m(0)=0$ [7,14].

In the following definition, we introduce the definition of the generalized Mittag-Leffler input stability in the context of the fractional differential equation described by the left generalized fractional derivative operator.

Definition 13. *The fractional differential equation described by the left generalized fractional derivative defined by Equation (24) is said to be generalized Mittag-Leffler input stable if, there exist a class $\gamma\in\mathcal{K}_\infty$ function such that for any initial condition $\|x_0\|$, its solution satisfies the following condition*

$$\|x(t,\|x_0\|)\| \le \left[m(\|x_0\|)\left(\frac{t^\rho-t_0^\rho}{\rho}\right)^{\alpha-1}E_{\alpha,\alpha}\left(\eta\left(\frac{t^\rho-t_0^\rho}{\rho}\right)^{\alpha}\right)\right]^a + \gamma\left(\|u\|_\infty\right), \tag{26}$$

where $a>0$ and $\eta<0$.

From the condition $\gamma\in\mathcal{K}_\infty$, we get $\gamma(0)=0$. We recover Definition 13. That is, the generalized Mittag-Leffler input stability of the fractional differential given in Equation (24) implies the generalized Mittag-Leffler stability of the trivial solution of the fractional differential equation with no input defined by $D^{\alpha,\rho}x=f(t,x,0)$. From the fact $\gamma\in\mathcal{K}_\infty$, when the input is bounded implies the function $\gamma\left(\|u\|_\infty\right)$ is bounded as well. Thus the state of the fractional differential Equation (24) is bounded too. We thus recover BIBS. From the fact $\gamma\in\mathcal{K}_\infty$, a converging input causes $\gamma\left(\|u\|_\infty\right)$ to converge. Thus the state of the fractional differential Equation (24) is converging as well. We thus recover CICS. In conclusion we can say that Definition 12 is well posed.

4. Lyapunov Characterizations of the Generalized Mittag-Leffler input Stability

In this section, we give the Lyapunov characterization of the generalized Mittag-Leffler input stability of the fractional differential equation. We know, it is not always trivial to get the analytical solution of the fractional differential equation with exogenous inputs. An alternative is to propose a method of analyzing the Mittag-Leffler input stability. The method consist of calculating the fractional energy of the fractional differential equation along the trajectories. In other words, we use the Lyapunov direct method.

Theorem 1. *Let us consider that there exists a positive function* $V : \mathbb{R}^+ \times \mathbb{R}^n \longrightarrow \mathbb{R}$ *continuous and differentiable, and a class* $\mathcal{K}_\infty$ *function* χ_1 *and class* $\mathcal{K}$ *functions* χ_2, χ_3 *satisfying the following assumptions:*

1. $\|x\|^a \leq V(t,x) \leq \chi_1(\|x\|)$.
2. *If for any* $\|x\| \geq \chi_2((|u|)) \Longrightarrow D_c^{\alpha,\rho}V(t,x) \leq -\chi_3((\|x\|))$.

Then the fractional differential Equation (24) described by the left generalized fractional derivative is generalized Mittag-Leffler input stable.

Proof. Summarizing [18], combining Assumption (1) and Assumption (2), we have

$$\begin{aligned} \|x\|^a &\leq V(x,t) \leq \alpha_1 \circ \alpha_2(\|u\|) \\ \|x\| &\leq (\alpha_1 \circ \alpha_2(\|u\|))^{1/a} \\ \|x\| &\leq \gamma(\|u\|), \end{aligned} \tag{27}$$

where the function $\gamma(\|u\|) = (\alpha_1 \circ \alpha_2(\|u\|))^{1/a} \in \mathcal{K}_\infty$.

From Assumption (2), using an exponential form of the Lyapunov function in, there exist positive constant such that, we have

$$\begin{aligned} \|x\| \geq \chi_2((|u|)) &\Longrightarrow D_c^{\alpha,\rho}V(t,x) \leq -\chi_3((\|x\|)) \\ &\Longrightarrow D_c^{\alpha,\rho}V(t,x) \leq -\chi_3((\|x\|)) \leq -kV(x,t). \end{aligned} \tag{28}$$

It follows from Equation (28), the following inequality

$$\begin{aligned} \|x\|^a &\leq V(t,x) \leq V(\|x_0\|)\left(\frac{t^\rho - t_0^\rho}{\rho}\right)^{\alpha\_1} E_{\alpha,\alpha}\left(-k\left(\frac{t^\rho - t_0^\rho}{\rho}\right)\right) \\ \|x\| &\leq \left\{V(\|x_0\|)\left(\frac{t^\rho - t_0^\rho}{\rho}\right)^{\alpha-1} E_{\alpha,\alpha}\left(-k\left(\frac{t^\rho - t_0^\rho}{\rho}\right)\right)\right\}^{1/a}. \end{aligned} \tag{29}$$

Finally, combining Equations (27) and (29), we obtain

$$\|x\| \leq \max\left\{\left\{V(\|x_0\|)\left(\frac{t^\rho - t_0^\rho}{\rho}\right)^{\alpha-1} E_{\alpha,\alpha}\left(-k\left(\frac{t^\rho - t_0^\rho}{\rho}\right)\right)\right\}^{1/a}; \gamma(\|u\|)\right\}. \tag{30}$$

Thus the fractional differential equation defined by Equation (24) is generalized Mittag-Leffler input stable. □

The second characterization is a consequence of the first theorem. It is more simplest to be applied in many cases. We have the following characterization.

Theorem 2. *Let there exist a positive function* $V : \mathbb{R}^+ \times \mathbb{R}^n \longrightarrow \mathbb{R}$ *continuous and differentiable, and a class* $\mathcal{K}_\infty$ *of functions* χ_1 *and class* $\mathcal{K}_\infty$ *function* γ *satisfying the following assumption:*

1. $\|x\|^a \leq V(t,x) \leq \chi_1(\|x\|)$.
2. $D_c^{\alpha,\rho}V(t,x) \leq -kV(x,t) + \gamma(\|u\|)$.

Then fractional differential Equation (24) described by the left generalized fractional derivative is generalized Mittag-Leffler input stable stable.

Proof. From Assumption (2), we have the following relationships

$$\begin{aligned} D_c^{\alpha,\rho}V(t,x) &\leq -kV(x,t)+\gamma(\|u\|) \\ D_c^{\alpha,\rho}V(t,x) &\leq -(1-\theta)kV(x,t)-\theta kV(x,t)+\gamma(\|u\|), \end{aligned} \tag{31}$$

where $\theta \in (0,1)$. We have

$$\begin{aligned} -\theta kV(x,t)+\gamma(\|u\|) \leq 0 &\implies D_c^{\alpha,\rho}V(t,x) \leq -(1-\theta)kV(x,t) \\ V(x,t) \geq \frac{\gamma(\|u\|)}{\theta k} &\implies D_c^{\alpha,\rho}V(t,x) \leq -(1-\theta)kV(x,t). \end{aligned} \tag{32}$$

From first assumption, it yields that

$$\theta k\chi_1(\|x\|) \geq \gamma(\|u\|) \implies D_c^{\alpha,\rho}V(t,x) \leq -(1-\theta)kV(x,t).$$

Thus the fractional differential equation described by the left generalized fractional derivative is Mittag-Leffler input stable. □

5. Practical Applications

In this section, we give many practical applications of the Mittag-Leffler input stability of the fractional differential equation described by the generalized fractional derivative using the Lyapunov characterizations.

Let us consider the fractional differential equation described by the left generalized fractional differential equation defined by

$$\begin{cases} D_c^{\alpha,\rho}x_1 = -x_1+\frac{1}{2}x_2+\frac{1}{2}u_1 \\ D_c^{\alpha,\rho}x_2 = -x_2+\frac{1}{2}u_2 \end{cases} \tag{33}$$

where $x = (x_1, x_2) \in \mathbb{R}^2$ and $u = (u_1, u_2) \in \mathbb{R}^2$ represents the exogenous input. Let us take the Lyapunov function defined by $V(x) = \frac{1}{2}\left(x_1^2+x_2^2\right)$. The left generalized fractional derivative of the Lyapunov function along the trajectories is given by

$$\begin{aligned} D_c^{\alpha,\rho}V(t,x) &= -x_1^2+\frac{1}{2}x_1x_2+\frac{1}{2}x_1u_1-x_2^2+\frac{1}{2}x_2u_2 \\ &\leq -\frac{1}{2}x_1^2-\frac{1}{2}x_2^2+\frac{1}{4}\|u\|^2 \\ &\leq -V(x)+\frac{1}{4}\|u\|^2 \end{aligned} \tag{34}$$

Consider $\gamma(\|u\|) = \frac{1}{4}\|u\|^2 \in \mathcal{K}_\infty$. It follows from Theorem 2, the fractional differential equation described by the left generalized fractional derivative given in Equation (33) is Mittag-Leffler input stable. Thus, the origin of the unforced fractional differential equation obtained with $u = (u_1, u_2) = (0, 0)$

$$\begin{cases} D_c^{\alpha,\rho}x_1 = -x_1+\frac{1}{2}x_2 \\ D_c^{\alpha,\rho}x_2 = -x_2 \end{cases} \tag{35}$$

where $x = (x_1, x_2) \in \mathbb{R}^2$, is Mittag-Leffler stable.

Let us consider the fractional differential equation described by the left generalized fractional differential equation defined by

$$\begin{cases} D_c^{\alpha,\rho}x_1 = -x_1+x_2+u_1 \\ D_c^{\alpha,\rho}x_2 = -x_2+u_2 \end{cases}, \tag{36}$$

where $x = (x_1, x_2) \in \mathbb{R}^2$ and $u = (u_1, u_2) \in \mathbb{R}^2$ represents the exogenous input. Let the Lyapunov function defined by $V(x) = \frac{1}{2}\left(x_1^2 + x_2^2\right)$. The left generalized fractional derivative of the Lyapunov function along the trajectories is given by

$$\begin{aligned} D_c^{\alpha,\rho} V(t,x) &= -x_1^2 + x_1 x_2 + x_1 u_1 - x_2^2 + x_2 u_2 \\ &\leq -x_1^2 + \frac{1}{2}x_1^2 + \frac{1}{2}x_1^2 - x_2^2 + \frac{1}{2}x_2^2 + \frac{1}{2}x_2^2 + \|u\|^2 \\ &\leq \|u\|^2 . \end{aligned} \tag{37}$$

Let $\gamma(\|u\|) = \frac{1}{4}\|u\|^2 \in \mathcal{K}_\infty$. It follows from Theorem 2, the fractional differential equation described by the left generalized fractional derivative in Equation (35) is bounded as well [36].

Let us consider the electrical RL circuit described by the left generalized fractional differential equation defined by

$$D_c^{\alpha,\rho} x = -\frac{\sigma^{1-\alpha} R}{L} x + u \tag{38}$$

with the initial boundary condition defined by $x(0) = x_0$. The parameter σ is associated with the temporal components in the differential equation. u represents the exogenous input. Let us take the Lyapunov function defined by $V(x) = \frac{1}{2}\|x\|^2$. The left generalized fractional derivative of the Lyapunov function along the trajectories is given by

$$\begin{aligned} D_c^{\alpha,\rho} V(t,x) &= \frac{\sigma^{1-\alpha} R}{L} x^2 + xu \\ &\leq -\frac{\sigma^{1-\alpha} R}{L}\|x\|^2 + \frac{1}{2}\|x\|^2 + \frac{1}{2}\|u\| \\ &\leq -\left(\frac{\sigma^{1-\alpha} R}{L} - \frac{1}{2}\right)\|x\|^2 + \frac{1}{2}\|u\| . \end{aligned} \tag{39}$$

Let us consider $k = \frac{\sigma^{1-\alpha} R}{L} - \frac{1}{2}$ and $\theta \in (0,1)$. We have the following relationship

$$D_c^{\alpha,\rho} V(t,x) \leq -(1-\theta)k\|x\|^2 + k\theta\|x\|^2 + \frac{1}{2}\|u\| \tag{40}$$

From Theorem 1, if $\|x\| \geq \frac{\|u\|}{2k\theta}$, we have $D_c^{\alpha,\rho} V(t,x) \leq -(1-\theta)k\|x\|^2$. Thus, the electrical RL circuit (36) is Mittag-Leffler input stable form.

Let us consider the fractional differential equation described in [4] by the left generalized fractional differential equation defined by

$$D_c^{\alpha,\rho} x = -x + xu, \tag{41}$$

where $x \in \mathbb{R}^n$ is a state variable. u represents the exogenous input. Let's the Lyapunov function defined by $V(x) = \frac{1}{2}\|x\|^2$. The left generalized fractional derivative of the Lyapunov function along the trajectories is given by

$$\begin{aligned} D_c^{\alpha,\rho} V(t,x) &= -x^2 + x^2 u \\ &\leq -\|x\|^2 + \|x\|^2\|u\| \\ &\leq -(1 - \|u\|)\|x\|^2 . \end{aligned} \tag{42}$$

We can observe, when we pick $\|u\| > 1$, using α-integration, the state x of Equation (42) diverge as t tends to infinity. Then the fractional differential equation is not BIBS. Thus, the fractional differential Equation (41) is not, in general, Mittag-Leffler input stable.

6. Conclusions

In this paper, the Mittag-Leffler input stability has been thoroughly investigated. We have tried to motivate this study with its connection to many real world applications known to use Mittag-Leffler functions. We also address the Lyapunov characterization of the fractional differential equations. In doing so, we have created a further Lyapunov characterization which is more useful. Finally, we give some numerical examples to help illustrate the work that was accomplished in this paper. Analyzing the generalized Mittag-Leffer input stability of the fractional differential equations without decomposing it can be non trivial. The possible issue is to decompose it as a cascade of triangular equations and to find a method to analyze the generalized Mittag-Leffer input stability of the obtained fractional differential equation. In other words, finding the conditions for the generalized Mittag-Leffer input stability of the fractional differential cascade equations will be subject of future works.

Author Contributions: N.S. was responsible for methodology, validation, conceptualization and formal analysis. G.S. was responsible for analysis, writing, review and draft preparation.

References

1. Kilbas, A.A.; Srivastava, H.M.; Trujillo, J.J. *Theory and Applications of Fractional Differential Equations*; North-Holland Mathematical Studies; Elsevier(North-Holland) Science Publishers: Amsterdam, The Nerherlands; London, UK; New York, NY, USA, 2006; Volume 204.
2. Fahd, J.; Abdeljawad, T. A modified Laplace transform for certain generalized fractional operators. *Results Nonlinear Anal.* **2018**, *2*, 88–98.
3. Souahi, A.; Makhlouf, A.B.; Hammami, M.A. Stability analysis of conformable fractional-order nonlinear systems. *Indagationes Math.* **2018**, *28*, 1265–1274. [CrossRef]
4. Sene, N. Lyapunov Characterization of the Fractional Nonlinear Systems with Exogenous Input. *Fractal Fract.* **2018**, *2*, 17. [CrossRef]
5. Sene, N. Fractional input stability for electrical circuits described by the Riemann–Liouville and the Caputo fractional derivatives. *Aims Math.* **2019**, *4*, 147–165. [CrossRef]
6. Sene, N. Stokes' first problem for heated flat plate with Atangana–Baleanu fractional derivative. *Chaos Solit. Fract.* **2018**, *117*, 68–75. [CrossRef]
7. Li, Y.; Chen, Y.Q.; Podlubny, I.; Cao, Y. Mittag-Leffler Stability of Fractional Order Nonlinear Dynamic Systems. *Automatica* **2009**, *45*, 1965–1969. [CrossRef]
8. Jarad, F.; Abdeljawad, T.; Baleanu, D. On the generalized fractional derivatives and their Caputo modification. *J. Nonlinear Sci. Appl.* **2017**, *10*, 2607–2619. [CrossRef]
9. Song, Q.; Yang, X.; Li, C.; Huang, T.; Chen, X. Stability analysis of nonlinear fractional-order systems with variable-time impulses. *J. Fran. Inst.* **2017**, *354*, 2959–2978. [CrossRef]
10. Tuan, H.T.; Trinhy, H. Stability of fractional-order nonlinear systems by Lyapunov direct method. *arXiv* **2018**, arXiv:1712.02921v2.
11. Makhlouf, A.B. Stability with respect to part of the variables of nonlinear Caputo fractional differential equations. *Math. Commun.* **2018**, *23*, 119–126.
12. Alidousti, J.; Ghaziani, R.K.; Eshkaftaki, A.B. Stability analysis of nonlinear fractional differential order systems with Caputo and Riemann–Liouville derivatives. *Turk. J. Math.* **2017**, *41*, 1260–1278. [CrossRef]
13. Guo, S.; Mei, L.; Li, Y. Fractional variational homotopy perturbation iteration method and its application to a fractional diffusion equation. *Appl. Math. Comput.* **2013**, *219*, 5909–5917. [CrossRef]
14. Sene, N. Exponential form for Lyapunov function and stability analysis of the fractional differential equations. *J. Math. Comp. Sci.* **2018**, *18*, 388–397. [CrossRef]
15. Sene, N. Mittag-Leffler input stability of fractional differential equations and its applications. *Discrete Contin. Dyn. Syst. Ser. S* **2019**, *13*. [CrossRef]
16. Shang, Y. Global stability of disease-free equilibria in a two-group SI model with feedback control. *Nonlinear Anal. Model Control* **2015**, *20* 501–508. [CrossRef]
17. Sene, N. Stability analysis of the generalized fractional differential equations with and without exogenous inputs. *J. Nonlinear Sci. Appl.* **2019**, *12*, 562–572. [CrossRef]

18. Sene, N. Fractional input stability of fractional differential equations and its application to neural network. *Discrete Contin. Dyn. Syst. Ser. S* **2019**, *13*. [CrossRef]
19. Li, C.; Chen, K.; Lu, J.; Tang, R. Stability and Stabilization Analysis of Fractional-Order Linear Systems Subject to Actuator Saturation and Disturbance. *IFAC-Papers On Line* **2017**, *50*, 9718–9723. [CrossRef]
20. Rezazadeh, H.; Aminikhah, H.; Sheikhani, A.H. Stability analysis of Hilfer fractional differential systems. *Math. Commun.* **2016**, *21*, 45–64.
21. Pagnini, G. Erdélyi-Kobar fractional diffusion. *J. Fract. Calc. Appl. Analy* **2012**, *15*, 117–127.
22. Atangana, A.; Baleanu, D. New fractional derivatives with nonlocal and non-singular kernel: Theory and application to heat transfer model. *arXiv* **2016**, arXiv:1602.03408.
23. Caputo, M.; Fabrizio, M. A new definition of fractional derivative without singular kernel. *Prog. Fract. Differ. Appl.* **2015**, *1*, 1–15.
24. Santos, M.A.F.D. Fractional Prabhakar derivative in diffusion equation with non-stochastic resetting. *Physica* **2019**, *1*, 5. [CrossRef]
25. Katugampola, U.N. New approach to a generalized fractional integral. *Appl. Math. Comput.* **2011**, *218*, 860–865. [CrossRef]
26. Gambo, Y.Y.; Ameen, R.; Jarad, F.; Abdeljawad, T. Existence and uniqueness of solutions to fractional differential equations in the frame of generalized Caputo fractional derivatives. *Adv. Diff. Equ.* **2018**, *2018*, 134 . [CrossRef]
27. Priyadharsini, S. Stability of fractional neutral and integrodifferential systems. *J. Fract. Calc. Appl.* **2016**, *7*, 87–102.
28. Jarad, F.; Gurlu, E.U.; Abdeljawad, T.; Baleanu, D. On a new class of fractional operators. *Adv. Diff. Equ.* **2017**, *2017*, 247. [CrossRef]
29. Adjabi, Y.; Jarad, F.; Abdeljawad, T. On generalized fractional operators and a Gronwall type inequality with applications. *Filo* **2017**, *31*, 5457–5473. [CrossRef]
30. Sene, N. Analytical solutions of Hristov diffusion equations with non-singular fractional derivatives. *Chaos* **2019**, *29*, 023112. [CrossRef] [PubMed]
31. Abdeljawad, T.; Mert, R.; Peterson, A. Sturm Liouville equations in the frame of fractional operators with exponential kernels and their discrete versions. *Quest. Math.* **2018**, 1–19. [CrossRef]
32. Shang, Y. Vulnerability of networks: Fractional percolation on random graphs. *Phys. Rev. E* **2015**, *89*, 812–813. [CrossRef]
33. Chaillet, A.; Loria, A. Uniform global practical asymptotic stability for non-autonomous cascaded systems. *Eur. J. Contr.* **2006**, *12*, 595–605. [CrossRef]
34. Qian, D.; Li, C.; Agarwal, R.P.; Wong, P.J. Stability analysis of fractional differential system with Riemann–Liouville derivative. *Math. Comput. Model.* **2010**, *52*, 862–874. [CrossRef]
35. Shang, Y. Fixed-time group consensus for multi-agent systems with non-linear dynamics and uncertainties. *IET Control Theory Appl.* **2017**, *12*, 395–404. [CrossRef]
36. Khader, M.M.; Sweilam, N.H.; Mahdy, A.M.S. An Efficient Numerical Method for Solving the Fractional Diffusion Equation. *J. Appl. Math. Bioinf.* **2011**, *1*, 1–12.

5

Collaborative Content Downloading in VANETs with Fuzzy Comprehensive Evaluation

Tianyu Huang [1], Xijuan Guo [1,*], Yue Zhang [1] and Zheng Chang [2]

[1] Colleage of Information Science and Engineering, Yanshan University, Qinhuangdao 066004, China; htyu3000@163.com (T.H.); zhangyue.1224@foxmail.com (Y.Z.)

[2] Department of Mathematical Information Technology, University of Jyväskylä, P.O. Box 35, FIN-40014 Jyväskylä, Finland; zheng.chang@jyu.fi

* Correspondence: xjguo@ysu.edu.cn.

Abstract: Vehicle collaborative content downloading has become a hotspot in current vehicular ad-hoc network (VANET) research. However, in reality, the highly dynamic nature of VANET makes users lose resources easily, and the transmission of invalid segment data also wastes valuable bandwidth and storage of the users' vehicles. In addition, the individual need of each customer vehicle should also be taken into consideration when selecting an agent vehicle for downloading. In this paper, a novel scheme is proposed for vehicle selection in the download of cooperative content from the Internet, by considering the basic evaluation information of the vehicle. To maximize the overall throughput of the system, a collaborative content downloading algorithm is proposed, which is based on fuzzy evaluation and a customer's own expectations, in order to solve the problems of agent vehicle selection. With the premise of ensuring successful downloading and the selection preferences of customer vehicles, linear programming is used to optimize the distribution of agent vehicles and maximize customer's satisfaction. Simulation results show that the proposed scheme works well in terms of average quality of service, average bandwidth efficiency, failure frequency, and average consumption.

Keywords: vehicle collaborative content downloading; fuzzy comprehensive evaluation; VANET

---

1. Introduction

With the rapid development of the network, the demand of the network extends to all aspects of people's lives. As a platform, which provides a specific network service, the vehicular ad-hoc network (VANET) brings new technical challenges to the transmission of information while providing information services, including: how to improve the efficiency of the vehicle network and how to meet the continuous improvement of the users' needs [1,2].

Scholars have discussed different ways to modify VANET, in order to improve the performance of vehicle networking and meet the growing demand of users. In terms of enhancing the performance of the vehicle-to-infrastructure (V2I) connection in VANET, the possibility of constructing a network with the TV white space geolocation database for vehicle networking was discussed by some scholars. Then vehicular communication architectures were proposed to mitigate the resulting high spectrum demands and provide vehicular connectivity with wider communication range, higher transmission rate, and lower data transfer cost [3,4]. By analyzing the end-to-end transmission performance from individual vehicles to a road side unit (RSU), an efficient message routing scheme was put forward to balance the data traffic across the network and improve the network throughput [5]. In Reference [6], a collaborative download algorithm, namely maximum throughput and minimum delay collaborative download (MMCD) was proposed, which minimizes the average transmission delay of each user's request and maximizes the number of packets downloaded from an RSU. Reference [7] mainly studies the cost of minimizing the download of a hybrid vehicle ad hoc network, and proposes the basic satisfaction algorithm (BMA) and heuristic algorithm (TSA) to solve the huge download delay caused by vehicle mobility in VANET. In order to solve the frequent collision among agent vehicles and customer vehicles, a transmission scheduling method was put forward to adjust the relationship between link routing and transmission time [8].

In terms of enhancing the vehicle-to-vehicle (V2V) collaborative download performance in VANET: ECDS gives an efficient collaborative downloading solution to popular content distribution in urban vehicle networks. Furthermore, a cross-domain relay selection strategy was proposed to build a peer-to-peer (P2P) network, which helps strengthen information dissemination [9]. In Reference [10], to solve the problems of popular content distribution (PCD) in a highway scene, the author modeled the problem as a coalition formation game with transferable utilities, and proposed a coalition formation algorithm that converges into a Nash-stable partition, adapting to environmental changes as a result of the VANET's rapid and unpredictable topological changes. In Reference [11], the design incentive mechanism is employed to propose a collaborative downloading method, which encourages cooperation between vehicles and helps users effectively obtain the required resources. The author designed a server-assisted key management scheme that promotes cooperation and ensures fairness and efficiency. In the scheme, vehicles with common interests form a cluster and take turns as the cluster head, which directly downloads data packets from the Internet and V2V shares the content with surrounding vehicles [12]. A delicate linear cluster formation scheme is proposed and applied to significantly enhance the probability of a successful file download in VANET [13]. In Reference [14], the author proposed a security incentive program (SIRC) to achieve reliable, fair, and secure collaborative downloading in VANET. SIRC stimulates vehicle users to help each other download and forward packets, encourages cooperation between users, and also punishes malicious vehicles to ensure the safety of vehicles. Efficient privacy-preserving cooperative data downloading for value-added services is used to solve the problems of limited communication range and high dynamics, which gains the access control in VANET [15].

The methods of improving the performance of the vehicle network are also discussed from other aspects. Digital fountain code (DFC) is proposed and applied in the field of cooperative downloading for VANET. As long as enough data packets encoded by hierarchical fountain code are available, the client can recover the raw data and avoid data transmission interruption [16,17]. In Reference [18], a fuzzy logic-based resource management (FLRM) scheme was proposed, and the lifetime of each storage resource was defined by the proposed fuzzy logic-based popularity evaluation algorithm.

1.1. Related Work

The agent vehicle selection method and vehicle distribution scheme are important links to achieve collaborative downloading. A fuzzy logic-based cooperative file transfer scheme (FL-CFT) was proposed to optimally select relays to help transfer the file and ensure the file integrity, in which the relative velocity, distance, and predicted connection time among vehicles were considered [19]. To solve the problems of the low utilization of spatiotemporal resources in DA and an unbalanced service of cooperative downloading, a balanced cooperative downloading method was proposed, which dynamically uses the Euclidean and Manhattan distances in order to select the vehicles according to the number of clients [20]. In Reference [21], a k-hop bandwidth aggregation scheme was proposed to select agent vehicles, to help download and forward videos and more effectively send video streams to requesters through DSRC VANET. In Reference [22], a preferential response incentive mechanism (PRIM) was proposed to motivate vehicles to participate in collaborative downloading, and game theory was used to analyze a vehicle's behavior in order to find the optimal strategy for each collaborator, reduce repeated downloads, and promote V2I cooperation to reduce delays and expenses. In Reference [23], a security collaboration data download framework for paid services in VANET was proposed. An application layer data sharing protocol was developed to coordinate vehicle data sharing according to its location. The seed screening scheme SIEVE was proposed in

Reference [24], using users' interest and near-term contact predictions to select the best vehicle node (vehicle) to download the object (via the cellular network) and propagate the object (via the RSU). In order to effectively characterize users' preferences and network performance, previous authors use parameters such as energy efficiency, signal intensity, network cost, delay, and bandwidth to construct utility functions. Then, these utility functions and multi-criteria utility theory are used to construct an energy-efficient network selection approach and a joint multi-criteria utility function for network selection of the appropriate access network [25].

1.2. Motivation and Contributions

In fact, the goal of a cooperative downloading method is to ensure more efficient data transmission, provide balanced services, and meet the requirements of all customers on the agent vehicles, so that the customers' cooperative unloading requirements can be satisfied. Based on the ideas above, this paper proposes a vehicle selection algorithm for the vehicle network agent based on fuzzy comprehensive evaluation. This algorithm takes the basic parameters of the customer vehicle, the agent vehicle, and the relationship between them into account, and improves the average throughput and customer satisfaction under the condition of satisfying a customer vehicle's information data requests. Compared with the previous articles, the contributions of this paper are in four aspects:

- We provide a fuzzy evaluation method based on the relationship between the agent vehicle and the customer vehicle, and evaluate the agent vehicle synthetically. In our opinion, we can judge whether the vehicle is suitable for cooperation by its relevant attributes. These attributes include computing capability, bandwidth, unit cost, credibility, and path consistency between vehicles, which are meaningful data for vehicle selection. Therefore, using this information as the evaluation factor for the fuzzy comprehensive evaluation, corresponding agent vehicles for each customer vehicle are scored, and the vehicles with higher scores are selected as the priority.
- In order to satisfy the requests of more customer vehicles and maximize resource utilization, this paper proposes an agent vehicle distribution strategy based on the maximization of service quality. Our approach allocates a certain number of agent vehicle resources to each customer vehicle, and takes the bandwidth limitation of the agent vehicle into consideration, so as to select the most suitable agent vehicle for the customer vehicle and maximize overall resource utilization.
- In a simulation, the performance of the proposed algorithm is compared with other schemes. The simulation results show that the proposed algorithm can gain significant performance achievements, which demonstrates the superiority of the scheme.
- By analyzing the fuzzy relationship between multiple constraints on the target, the fuzzy comprehensive evaluation method quantifies and unifies the relationship as an index to realize vehicle selection. This method woks well in dealing with the problems of fuzzification that are constrained by many factors. Additionally, it can be used as a reference for the solutions of multi-factor constraint model problems such as mobile vehicle network selection problems, vehicle routing problems in complex environments, and so on.

The organization of this paper is as follow. Section 2 describes the system model used by this scenario. Section 3 explains, in detail, the vehicle network cooperation content downloading method, based on fuzzy comprehensive evaluation, proposed in this paper. Section 4 shows our simulation results and discussion. Finally, Section 5 summarizes the method of this paper and points out the future work.

2. System Model

In this section, the model is first introduced, and some parameters are defined, including the vehicle evaluation index and the format of the data packet.

2.1. System Model

In the system model shown in Figure 1, the vehicles are grouped as customer vehicles and agent vehicles. Customer vehicles send download requests to the local server (RC), and agent vehicles are responsible for helping them download the requested content. The customer and agent vehicles together form a VANET.

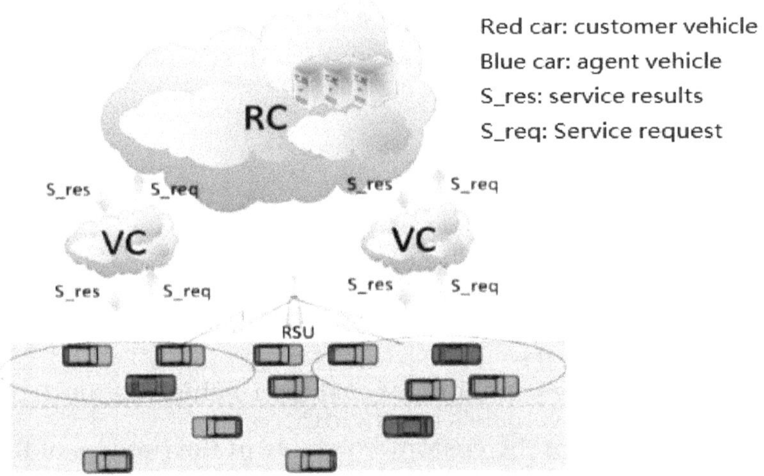

Figure 1. Comprehensive evaluation index system.

When passing by an RSU, a vehicle in the vehicle cloud downloads the response file (vehicle-to-infrastructure, V2I); when leaving the RSU affected area, the vehicles in the VANET share the downloaded files (vehicle-to-vehicle, V2V). The V2I and V2V process forms a circulation, and many such circulations have to be gone through to complete the download of a large file.

In this paper, we make the following assumptions:

- The customer vehicle which requests cooperation, selects an agent vehicle only once every period. When the agent vehicle is selected, its map route must be consistent with the customer vehicle.
- The local server can obtain the vehicle's navigation information (that is, the driving route of each vehicle on a map), in order to allocate an agent vehicle traveling on the same road section as the customer vehicle and reduce the waste of resources. The vehicle uploads any relevant information to the local server. The local server selects the agent vehicle for the customer vehicle according to the scheme proposed herein.
- On the basis of content consistency, the local server counts the request information of the customer vehicle and the service information of the agent vehicle. There are two forms of vehicle computing capability. The first form is collaborative computing. In this case, the computing capability of the agent vehicle is determined by the hardware of the vehicle itself, which represents the total amount of data that the agent vehicle needs to receive and send in the service. In the second form, the request of the customer vehicle is content downloading. It is set that the storage and removal of files in the vehicle are in chronological order. In this case, the computing capability of the customer vehicle is a request for the files that have not been downloaded yet, which can be part of a file or an entire file. The computing capability of the agent vehicle is the part of the file reserved in the current storage, which can be part of the file or the entire file. In this paper, the method for obtaining data from the agent vehicle will not be discussed and we assume that the agent vehicle has had the corresponding computing capability before providing services to the customer vehicle.

Failure of a customer vehicle's request will occur due to the following:

- If there is only one agent vehicle for more than one customer vehicle, the selected agent vehicle cannot serve more than one vehicle, and it will be allocated to the customer vehicle with the greatest satisfaction. The other customer vehicles' requests will fail.
- According to Algorithm 1, a request fails if the customer vehicle cannot find an agent vehicle that meets its requirements.

This paper mainly studies the content downloading through V2V in VANET, and focuses on how to choose the best cooperators for a customer vehicle. According to the relevant data of the vehicles, under the premise of satisfying the cooperation standard expected by the customer vehicle, the overall rating of the agent vehicle is maximized, so each customer obtains a satisfactory downloading experience.

2.2. *Definitions*

In order to record the data set, the packet format for the customer vehicle (CV) and the agent vehicle (AV) are defined respectively as:

- Request package of customer vehicle:
  - CV-ID: Customer vehicle's ID;
  - CV-computing: Computing capability of customer vehicle request;
  - CV-bandwidth: Customer vehicle's bandwidth;
  - CV-path: The travel route of the customer vehicle in the process of the data request;
  - CV-position: Customer vehicle's position;
  - CV-speed: Customer vehicle's speed;
- Service package of agent vehicle:
  - AV-ID: Agent vehicle's ID;
  - AV-computing: Agent vehicle's computing capability;
  - AV-bandwidth: Agent vehicle's bandwidth;
  - AV-path: The travel route of the agent vehicle in the process of the data service;
  - AV-credit: Agent vehicle's credit;
  - AV-position: Agent vehicle's position;
  - AV-cost: Service cost of agent vehicle in unit time;
  - AV-speed: Agent vehicle's speed;
- The format of the reply message of the local server is as follow:
  - Server-ID: ID of the local server that communicates with the current vehicle;
  - Reply (N = AV-ID): if the reply message is 0, the local server finds the agent vehicle. If the reply message is a series of numbers (which are defined as positive integers), they represent the IDs of all the agent vehicles assigned to it by the local server;

Therefore, the vehicle and the local server use the information as an evaluation factor in the communication process to complete the evaluation of the vehicle. A comprehensive evaluation index system is designed, as shown in Table 1.

Table 1. Comprehensive evaluation index system.

Target Layer	Factor Layer
Vehicle selection result	Computing capability Bandwidth Unit cost Credibility Path consistency

The computing capability is determined by the customer vehicle's requirement data and agent vehicle's service data. Bandwidth is determined by the hardware properties of the vehicle. Unit cost is the remuneration to be paid per unit time when the service is provided by the agent vehicle. Credibility is the score given on cooperation in the vehicle's historical records, which is evaluated in VANET. If it can serve the customer vehicle very well every time, the score will be high; if there is a malicious termination of the cooperation, the behavioral reputation value will be correspondingly reduced. Path consistency represents the proportion of path that maintains communication between an agent and customer vehicle in the total path.

In addition, due to the mobility of the vehicle, datagrams will be updated every time period to ensure good transmission. In the next section, the vehicle selection method based on fuzzy comprehensive evaluation will be introduced in detail.

3. Vehicle Network Collaborative Content Downloading Method Based on Fuzzy Comprehensive Evaluation

In this section, we describe the specific method for the local server to select an agent vehicle for a customer vehicle, in detail. The fuzzy comprehensive evaluation model is also introduced to make a fuzzy comprehensive evaluation of the factors affecting the vehicle selection in agent vehicle unloading. The choice of vehicles tends to be optimal.

The detailed communication process of finding agent vehicles is as follow:

- Several customer vehicles send request packets to a local server. A request packet contains the requirements for an agent vehicle and the relevant information of the customer vehicle itself.
- After the local server receives the message, it uses the fuzzy comprehensive evaluation method proposed in this paper to analyze the request packet of the customer vehicle and the service packet of an agent vehicle. Then it forms the distribution plan of the agent vehicle for the customer vehicle, and sends a response message back to them.
- Response message. If the message is 0, it means that the local server did not find an agent vehicle and the customer vehicle needs to wait for the next assignment. If the message is a series of numbers (which are defined as positive integers, indicating the IDs of all the agent vehicles assigned by the local server), it means that the distribution of agent vehicles was successful, and the local server notifies the agent vehicle to serve the corresponding customer vehicle according to the allocation plan.
- After the entire communication is over, the local server records the evaluation of the agent vehicle, to update the credibility of the agent vehicle. A penalty mechanism is established to punish a vehicle which is rated poorly by the customer vehicle in this cooperation. A punished vehicle is unable to participate in the next cooperation and cannot obtain the expected rewards.

This section mainly evaluates objective ratings and customers' satisfaction for agent vehicles in the decision domain based on certain fuzzy constraints. Agent vehicles with higher scores in comprehensive evaluation should be given priority, while those with lower scores should be given a second thought, when selecting vehicles based on the demand.

3.1. Pre-Selection of Agent Vehicles

To find an appropriate agent vehicle for the customer vehicle from a large number of vehicles, in order to meet their information requests in the process of routing, we need to establish an information selecting mechanism. In the mechanism, the relationships between a customer vehicle's and an agent vehicle's information are compared and analyzed, to meet the customer's data requests. Alternative vehicles should meet the following requirements:

- Computing capability c: Computing capability is the main content of requests for customer vehicles. For the agent vehicle, it decides whether it can serve the customer vehicle or not. Computing capability c_j provided by the agent vehicle j should be better than or equal to the computing capability c_i requested by the customer vehicle i; so as to meet the demand of the customer vehicle:

$$c_j \geq c_i.$$

- Bandwidth b: Bandwidth determines the fluency of a customer vehicle's data request. The bandwidth b_j provided by the agent vehicle j should be better than or equal to the bandwidth b_i requested by the customer vehicle i, so as to meet the need of the customer vehicle:

$$b_j \geq b_i.$$

- Agent vehicle j should satisfy customer i's requests for computing capability within the time of collaboration between the two vehicles. L_a is the effective distance between the customer and agent vehicles, and if the distance between them exceeds L_a, then the connection will fail. L is the path length. v_i is the average speed of the customer vehicle. v_j is the average driving speed of the agent vehicle. Thus:

$$\frac{c_i}{b_i} \leq \min\left(\frac{L_a}{|v_i - v_j|}, \frac{L}{v_i}\right).$$

- Path consistency determines the time length of the service that a customer vehicle obtains from the agent vehicle. It indicates whether the customer vehicle can get complete service from the agent vehicle or not. The path consistency pc_{ij} is calculated to express the ratio of the effective signal path to the whole path when the agent vehicle provides data service to the customer vehicle:

$$pc_{ij} = \frac{L_a v_i}{|v_i - v_j| L}.$$

Based on the requirements above, we filter the agent vehicles according to Algorithm 1, and record the information of the selected agent vehicles for each customer vehicle.

Algorithm 1 Attaining the Available Agent Vehicle List

Input: Customer vehicle request package; agent vehicle service package; signal effective distance L_a; path length L;
Output: Available agent vehicle (AV) list N_i and path consistency pc_{ij}, for each customer vehicle CV_i

for each $CV_i, i \in [1, n]$ **do**
    for each $AV_j, j \in [1, m]$ **do**
        if $c_i \leq c_j$ & $b_i \leq b_j$ & $\frac{c_i}{b_i} \leq \min\left(\frac{L_a}{|v_i - v_j|}, \frac{L}{v_i}\right)$ **then**
            write AV_j into the list N_i
            $pc_{ij} = \frac{L_a v_i}{|v_i - v_j| L}$
        end if
    end for
end for

3.2. *Comprehensive Evaluation of Customer Satisfaction*

3.2.1. The Determination of the Domain and Various Factors of Agent Vehicles:

Based on the illustration above, the factor domain of agent vehicles is recorded as: $U = \{u_1, u_2, u_3, u_4, u_5\}$,where u_1 is the computing capability; u_2 is the bandwidth; u_3 is the unit cost; u_4 is the credibility; u_5 is the path consistency;

Among them, each factor belongs to a different domain, i.e.,

$$\mu_1 \in D_1 = (\chi_1, \psi_1);$$
$$\mu_2 \in D_2 = (\chi_2, \psi_2);$$
$$\mu_3 \in D_3 = (\chi_3, \psi_3);$$
$$\mu_4 \in D_4 = (\chi_4, \psi_4);$$
$$\mu_5 \in D_5 = (\chi_5, \psi_5);$$

The local server divides the data sets of factors u_1, u_2, u_3, u_4, and u_5 into three categories: low, medium, and high. These are represented by V_1, V_2, and V_3, respectively, and the level domain of each factor is $V = \{V_1, V_2, V_3\}$, which corresponds to the numerical values $\{1, 2, 3\}$, in order.

If the fuzzy experiment determines the first division of the factor u_i on the domain (χ_i, ψ_i), one pair can be determined for each division: (ξ_{u_i}, η_{u_i}), where ξ_{u_i} is the demarcated point between $V_{1\_u_i}$ and $V_{2\_u_i}$ and η_{u_i} is the demarcated point between $V_{2\_u_i}$ and $V_{3\_u_i}$.

On the contrary, if (ξ, η) is given, the mapping e is also determined, and $V_{1\_u_i}, V_{2\_u_i}, V_{3\_u_i}$ are separated, thus the fuzzy concept is clarified.

The interval of $V_{1\_u_i}, V_{2\_u_i}, V_{3\_u_i}$ is a random interval, and so ξ_{u_i} and η_{u_i} are random variables. They follow characteristic normal distributions, as shown in Figure 2, namely: $\xi_{u_i} : N(\alpha_{1\_u_i}, \sigma^2_{1\_u_i})$; $\eta_{u_i} : N(\alpha_{2\_u_i}, \sigma^2_{2\_u_i})$

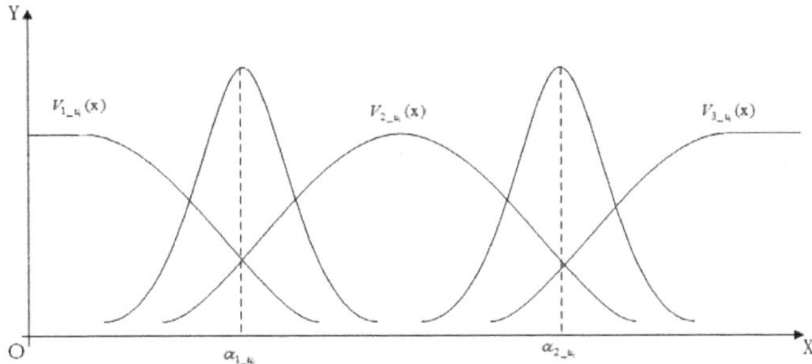

Figure 2. The normal distribution properties.

Based on the definition of each factor, the values $\alpha_{1\_u_i}, \alpha_{2\_u_i}, \sigma^2_{1\_u_i}$, and $\sigma^2_{2\_u_i}$ are determined.

For factor u_1, u_2, u_4, and u_5 (i.e., computing capability, bandwidth, credibility, and path consistency), the bigger they are, the better it is for the vehicle cooperative downloading. Thus they are defined as:

$$\alpha_{1\_u_i} = \omega(u_{i\_\min} + u_{i\_ave}), 0 < \omega < 1$$

$$\alpha_{2\_u_i} = \vartheta(u_{i\_\max} + u_{i\_ave}), 0 < \vartheta < 1$$

For factor u_3 (unit cost), the smaller the better in consideration of a user's benefits. Thus, they are defined as:

$$\alpha_{1\_u_i} = \omega(u_{i\_\max} + u_{i\_ave}), 0 < \omega < 1$$

$$\alpha_{2\_u_i} = \vartheta(u_{i\_\min} + u_{i\_ave}), 0 < \vartheta < 1$$

In order to make the distribution of demarcated points relatively centralized:

$$0 < \sigma^2_{1\_u_i} \leq 1, 0 < \sigma^2_{2\_u_i} \leq 1$$

$$u_{i\_\max} = \max(u_{i1}, u_{i2}, \ldots, u_{in}), u_{il} \in u_i$$

$$u_{i\_\min} = \min(u_{i1}, u_{i2}, \ldots, u_{in}), u_{il} \in u_i$$

$$u_{\text{ave}} = \frac{u_{i1}, u_{i2}, \ldots, u_{in}}{n}, u_{il} \in u_i$$

3.2.2. Determination of Membership Functions

The membership function of factors $u_i \in (\chi_i, \psi_i)$ with the three levels $V_{1\_u_i}$, $V_{2\_u_i}$, and $V_{3\_u_i}$ is determined by the three division method. The three division method is a fuzzy statistical method that determines the membership function with three levels of fuzzy concepts. The basic principles of this method are as follow.

From above, we know that the partition (ξ_{u_i}, η_{u_i}) of the factor u_i on the universe (χ_i, ψ_i) obeys a normal distribution: ξ_{u_i} obeys $N(\alpha_{1\_u_i}, \sigma^2_{1\_u_i})$, and η_{u_i} obeys $N(\alpha_{2\_u_i}, \sigma^2_{2\_u_i})$.

Furthermore, the number of (ξ, η) determines the mapping $e\,(\xi_i, \eta_i) : U \to \{V_{1\_u_1}, V_{2\_u_2}, V_{3\_u_3}\}$, which is:

$$e\,(\xi_{u_i}, \eta_{u_i})\,(x) = \begin{cases} V_{1\_u_i}(x) & , x \le \xi_{u_i} \\ V_{2\_u_i}(x), \xi_{u_i} < x < \eta_{u_i} & \\ V_{3\_u_i}(x) & , \eta_{u_i} < x \end{cases} \tag{1}$$

The value $P\left(x \le \xi_{u_i}\right)$ is the probability that the random variable x falls in the interval $[x, b)$. If x increases, $[x, b)$ becomes smaller, and the probability of falling in the interval $[x, b)$ also becomes smaller. This character of probability $P\left(x \le \xi_{u_i}\right)$ is the same as the "low" fuzzy set $V_{1\_u_i}$, so $V_{1\_u_i}(x) = P\{x \le \xi_{u_i}\} = \int_x^{\infty} P_{\xi_{u_i}}(x)dx$. Similarly $V_{3\_u_i}(x) = P\{\eta_{u_i} < x\} = \int_x^{\infty} P_{\eta_{u_i}}(x)dx$. In these expressions $P_{\xi_{u_i}}(x)$ and $P_{\eta_{u_i}}(x)$ are the probability densities of the random variable ξ_{u_i} and η_{u_i} respectively, and $V_{2\_u_i}(x) = 1 - V_{1\_u_i}(x) - V_{3\_u_i}(x)$.

Calculated in the probabilistic method, the membership function of each level can be obtained:

$$V_{1\_u_i}(x) = 1 - \Phi\left(\frac{x - a_{1\_u_i}}{\sigma_{1\_u_i}}\right) \tag{2}$$

$$V_{3\_u_i}(x) = \Phi\left(\frac{x - a_{2\_u_i}}{\sigma_{2\_u_i}}\right) \tag{3}$$

$$V_{2\_u_i}(x) = \Phi\left(\frac{x - a_{1\_u_i}}{\sigma_{1\_u_i}}\right) - \Phi\left(\frac{x - a_{2\_u_i}}{\sigma_{2\_u_i}}\right), \tag{4}$$

where $\Phi(x) = \int_{-\infty}^{x} \frac{1}{\sqrt{2\pi}} e^{-\frac{t^2}{2}} dt$.

However, for the convenience of presentation we still use $V_{1\_u_i}$, $V_{2\_u_i}$, and $V_{3\_u_i}$ to represent the three level membership function of a factor $u_i \in (\chi_i, \psi_i)$.

3.2.3. Constructing the Fuzzy Evaluation Matrix

From the above, the membership function of each factor u_i can be obtained. Bringing the data of the five factors of an agent vehicle j into the corresponding membership functions, the relationship between the five factors u_i and the grading of V can be obtained as:

$$(V^j_{1\_u_i}(x), V^j_{2\_u_i}(x), V^j_{3\_u_i}(x))$$

Thus a fuzzy relation matrix for vehicle j can be obtained:

$$\mathrm{R}^j = \begin{bmatrix} V^j_{1\_u_1}(x) & V^j_{2\_u_1}(x) & V^j_{3\_u_1}(x) \\ V^j_{1\_u_2}(x) & V^j_{2\_u_2}(x) & V^j_{3\_u_2}(x) \\ V^j_{1\_u_3}(x) & V^j_{2\_u_3}(x) & V^j_{3\_u_3}(x) \\ V^j_{1\_u_4}(x) & V^j_{2\_u_4}(x) & V^j_{3\_u_4}(x) \\ V^j_{1\_u_5}(x) & V^j_{2\_u_5}(x) & V^j_{3\_u_5}(x) \end{bmatrix} = \begin{bmatrix} r^j_{1,1} & r^j_{1,2} & r^j_{1,3} \\ r^j_{2,1} & r^j_{2,2} & r^j_{2,3} \\ r^j_{3,1} & r^j_{3,2} & r^j_{3,3} \\ r^j_{4,1} & r^j_{4,2} & r^j_{4,3} \\ r^j_{5,1} & r^j_{5,1} & r^j_{5,3} \end{bmatrix}. \tag{5}$$

Assuming that there are m agent vehicle participating in the evaluation, m fuzzy relation matrices will be obtained for each customer vehicle i: $R^1{}_i, R^2{}_i, \cdots, R^m{}_i$. In Algorithm 1, we select the agent vehicles that meet the customer vehicle's needs.

3.2.4. Determination of the Weight a_k

The importances of the five factors in the comprehensive evaluation system are not the same. If the status is important, it should be given a greater weight; otherwise, it should be given a smaller weight.

Assume that the weight set is $A = \{a_1, a_2, ..., a_5\}$, where $\sum_{i=1}^{5} a_i = 1$.

3.2.5. Fuzzy Comprehensive Evaluation

A and R^j are used in a fuzzy synthesis operation: $A \circ R^j = B^j$ to obtain a comprehensive evaluation $B^j = (b_1^j, b_2^j, b_3^j)$ for the agent vehicle j. Here, $B^j = A^\circ R^j = \min(1, \sum_{i=1}^{n} a_i \cdot r_{ij})$ considers the degree of subordination of the agent vehicle j. Then according to the principle of maximum subordination, we can get the evaluation level of agent vehicle j:

$$B^j = A \circ R^j = (a_1, a_2, a_3, a_4, a_5) \begin{bmatrix} r^j_{1,1} & r^j_{1,2} & r^j_{1,3} \\ r^j_{2,1} & r^j_{2,2} & r^j_{2,3} \\ r^j_{3,1} & r^j_{3,2} & r^j_{3,3} \\ r^j_{4,1} & r^j_{4,2} & r^j_{4,3} \\ r^j_{5,1} & r^j_{5,1} & r^j_{5,3} \end{bmatrix} = (b_1^j, b_2^j, b_3^j). \tag{6}$$

Here $b_k^j = (a_1 \bullet r_{1k}) \oplus (a_2 \bullet r_{2k}) \oplus ... \oplus (a_5 \bullet r_{5k})$. Additionally, the fuzzy synthesis operator "$\circ$" selects the fuzzy operator $M(\bullet, \oplus)$. In the fuzzy operator $M(\bullet, \oplus)$, $\bullet$ is defined as multiplication, and $\oplus$ is defined as the operation $x \oplus y = \min(1, x + y)$. Thus, $B^j = A \circ R^j = \min(1, \sum_{i=1}^{n} a_i \cdot r_{ij})$.

The following normalization is performed on B:

$$B^j = \left(\frac{b_1^j}{\sum_{i=1}^{3} b_i^j}, \frac{b_2^j}{\sum_{i=1}^{3} b_i^j}, \frac{b_3^j}{\sum_{i=1}^{3} b_i^j} \right) \triangleq (C_1^j\%, C_2^j\%, C_3^j\%) \tag{7}$$

An understanding of B can be achieved through the following example: for an agent vehicle j, its comprehensive evaluation $B^j = (10\%, 50\%, 40\%)$ indicates that taking the five factors of the agent vehicle j into consideration, 10% of vehicles evaluate it as "low", 50% of vehicles evaluate it as "medium", and 40% of vehicles evaluate it as "high". According to the principle of maximum degree of membership, the evaluation level of agent vehicle j is "medium".

Next, based on the quantized value of the fuzzy comment set, that is:

$$V = \{V_1, V_2, V_3\} = \{1, 2, 3\},$$

the overall rating of the agent vehicle j is:

$$E_j = B^j V^T = (B^1, B^2, B^3, B^4, B^5)(V_1, V_2, V_3)^T = (B^1, B^2, B^3, B^4, B^5)(1, 2, 3)^T. \tag{8}$$

In this way, we can get the comment sets of several agent vehicles from the customers who participate in the evaluation:

$$E_i = (E_{i1}, E_{i2}, \cdots, E_{im}). \tag{9}$$

Then the comments on agent vehicles from customers are expressed as follows, where n represents the number of customer vehicles, and m represents the number of agent vehicles:

$$E = (E_1, E_2, \cdots, E_n) = \begin{bmatrix} E_{11}E_{12}\cdots\cdots E_{1m} \\ E_{21}E_{22}\cdots\cdots E_{2m} \\ \cdots\cdots\cdots\cdots\cdots \\ \cdots\cdots\cdots\cdots\cdots \\ E_{n1}E_{n2}\cdots\cdots E_{nm} \end{bmatrix}. \tag{10}$$

Algorithm 2 Fuzzy Comprehensive Evaluation Algorithm

Input: Customer vehicle request package; agent vehicle service package; signal effective distance L_a; path length L; available agent vehicle list N_i for each CV_i

Output: Available QoS (quality of service) for each agent vehicle j; list E_{ij} for customer vehicle i

for each CV_i $i \in [1, n]$ **do**
  for each AV_j in N_i **do**
    compute path consistency $pc_{ij} = \frac{L_a v_i}{|v_i - v_j| L}$
  end for
  for each AV_j in N_i **do**
    for each element k of the AV_j **do**
      $R(j,k,1) = 1 - \Phi\left(\frac{x - a_{1\_u_1}}{\sigma_{1\_u_1}}\right)$; $R(j,k,2) = \Phi\left(\frac{x - a_{1\_u_i}}{\sigma_{1\_u_i}}\right) - \Phi\left(\frac{x - a_{2\_u_i}}{\sigma_{2\_u_i}}\right)$; $R(j,k,3)(x) = \Phi\left(\frac{x - a_{2\_u_i}}{\sigma_{2\_u_i}}\right)$;
    end for
  end for
  for each AV_j in N_i **do**
    $B(j) = A \circ R(j,:,:)$; $B(j) = \frac{1}{\sum_{i=1}^{3} b_i^j} B(j)$; $E_{ij} = B(j) \cdot (1,2,3)^T$;
  end for
end for
return E

3.3. Optimization

3.3.1. Comprehensive Vehicle Evaluation

To satisfy the requests of more customer vehicles and enable the agent vehicles to provide more effective service, taking the bandwidth limitation of the agent vehicles and the comprehensive scores given by the customer vehicles into account, this section distributes the agent vehicle resources and chooses the most suitable vehicle for customers. According to the discussion above, we propose the following access selection model:

$$\text{QoS} = \max\left(\sum_{i=1}^{n}\sum_{j=1}^{m} E_{ij} x_{ij}\right) \tag{11}$$

s.t.

$$\begin{aligned} &\sum_{i=1}^{n} x_{ij} = 1 \\ &\sum_{i=1}^{n} x_{ij} \times b_i \le b_j \\ &c_i \le c_j \\ &\frac{c_i}{bi} \le \min\left(\frac{L_a}{|v_i - v_j|}, \frac{L}{v_i}\right), \end{aligned}$$

where: $\sum_{i=1}^{n} x_{ij} = 1$ means that each customer vehicle can only be connected to one agent vehicle at the same time; $\sum_{i=1}^{n} x_{ij} \times b_i \leq b_j$ means that the bandwidth sum of customer vehicles served by the same proxy vehicle should not exceed its bandwidth capacity; $c_i \leq c_j$ means that the computing capability provided by the agent vehicle shall be no less than the computing capability of the customer vehicle's requirements; and $\frac{c_i}{b_i} \leq \min\left(\frac{L_a}{|v_i - v_j|}, \frac{L}{v_i}\right)$ indicates the computing capability that the customer vehicle should meet to satisfy the requirement of collaborative download within the service time of the agent vehicle in the path, so as to ensure the integrity of data transmission.

3.3.2. Agent Vehicle Resource Allocation Algorithm

Algorithm 3 Agent Vehicle Distribution Optimization Algorithm

Input: The comments on agent vehicles for each customer vehicle E

Output: Collaborative offload distribution scheme X

1: Construct the formula of overall customer satisfaction by maximizing the customer satisfaction: $\text{QoS} = \max\left(\sum_{i=1}^{n}\sum_{j=1}^{m} E_{ij}x_{ij}\right)$
2: The bandwidth sum of customer vehicles served by the same proxy vehicle should not exceed its bandwidth capacity: $\sum_{i=1}^{n} x_{ij} \times b_i \leq b_j$
3: Each customer vehicle can only be connected to one agent vehicle at the same time: $\sum_{i=1}^{n} x_{ij} = 1$
4: Determine the distribution of agent vehicles that maximize customer satisfaction by using linear programming.
5: Output the agent vehicles' distribution X.

4. Performance Evaluation

In this section, we use the proposed FCE (fuzzy comprehensive evaluation) algorithm to construct a series of experiments for the V2V agent vehicle selection problem based on the MATLAB platform. The experimental parameters are shown in Table 2. We compare the performance of the FCE algorithm with the FL-CFT [19] and RSB (random selection based on computing capability and bandwidth) algorithms under different numbers of customer requests. In order to realize the comparison between the FCE algorithm and the FL-CFT algorithm, we quantify the index obtained by FL-CFT using the process after the second step 13 of the FL-CFT algorithm. The comparative performance is as follows.

Table 2. The basic parameters of the simulation.

Parameter	Number	Unit	Information Description
L	3000	m	Path length
L_a	200	m	Effective distance between the customer vehicle and the agent vehicle
n	50		Number of customer vehicles requesting data
m	300		Number of agent vehicles providing data services
c_i	20–80	Mb	Computing capability of customer vehicle i's request
b_i	3–12	Mbps	Bandwidth of customer vehicle i
v_i	20-35	m/s	Speed of customer vehicle i
c_j	20–80	Mb	Computing capability of agent vehicle j
b_j	3–12	Mbps	Bandwidth of agent vehicle j
co_j	0–3		Service cost of agent vehicle j in unit time
cr_j	0–1		Accumulated credit ratio of agent vehicle j
v_j	20–35	m/s	Speed of agent vehicle j
pc_{ij}	0–1		Path consistency between customer vehicle i and agent vehicle j
N_i			The list of agent vehicles available for customer vehicle i
x_{ij}			Connection status between customer vehicle i and agent vehicle j
E_{ij}			Available QoS list of agent vehicle j to customer vehicle i

4.1. Experimental Setup

In the experiment, we consider that cooperative uninstallation occurs in the area without network coverage between two RSUs. The information requested between vehicles can only be shared through the information sharing mechanism between V2V. Vehicles apply to the vehicle cloud (VC) before arriving in the region. The vehicle cloud aggregates vehicle information, and uses the FCE algorithm proposed in this paper to analyze the information of customer vehicles and agent vehicles, so as to provide an agent vehicle allocation scheme that maximizes customer satisfaction. The following assumptions are employed in our simulations:

- Set the same driving path between the customer vehicle and the agent vehicle.
- Equip each vehicle (including the customer and agent vehicles) with an OBU, which can receive information and transmit information to the surrounding vehicles, and set the effective communication range of the vehicle.
- There are only two forms of data transmission between a customer vehicle and an agent vehicle: completion and failure.
- Each vehicle can act as a customer vehicle when requesting data and an agent vehicle when providing data service, but it can only be one in a period.

4.2. Performance Analysis

In this paper, we analyze the performance of the algorithm in four aspects: quality of service, average throughput, number of request failures, and average consumption. Quality of service is a comprehensive evaluation index under multi-factor consideration. It is the standard to verify the performance of the algorithm. Average throughput is the data transmission volume per unit time, which is the main factor to ensure the fluency of a customer vehicle's data requests. The number of request failures is the number of times that the agent vehicle cannot provide complete data transmission for the customer vehicle, which shows the stability of data transmission. Cost is an important reference factor for each customer vehicle in choosing agent vehicle service. We discuss the impact of the number of customer vehicles on the quality of service, average throughput, number of request failures, and average consumption in the process of collaboration between customer and agent vehicles.

Figure 3 shows that the average customer satisfaction curve obtained by the FCE algorithm is higher than that of the FL-CFT and RSB algorithms when changing the number of customer vehicle requests. The RSB curve has the worst performance. This is because the FCE algorithm considers the computing capability, bandwidth, unit cost, credibility, and path consistency of the agent vehicle in the process of selection; while FL-CFT only considers the velocity, distance, and connection of the agent vehicle; and RSB only considers the bandwidth and path consistency. Figure 3 shows that the FCE algorithm has better average customer satisfaction performance than the FL-CFT and RSB algorithms.

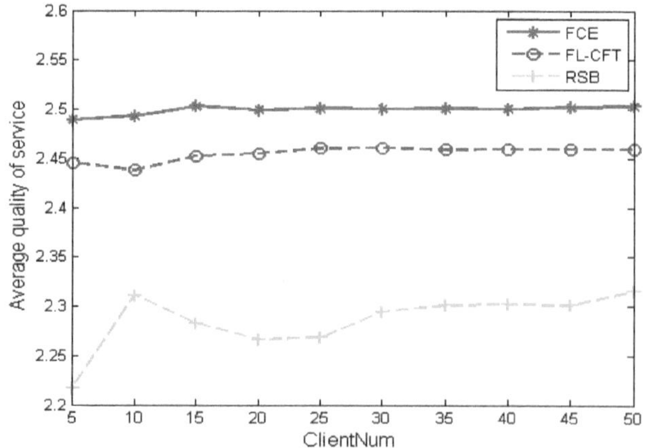

Figure 3. Average quality of service of the three algorithms.

Figure 4 shows that in the process of changing the number of customer vehicle requests, the effective bandwidth ratio of the FCE algorithm to the RSB Algorithm is 1, and the performance of the FL-CFT method is the worst. This is because the FCE and RSB algorithms take the bandwidth as an important index to evaluate the agent vehicle selection process, while FL-CFT does not consider this index. Figure 4 shows that the FCE and RSB algorithms have better average bandwidth utilization than the FL-CFT algorithm. This index also shows whether the selected agent vehicle can meet the customer's bandwidth requirements. The FCE and RSB algorithms can provide a better data fluency experience for a customer vehicle.

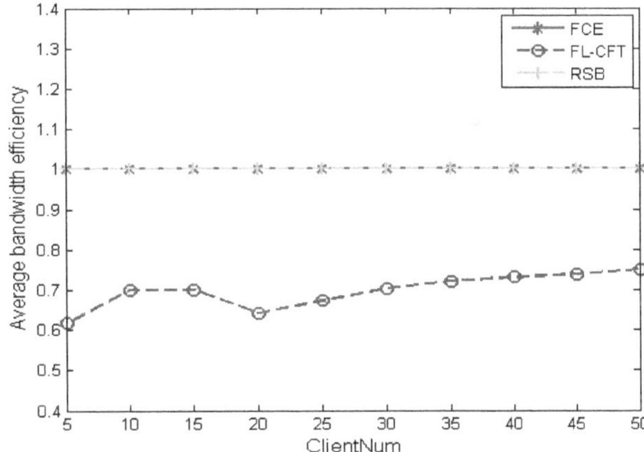

Figure 4. Average bandwidth efficiency of the three algorithms.

In order to verify the correctness and stability of the algorithm, we run a model experiment with 500 customer vehicles and 3000 agent vehicles, and count the failure times of customer requests under the experimental conditions. Figure 5 shows that in the process of changing the number of customer vehicle requests, the FCE algorithm does not fail, while the failure rates of the FL-CFT and RSB algorithms increase with an increase in the number of customer requests. This is because the FCE algorithm takes into account the interaction of many factors, and takes the path matching degree of customer vehicles and agent vehicles and the reputation of customer vehicles as important indicators. Figure 5 shows that the FCE algorithm has better selectivity and stability than the FL-CFT and RSB algorithms.

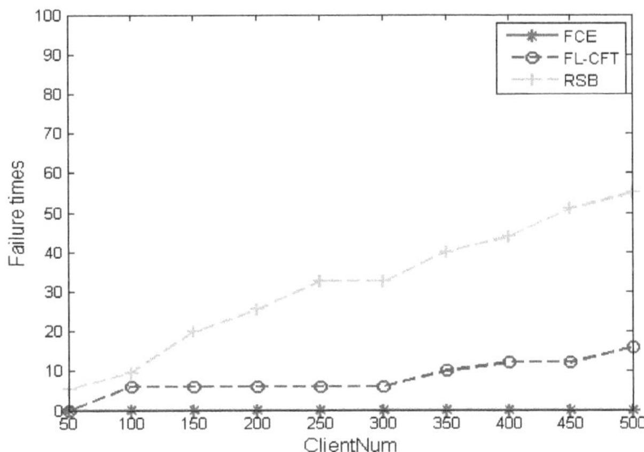

Figure 5. Number of failures for the three algorithms.

Customer consumption is always an important indicator of customer vehicles in the selection of agent vehicles. Figure 6 shows that the average consumption of the FCE algorithm is less than that of the FL-CFT and RSB algorithms with increasing customer vehicle requests. The main reason is that the FCE algorithm considers all the required vehicle information when selecting the agent vehicle, and seeks an optimal selection strategy. The FL-CFT and RSB algorithms ignore the interaction of these factors. Figure 6 shows that the FCE algorithm is more economical than the FL-CFT and RSB algorithms.

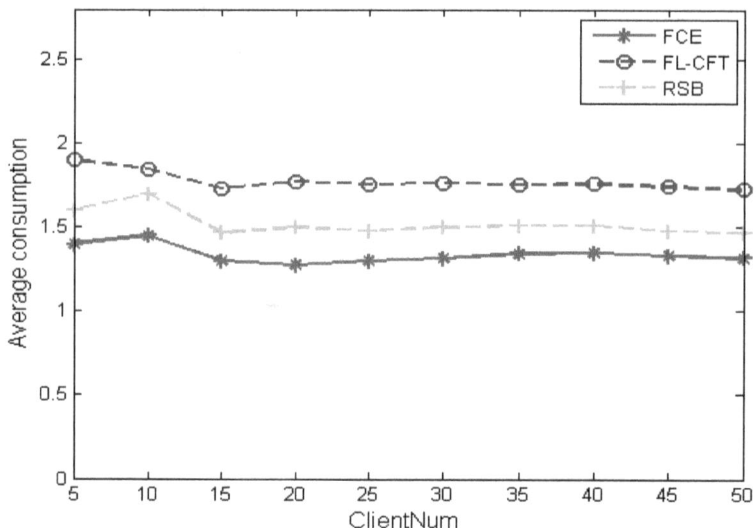

Figure 6. Average consumption of the three algorithms.

5. Conclusions

In this paper, the problem of agent vehicle selection for V2V collaborative unloading in a vehicle network is studied. A method of joint selection in a V2V network, based on fuzzy comprehensive evaluation is proposed. Different from the previous articles, many factors such as computing capability, bandwidth, consumption index, reputation, and path matching of the vehicles are considered in this paper. At the same time, we take the relationship between the customer vehicle demand and the agent vehicle as tan index item. In this paper, we propose a fuzzy comprehensive evaluation method to evaluate customer and agent vehicles, and give a multi-constrained optimization model to describe the agent vehicle allocation scheme. The simulation results show that the proposed vehicle selection algorithm has good prospects of usability and application.

Author Contributions: These authors contributed equally to this work. All authors discussed the required algorithm to complete the manuscript. T.H., X.G. and Z.C. conceived the paper. T.H. and Y.Z. did the experiments and wrote the paper. X.G. and Z.C. checked typographical errors.

Acknowledgments: The authors are very grateful to the reviewers for their valuable comments on the revision of the paper. These suggestions play an important role in the construction of Algorithm 3 and the simulations.

References

1. Lu, N.; Cheng, N.; Zhang, N.; Shen, X.; Mark, J.W. Connected Vehicles: Solutions and Challenges. *IEEE Internet Things J.* **2014**, *1*, 289–299. [CrossRef]
2. Amadeo, M.; Campolo, C.; Molinaro, A. Information-centric networking for connected vehicles: A survey and future perspectives. *IEEE Commun. Mag.* **2016**, *54*, 98–104. [CrossRef]
3. Zhou, H.; Zhang, N.; Bi, Y.; Yu, Q.; Shen, X.S.; Shan, D.; Bai, F. TV White Space Enabled Connected Vehicle Networks: Challenges and Solutions. *IEEE Netw.* **2017**, *31*, 6–13. [CrossRef]
4. Lim, J.H.; Kim, W.; Naito, K.; Yun, J.H.; Cabric, D.; Gerla, M. Interplay Between TVWS and DSRC: Optimal Strategy for Safety Message Dissemination in VANET. *IEEE J. Sel. Areas Commun.* **2014**, *32*, 2117–2133. [CrossRef]

5. Wang, M.; Shan, H.; Luan, T.H.; Lu, N.; Zhang, R.; Shen, X.; Bai, F. Asymptotic Throughput Capacity Analysis of VANETs Exploiting Mobility Diversity. *IEEE Trans. Veh. Technol.* **2015**, *64*, 4187–4202. [CrossRef]
6. Ota, K.; Dong, M.; Chang, S.; Zhu, H. MMCD: Cooperative downloading for highway VANETs. *IEEE Trans. Emerg. Top. Comput.* **2015**, *3*, 34–43. [CrossRef]
7. Yue, D.; Li, P.; Zhang, T.; Cui, J.; Jin, Y.; Liu, Y.; Liu, Q. Cooperative Content Downloading in Hybrid VANETs: 3G/4G or RSUs Downloading. In Proceedings of the 2016 IEEE International Conference on Smart Cloud (SmartCloud), New York, NY, USA, 18–20 November 2016.
8. Liu, J.; Ge, Y.; Li, S.; Shu, R.; Ding, S. A transmission scheduling method of cooperative downloading for Vehicular Networking. In Proceedings of the 2014 14th International Symposium on Communications and Information Technologies (ISCIT), Incheon, Korea, 24–26 September 2015.
9. Huang, W.; Wang, L. ECDS: Efficient collaborative downloading scheme for popular content distribution in urban vehicular networks. *Comput. Netw.* **2016**, *101*, 90–103. [CrossRef]
10. Wang, T.; Song, L.; Han, Z.; Lu, Z.; Hu, L. Popular content distribution in vehicular networks using coalition formation games. In Proceedings of the 2013 IEEE International Conference on Communications (ICC), Budapest, Hungary, 9–13 June 2013.
11. Wang, J.; Wang, S.; Sun, Y.; Changchun, Y.; Lu, W.; Wu, D. An incentive mechanism for cooperative downloading method in VANET. In Proceedings of the 2013 IEEE International Conference on Vehicular Electronics and Safety, Dongguan, China, 28–30 July 2013.
12. Wu, C.; Gerla, M.; Mastronarde, N. Incentive driven LTE content distribution in VANETs. In Proceedings of the 2015 14th Annual Mediterranean Ad Hoc Networking Workshop (MED-HOC-NET), Vilamoura, Portugal, 17–18 June 2015.
13. Zhou, H.; Liu, B.; Luan, T.H.; Hou, F.; Gui, L.; Li, Y.; Yu, Q.; Shen, X.S. ChainCluster: Engineering a Cooperative Content Distribution Framework for Highway Vehicular Communications. *IEEE Trans. Intell. Transp. Syst.* **2014**, *15*, 2644–2657. [CrossRef]
14. Lai, C.; Zhang, K.; Cheng, N.; Li, H.; Shen, X. SIRC: A Secure Incentive Scheme for Reliable Cooperative Downloading in Highway VANETs. *IEEE Trans. Intell. Transp. Syst.* **2017**, *18*, 1559–1574. [CrossRef]
15. Zhang, W.; Jiang, S.; Zhu, X.; Wang, Y. Privacy-Preserving Cooperative Downloading for Value-Added Services in VANETs. In Proceedings of the International Conference on Intelligent Networking & Collaborative Systems, Xi'an, China, 9–11 September 2013.
16. Liu, J.; Zhang, W.; Wang, Q.; Li, S.; Chen, H.; Cui, X.; Sun, Y. A Cooperative Downloading Method for VANET Using Distributed Fountain Code. *Sensors* **2016**, *16*, 1685. [CrossRef] [PubMed]
17. Molisch, A.F.; Mehta, N.B.; Yedidia, J.S.; Zhang, J. Performance of Fountain Codes in Collaborative Relay Networks. *IEEE Trans. Wirel. Commun.* **2007**, *6*, 4108-4119. [CrossRef]
18. Lin, C.C.; Deng, D.J.; Yao, C.C. Resource allocation in vehicular cloud computing systems with heterogeneous vehicles and roadside units. *IEEE Internet Things J.* **2017**, *5*, 3692–3700. [CrossRef]
19. Luo, Q.; Cai, X.; Luan, T.H.; Ye, Q. Fuzzy logic-based integrity-oriented file transfer for highway vehicular communications. *EURASIP J. Wirel. Commun. Netw.* **2018**, *2018*, 3. [CrossRef]
20. Liu, J.; Zhai, H.; Jia, Z.; Li, S.; Chen, H.; Cui, X. A balanced cooperative downloading method for VANET. In Proceedings of the 2016 16th International Symposium on Communications and Information Technologies (ISCIT), Qingdao, China, 26–28 September 2016.
21. Huang, C.M.; Yang, C.C.; Lin, H.Y. A bandwidth aggregation scheme for member-based cooperative networking over the hybrid VANET. In Proceedings of the 2011 IEEE 17th International Conference on Parallel and Distributed Systems (ICPADS), Tainan, Taiwan, 7–9 December 2011.
22. Deng, G.; Li, F.; Wang, L. Cooperative downloading in vanets-lte heteroge- neous network based on named data. In Proceedings of the 2016 IEEE Conference on Computer Communications Workshops (INFOCOM WKSHPS), San Francisco, CA, USA, 10–14 April 2016.
23. Hao, Y.; Tang, J.; Cheng, Y. Secure cooperative data downloading in ve- hicular ad hoc networks. *IEEE J. Sel. Areas Commun.* **2013**, *31*, 523–537.
24. Mezghani, F.; Dhaou, R.; Nogueira, M.; Beylot, A.L. Offloading Cellular Networks Through V2V Communications—How to Select the Seed-Vehicles? In Proceedings of the 2016 IEEE International Conference on Communications (ICC), Kuala Lumpur, Malaysia, 22–27 May 2016.
25. Jiang, D.; Huo, L.; Lv, Z.; Song, H.; Qin, W. A Joint Multi-Criteria Utility-Based Network Selection Approach for Vehicle-to-Infrastructure Networking. *IEEE Trans. Intell. Transp. Syst.* **2018**, *19*, 3305–3319. [CrossRef]

Construction of Weights for Positive Integral Operators

Ron Kerman

Department of Mathematics, Brock University, St. Catharines, ON L2S 3A1, Canada; rkerman@brocku.ca

Abstract: Let (X, M, μ) be a σ-finite measure space and denote by $P(X)$ the μ-measurable functions $f\colon X \to [0,\infty]$, $f < \infty$ μ ae. Suppose $K\colon X \times X \to [0,\infty)$ is $\mu \times \mu$-measurable and define the mutually transposed operators T and T' on $P(X)$ by $(Tf)(x) = \int_X K(x,y)f(y)\,d\mu(y)$ and $(T'g)(y) = \int_X K(x,y)g(x)\,d\mu(x)$, $f, g \in P(X)$, $x, y \in X$. Our interest is in inequalities involving a fixed (weight) function $w \in P(X)$ and an index $p \in (1,\infty)$ such that: (*): $\int_X [w(x)(Tf)(x)]^p d\mu(x) \lesssim C \int_X [w(y)f(y)]^p d\mu(y)$. The constant $C > 1$ is to be independent of $f \in P(X)$. We wish to construct all w for which (*) holds. Considerations concerning Schur's Lemma ensure that every such w is within constant multiples of expressions of the form $\phi_1^{1/p-1}\phi_2^{1/p}$, where $\phi_1, \phi_2 \in P(X)$ satisfy $T\phi_1 \leq C_1\phi_1$ and $T'\phi_2 \leq C_2\phi_2$. Our fundamental result shows that the ϕ_1 and ϕ_2 above are within constant multiples of (**): $\psi_1 + \sum_{j=1}^{\infty} E^{-j}T^{(j)}\psi_1$ and $\psi_2 + \sum_{j=1}^{\infty} E^{-j}T'^{(j)}\psi_2$respectively; here $\psi_1, \psi_2 \in P(X)$, $E > 1$ and $T^{(j)}, T'^{(j)}$ are the jth iterates of T and T'. This result is explored in the context of Poisson, Bessel and Gauss–Weierstrass means and of Hardy averaging operators. All but the Hardy averaging operators are defined through symmetric kernels $K(x,y) = K(y,x)$, so that $T' = T$. This means that only the first series in (**) needs to be studied.

Keywords: weights; positive integral operators; convolution operators

MSC: 2000 Primary 47B34; Secondary 27D10

---

1. Introduction

Consider a σ-finite measure space (X, M, μ) and a positive integral operator T defined through a nonnegative kernel $K = K(x,y)$ which is $\mu \times \mu$ measurable on $X \times X$; that is, T is given on the class, $P(X)$, of μ-measurable functions $f : X \to [0,\infty]$, $f < \infty$ μ ae, by

$$(Tf)(x) = \int_X K(x,y)f(y)\,d\mu(y),\ x \in X.$$

The transpose, T', of T at $g \in P(X)$ is

$$(T'g)(y) = \int_X K(x,y)g(x)\,d\mu(x),\ y \in X;$$

it satisfies

$$\int_X gTf\,d\mu = \int_X fT'g\,d\mu,\ f, g \in P(X).$$

Our focus will be on inequalities of the form

$$\int_X [uTf]^p\,d\mu \leq B^p \int_X [vf]^p\,d\mu, \tag{1}$$

with the index p fixed in $(1,\infty)$ and $B > 0$ independent of $f \in P(X)$; here, $u, v \in P(X)$, $0 \le u, v < \infty$, μ *ae*, are so-called weights.

The equivalence need only be proved in one direction. Suppose, then, (1) holds and $g \in P(x)$ satisfies $\int_X [u^{-1}g]^p \, d\mu < \infty$. Then

$$\left[\int_X [v^{-1}T'g]^{p'} \, d\mu\right]^{\frac{1}{p'}} = \sup \int_X f v^{-1} T' g \, d\mu,$$

the supremum being take over $f \in P(X)$ with $\int_X f^p \, d\mu \le 1$. But, Fubini's Theorem ensures

$$\begin{aligned}\int_X f v^{-1} T' g \, d\mu &= \int_X g T(f v^{-1}) \, d\mu \\ &= \int_X (u^{-1}g) u T(f v^{-1}) \, d\mu \\ &\le \left[\int_X [u^{-1}g]^{p'} \, d\mu\right]^{\frac{1}{p'}} \left[B^p \int_X [v f v^{-1}]^p \, d\mu\right]^{\frac{1}{p}} \\ &\le \left[B^{p'} \int_X [u^{-1}g]^{p'} \, d\mu\right]^{\frac{1}{p'}}.\end{aligned}$$

Further, (1) holds if and only if the dual inequality

$$\int_X [v^{-1}T'g]^{p'} \, d\mu \le B^{p'} \int_X [u^{-1}g]^{p'} \, d\mu, \; p' = \frac{p}{p-1}, \tag{2}$$

does.

Inequality (1) has been studied for various operators T in such papers as [1–9].

In this paper, we are interested in constructing weights u and v for which (1) holds. We restrict attention the case $u = v = w$; the general case will be investigated in the future. Our approach is based on the observation that, implicit in a proof of the converse of Schur's lemma, given in [10], is a method for constructing w. An interesting application of Schur's lemma itself to weighted norm inequalities is given in Christ [11].

In Section 2, we prove a number of general results the first of which is the following one.

Theorem 1. *Let (X, M, μ) be a σ-finite measure space with $u, v \in P(X)$, $0 \le u, v < \infty$, μ ae. Suppose that T is a positive integral operator on $P(X)$ with transpose T'. Then, for fixed p, $1 < p < \infty$, one has (1), with $C > 1$ independent of $f \in P(X)$, if and only if them exists a function $\phi \in P(X)$ and a constant $C > 1$ for which*

$$T(v^{-1}\phi^{p'}) \le C u^{-1} \phi^{p'} \text{ and } T'(u\phi^p) \le C v \phi^p. \tag{3}$$

In this case, B_0, the smallest B possible in (1) and Co, the smallest possible C so that (3) holds for some ϕ, satisfy

$$B_0 \le C_0 = \max\left[B_1^p, B_1^{p'}\right],$$

where $B_1 = B_0^{1/p} + B_0^{1/p'}$.

Theorem 1 has the following consequence.

Corollary 1. *Under the condition of Theorem 1, (1) holds for $u = v = w$ if and only if $w = \phi_1^{-1/p'} \phi_2^{1/p}$, where ϕ_1, ϕ_2 are functions in $P(X)$ satisfying*

$$T\phi_1 \le C\phi_1 \text{ and } T'\phi_2 \le C\phi_2, \tag{4}$$

for some $C > 1$.

Though it is often possible to work with the inequalities (4) directly (see Remark 1) it is important to have a general method to construct the functions ϕ_1 and ϕ_2. This method is given in our principal result.

Theorem 2. *Suppose* X, μ *and* T *are as in Theorem 1. Let* $\phi \in P(X)$. *Then,* ϕ *satisfies an inequality of the form*

$$T\phi \leq C_1\phi,\ C_1 > 0 \text{ constant}, \tag{5}$$

if and only if there is a constant $C > 1$ *such that*

$$C^{-1}\phi \leq \psi + \sum_{j=1}^{\infty} C_2^{-j} T^{(j)}\psi \leq C\phi, \tag{6}$$

where $\psi \in P(X)$, $C_2 > 1$ *is constant and* $T^{(j)} = T \circ T \cdots \circ T$, *j times.*

The kernels of operator of the form

$$\sum_{j=1}^{\infty} C^{-j} T^{(j)} \text{ and } \sum_{j=1}^{\infty} C^{-j} T'^{(j)}$$

will be called the weight generating kernels of T. In Sections 3–6 these kernels will be calculated for particular T. All but the Hardy operators considered in Section 6 operate on the class $P(R^n)$ of nonnegative, Lebesgue-measurable functions on R^n.

The operators last referred to are, in fact, convolution operators

$$(T_k f)(x) = (k * f)(x) = \int_{R^n} k(x-y) f(y)\,dy,\ x \in R^n,$$

with even integrable kernels k, $\int_{R^n} k(y)\,dy = 1$. In particular, the kernel $k(x-y)$ is symmetric, so $T_k' = T_k$, whence only the first series in (**) need be considered.

Further, the convolution kernels are part of an approximate identity $\{k_t\}_{t>0}$ on

$$L^P(R^n) = \left\{ f \text{ Leb. meas: } \left[\int_{R_n} |f|^p\right]^{1/p} < \infty \right\},$$

see [12]. Thus, it becomes of interest to characterize the weights w for which $\{k_t\}_{l>0}$ is an approximate identity on

$$L^p(w) = L^p(R^n, w) = \left\{ f \text{ Leb. meas: } \|f\|_{p,w} = \left[\int_{R^n} |wf|^p\right]^{1/p} < \infty \right\};$$

that is $k_t * f \in L^p(w)$ and

$$\lim_{t \to 0+} \|k_t * f - f\|_{p,w} = 0$$

for all $f \in L^p(w)$. It is a consequence of the Banach-Steinhaus Theorem that this will be so if and only if

$$\sup_{0<t<a} \|k_t\| < \infty$$

for some fixed $a > 0$, where $\|k_t\|$ denotes the operator norm of T_{k_t} on $L^p(w)$. We remark here that the operators in Sections 3–5 are bounded on $L^p(w)$ and, indeed, form part of an approximate identity on $L^p(w)$, if w satisfies the A_p condition, namely,

$$\sup \left[\frac{1}{|Q|}\int_Q w^p\right]\left[\frac{1}{|Q|}\int_Q w^{-p'}\right]^{1/p'} < \infty,\ p' = \frac{p}{p-1}, \tag{7}$$

the supremum being taken over all cubes Q in R^n whose sides are parallel to the coordinate axes with $\infty > |Q| =$ Lebesgue measure of Q. See ([13], p. 62) and [14].

Finally, all the convolution operators are part of a convolution semigroup $(k_t)_{t>0}$; that is $k_t(x) = t^{-n}k\left(\frac{x}{t}\right)$ and $k_{t_1} * k_{t_2} = k_{t_1+t_2}$, $t_1, t_2 > 0$. The approximate identity result can thus be interpreted as the continuity of the semigroup.

We conclude the introduction with some remarks on terminology and notation. The fact that T is bounded on $L^p(w)$ if and only if T' is bounded on $L^{p'}(w^{-1})$ is called the principle of duality or, simply, duality. Two functions $f, g \in P(X)$ are said to be equivalent if a constant $C > 1$ exists for which

$$C^{-1}g \le f \le Cg. \tag{8}$$

We indicate this by $f \approx g$, with the understanding that C is independent of all parameters appearing, (except dimension) unless otherwise stated. If only one of the inequalities in (8) holds, we use the notation $f \succeq g$ or $f \preceq g$, as appropriate. Lastly, a convolution operator and its kernel are frequently denoted by the same symbol.

2. General Results

In this section we give the proofs of the results stated in the Introduction, together with some remarks.

Proof of Theorem 1. The conditions (3) are, respectively, equivalent to

$$T' : L^1(u^{-1}\phi^{p'}) \to L^1(v^{-1}\phi^{p'})$$
$$\text{i.e., } T : L^\infty(v\phi^{-p'}) \to L^\infty(u\phi^{-p'})$$

and

$$T : L^1(v\phi^p) \to L^1(u\phi^p).$$

It will suffice to deal with the first condition in (3). So, Fubini's Theorem yields

$$\int_X v^{-1}\phi^{p'} T'f\, d\mu \le C\int_X u^{-1}\phi^{p'} f\, d\mu$$

equivalent to

$$\int_X f\, T(v^{-1}\phi^{p'})\, d\mu \le C\int_X f u^{-1}\phi^{p'}\, d\mu, \quad f \in P(X),$$

and hence to

$$T(v^{-1}\phi^{p'}) \le Cu^{-1}\phi^{p'},$$

since f is arbitrary.

According to the main result of [15], then,

$$T : L^p\left((v\phi^p)^{1/p}(v\phi^{-p'})^{1/p'}\right) \to L^p\left((u\phi^p)^{1/p}(u\phi^{-p'})^{1/p'}\right)$$

i.e., $T : L^p(v) \to L^p(u)$, with norm $\le C$, so that (1) holds with $B \le C$.

Suppose now (1) holds. Following [10], choose $g \in P(X)$ with

$$\int_X g^{pp'}\,d\mu = 1.$$

Let $T_1 g = \left[uT(v^{-1}g^{p'})\right]^{1/p'}$ and $T_2 g = \left[v^{-1}T'(ug^p)\right]^{1/p}$. Set

$$S = T_1 + T_2,\ A = B_0 + \varepsilon \text{ and } \phi = \sum_{j=0}^{\infty} A^{-j} S^{(j)} g.$$

As in [10], conclude $T_1\phi \le A\phi$ and $T_2\phi \le A\phi$, so that (2) is satisfied for $C_0 \le \left[B_1^p, B_1^{p'}\right]$, where $B_1 = B_0^{1/p} + B_0^{1/p'}$. □

Proof of Corollary 1. Given (1), one has (2) and Theorem 1 then implies (3), with T replaced by T', namely for $u = v = w$,

$$T(w^{-1}\phi^{p'}) \le Cw^{-1}\phi^{p'} \text{ and } T(w\phi^p) \le Cw\phi^p,$$

whence the inequalities (4) are satisfied by $\phi_1 = w\phi^p$ and $\phi_2 = w^{-1}\psi^{p'}$. Conversely, given (4), and taking $u = v = w = \phi_1^{1/p-1}\phi_2^{1/p}$, one readily obtains (3), with $\psi = (\psi_1\psi_2)^{1/pp'}$. □

Proof of Theorem 2. Clearly, if (6) holds,

$$T\phi \le C\left[T\psi + \sum_{j=1}^{\infty} C_2^{-j} T^{(j+1)}\psi\right] = CC_2 \sum_{j=1}^{\infty} C_2^{-j} T^{(j)}\psi \le C^2 C_2 \phi.$$

Suppose $\phi \in P(X)$ satisfies (5). Then,

$$T^{(j)}\phi \le C_1^j \phi_1,\ j = 1, 2, \ldots.$$

It only remains to observe that

$$\left(1 + \frac{C_1}{\varepsilon}\right)^{-1} \phi \le \phi + \sum_{j=1}^{\infty} (C_1 + \varepsilon)^{-j} T^{(j)}\phi \le \phi + \sum_{j=1}^{\infty} \left(\frac{C_1}{C_1 + \varepsilon}\right)^j \phi \le \left(1 + \frac{C_1}{\varepsilon}\right)\phi,$$

for any $\varepsilon > 0$. □

Remark 1. *The class of functions ϕ determined by the weight-generating operators $\sum_{j=1}^{\infty} C^{-j}T^{(j)}$ effectively remains the same as C increases. Thus, suppose $0 < C_1 < C_2$, $\psi \in P(X)$ and $\phi = \psi + \sum_{j=1}^{\infty} C_1^{-j} T^{(j)}\psi$. Then, ϕ is equivalent to $\psi + \sum_{j=1}^{\infty} C_2^{-j} T^{(j)}\psi$, since*

$$\begin{aligned}
\phi \le \phi + \sum_{j=1}^{\infty} C_2^{-j} T^{(j)}\phi &= \sum_{j=0}^{\infty} C_2^{-j} \sum_{k=0}^{\infty} C_1^{-k} T^{(j+k)}\psi = \sum_{l=0}^{\infty} \left(\sum_{j+h=l} C_1^{-k} C_2^{-j}\right) T^{(l)}\psi \\
&= \sum_{l=0}^{\infty}\sum_{j=0}^{\infty} \left(\frac{C_1}{C_2}\right)^j C_1^{-l} T^{(l)}\psi = \frac{C_2}{C_2 - C_1} \sum_{l=0}^{\infty} C_1^{-l} T^{(l)}\psi \\
&= \frac{C_2}{C_2 - C_1}\phi.
\end{aligned}$$

This means that in dealing with weight-generating operators we need only consider $C > 1$.

We conclude this section with the following observations on approximate identities in weighted Lebesgue spaces.

Remark 2. *Suppose* $\{k_t\}_{t>0}$ *is an approximate identity in* $L^p(R^n), 1 < p < \infty$. *If the inequalities* (4) *involving* ϕ_1 *and* ϕ_2 *can be shown to hold for* $T_{kt}, t \in (0, a]$ *for some* $a > 0$, *with* $C > 1$ *independent of such* t, *then* $\{k_t\}_{t>0}$ *will also be an approximate identity in* $L^p(w) = L^p(R^n, w), w = \phi_1^{-1/p'} \phi_2^{1/p}$.

Example 1. *Let* $k = k(|x|)$ *be any bounded, nonnegative radial function on* R^n *which is a decreasing function of* $|x|$ *and suppose* $\int_{R^n} k(x)\,dx = 1$. *It is well-known, see ([13], p. 63), that* $k_t(x) = t^{-n}k(x/t), x \in R^n$, *is an approximate identity in* $L^p(R^n), 1 < p < \infty$.

The weight $w(x) = 1 + |x|^{-n/p}(1 + \log^+(1/|x|))^{-1}$, for fixed $p, 1 < p < \infty$, has the interesting properly that $T_{k_t} : L^p(w) \to L^p(w)$ for all $t > 0$, yet $\{k_t\}_{t>0}$ is never an approximate identity in $L^p(w)$.

To obtain the boundedness assertion take $\phi_1(x) = 1$ and $\phi_2(x) = 1 + |x|^{-n}\,(1 + \log^+(1/|x|))^{-p}$ in Corollary 1.

Arguments similar to those in [6] show that if $\{k_t\}_{t>0}$ is an approximate identity in $L^p(w)$, then w must satisfy the A_p condition for all cubes Q will sides parallel to the coordinate axes and $|Q| \le a$ for some $a > 0$. However, the weight w does not have this property.

3. The Poisson Integral Operators

We recall that for $t > 0$ and $y \in R^n$, the Poisson kernel, P_t, is defined by

$$P_t(y) = c_n t(t^2 + |y|^2)^{-(n+1)/2}, \; c_n = \pi^{-(n+1)/2}\Gamma((n+1)/2).$$

Theorem 3. *The weight-generating kernels for* $P_t, t > 0$, *are equivalent to* $P \equiv P_0$. *Indeed, given* $\psi \in P(R^n)$, *with* $P_\psi < \infty$ *a.e.,*

$$C_t^{-1}P_\psi \le \sum_{j=1}^{\infty} C^{-j}P_{jt}\psi \le C_t' P_\psi, \tag{9}$$

where $C > 1, C_t = C\max[t^{-1}, t^n]$ *and* $C_t' = C_t \sum_{j=1}^{\infty} C^{-j}\max[jt, (jt)^{-n}]$.

Proof. Observe that by the semigroup property $P_t^{(j)} = P_{jt}, j = 1, 2, \ldots$.

Also,

$$\min[t, t^{-n}]P \le P_t \le \max[t, t^{-n}]P.$$

Now, suppose

$$\psi + \sum_{j=1}^{\infty} C^{-j}P_{jt}\psi \text{ is in } P(R^n),$$

with $C > 1$. Then,

$$\begin{aligned} P_\psi &\le C_t P_t\psi + \sum_{j=1}^{\infty} C^{-j}P_{(j+1)t}\psi \le C_t \sum_{j=1}^{\infty} C^{-j}P_{jt}\psi \le C_t \sum_{j=1}^{\infty} C^{-j}\max[jt, (j)^{-n}]P_\psi \\ &\le C_t' P_\psi. \end{aligned}$$

□

As stated in Section 1, $w \in A_p$ is sufficient for $\{P_t\}_{t>0}$ to be an approximate identify in $L^P(w)$. Moreover, $w \in A_p$ is also necessary for this in the periodic case. See [6,8,16]. It is perhaps surprising then that the class of approximate identity weights is much larger than A_p, as is seen in the next result.

Proposition 1. *Let $w_\alpha(x) = [1+|x|]^\alpha$, $\alpha \in R$. Then, for any $t > 0$, P_t is bounded on $L^p(w_\alpha)$ if any only if $-\frac{n}{p} - 1 < \alpha < \frac{n}{p'} + 1$. Moreover, on that range of α one has*

$$\lim_{t\to 0+} \|P_t * f - f\|_{p,\omega_\alpha} = 0, \tag{10}$$

for all $f \in L^p(\omega_\alpha)$. The set of α for which $w_\alpha \in A_p$, however, is

$$-\frac{n}{p} < \alpha < \frac{n}{p'}.$$

Proof. We omit the easy proof of the assertion concerning the α for which $w_\alpha \in A_p$.

To obtain the "if" part of the other assertion we will show

$$P_t * w_\beta \le C w_\beta,\ t > 0, \tag{11}$$

if and only if $-n-1 \le \beta < 1$, with $C > 1$ independent of both s and t, if $t \in (0,1)$. Corollary 1 and Remark 2, then yield (10) when $-\frac{n}{p} - 1 < \alpha < \frac{n}{p'} + 1$.

Consider, then, fixed $x \in R^n$ and $0 < t < 1$. We have

$$\begin{aligned}(P_t * w_\beta)(x) &= \left(\int_{|y|\le \frac{|x|}{2}} + \int_{\frac{|x|}{2}<|y|<2|x|} + \int_{|y|\ge 2|x|}\right) P_t(y) w_\beta(s-t)\, dy \\ &= I_1 + I_2 + I_3.\end{aligned}$$

Now,

$$I_1 \le w_\beta(x) \int_{|y|<\frac{|x|}{2}} P_t(y)\, dy \le C w_\beta(x),$$

for all $\beta \in R$.

Again,

$$I_2 \ge c P_t(x) \int_{|x-y|\le 1} (1+|x-y|)^\beta\, dy \ge c P_t(x) \ge c|x|^{-n-1},$$

so we require $\beta > n-1$, if (11) is to hold.

Moreover, for $x \in R^n$ and $0 < t < 1$,

$$\begin{aligned} I_2 &\approx P_t(x)\left[|x|^n \chi_{|x|\le 1} + |x|^{\beta+n}\chi_{|x|>1}\right] \\ &\approx \left(\frac{|x|}{t}\right)^n \chi_{|x|\le 1} + \frac{t}{|x|}\chi_{t\le|x|\le 1} + \frac{t}{|x|}|x|^\beta \chi_{|x|\ge 1} \\ &\le C w_\beta(x). \end{aligned}$$

Next, for $|x| \gg 1$

$$I_3 = \int_{|y|>2|x|} P_t(y) w_\beta(y)\, dy \preceq t \int_{|y|>2|x|} |y|^{-n-1+\beta} dy$$

which requires $\beta < 1$ to have $I_3 < \infty$. In that case

$$I_3 \preceq \int_{r>2|x|} r^{-n-1+\beta} r^{n-1}\, dr \preceq |x|^{\beta-1} \preceq w_\beta(x).$$

That P_t is not bounded on $L^p(w_\alpha)$ when $\alpha \le -\frac{n}{p} - 1$ can be seen by noting that, for appropriate $\varepsilon > 0$, the function $f(x) = |x|[\log(1+|x|)]^{-(1+\varepsilon)/p}$ is in $L^p(w_\alpha)$, while $P_t f \equiv \infty$. The range $\alpha \ge n/p+1$ is then ruled out by duality. □

4. The Bessel Potential Operators

The Bessel kernel, $G_\alpha, \alpha > 0$, can be defined explicitly by

$$G_\alpha(y) = C_\alpha |y|^{(\alpha-n)/2} K_{(n-\alpha)/2}(|y|),\ y \in R^n,$$

where K_r is the modified Bessel function of the third kind and

$$C_\alpha^{-1} = \pi^{n/2} 2^{(n+\alpha-2)} \Gamma(\alpha/2).$$

It is, however, more readily recognized by its Fourier transformation

$$\hat{G}_\alpha(z) = (2\pi)^{-n/2}[1+|z|^2]^{-\alpha/2}.$$

Using the latter formula one picks out the special cases G_{n-1} and G_{n+1} which, except for constant multiplies, are, respectively, $|y|^{-1}e^{-|y|}$ and the Picard kernel $e^{-|y|}$.

The semigroup properly $G_\alpha * G_\beta = G_{\alpha+\beta}$ holds and so the jth convolution iterate has kernel $G_{j\alpha}$. Also, $\int_{R^n} G_\alpha(y)\,dy = 1$.

We use the integral representation

$$G_\alpha(y) = g_{\alpha,n}(|y|) = (4\pi)^{-n/2}\Gamma(\alpha/2)^{-1}\int_0^\infty e^{-|y|^2 t/4} e^{-1/t} t^{(n-2)/2}\frac{dt}{t} \tag{12}$$

to show in Lemma 1 below that known estimates [17], are in fact, sharp.

Lemma 1. *Suppose $n, \alpha > 0, n \in Z_+$. Set $m = n - \alpha$ and define r^{-m+} to be r^{-m}, $\log_+\left(\frac{2}{r}\right)$ or 1, according as $m > 0, m = 0$ or $m < 0$. Then, a constant $C > 1$ exists, depending on n, such that*

$$\begin{aligned} C^{-1}r^{-m+} \le g_{\alpha,n}(r) \le Cr^{-m+},\ 0 < r < 1, \\ C^{-1}r^{-(m+1)/2}e^{-r} \le g_{\alpha,n}(r) \le Cr^{-(m+1)/2}e^{-r},\ r \ge 1. \end{aligned} \tag{13}$$

Proof. As in [17], p. 296

$$g_{\alpha,n}(r) = C_\alpha e^{-r}(\alpha/r)^{m/2}\int_1^\infty e^{-\frac{r}{2}(x+\frac{1}{x}-2)}\left[x^{m/2}+x^{-m/2}\right]\frac{dx}{x}$$

with $C_\alpha = (4\pi)^{-n/2}\Gamma(\alpha/2)^{-1}$. Clearly,

$$g_{\alpha,n}(r) \approx r^{-m/n}e^{-r}\int_1^\infty e^{-\frac{r}{2}\left(\sqrt{x}-\frac{1}{\sqrt{x}}\right)^2} x^{|m|/2}\frac{dx}{x}. \tag{14}$$

Let $y = \sqrt{x} - 1/\sqrt{x}$, so that $x = \dfrac{2+y^2+\sqrt{(1+y^2)^2-4}}{2}$ which is essentially 1, when $0 < y < 2$ and y^2 when $y > 2$. The integral in (14) is thus equivalent to

$$\int_0^{\sqrt{2}} e^{-\frac{r}{2}y^2}dy + \int_{\sqrt{2}}^\infty e^{-\frac{r}{2}y^2}y^{|m|}\frac{dy}{y}. \tag{15}$$

Next, let $y = \sqrt{2z/t}$ to get (15) equivalent to

$$r^{-1/2}\int_0^r e^{-2}\frac{dz}{\sqrt{z}} + \int_r^\infty e^{-z}|z|^{|m|/2}\frac{dz}{z}. \tag{16}$$

Using L'Hospital's Rule and the asymptotic formula for the incomplete gamma function we find that the expression (16) is effectively $r^{-|m|/2}$ in $(0,0)$ and $r^{-1/2}$ in $(1,\infty)$. This completes the proof when $m \neq 0$. The case $m = 0$ is left to the reader. □

Remark 3. *For $p \in (1,\infty)$, let $W_{\alpha,p}$ denote the class of weights w for which G_α is bounded on $L^p(w)$. Then $W_{\alpha,p}$ increases with α and $W_{\alpha,p} = W_{p,p}$, whenever $\alpha, \beta > n$. These facts follow from the semigroup property, the estimates (13) and the inequality $G_{\alpha_t} \leq CG^{1-t}_{\alpha_1}, G^t_{\alpha_2}$ which holds for $\alpha_t = (1-t)_{\alpha_1} + t\alpha_2$, provided $0 < \alpha_1 < \alpha_2$, $0 < t < 1$ and either $\alpha_2 < n$ or $\alpha_1 > n$. However,no two classes $W_{\alpha,p}$ are identical, as is shown in the following proposition.*

Proposition 2. *Fix $p \in (1,\infty)$ and $\alpha, \beta \in (0,n)$, with $\alpha < \beta/p$. Then, there is a weight $w \in W_{\beta,p} - W_{\alpha,p}$.*

Proof. Let $\phi_\gamma(x) = 1 + \sum\limits_{k=1}^{\infty} |x - 4^{-k}|^{-\gamma}\chi_{E_k}(x)$, where

$$E_k = \left\{x \in R^n : |x - 4^{-k}| \leq \tfrac{1}{2}4^k\right\}.$$

One readily shows $G_\beta\phi_\gamma \leq C\phi_\gamma$, if $0 < \gamma < \beta$. Hence, taking $w_\gamma = \phi_\gamma^{1/p}$, we have $w_\gamma \in W_{\beta,p}$. For $0 < \delta < n$, $L^p(w_\gamma)$ contains the function

$$f(x) = \sum_{k=1}^{\infty} |x - x_k|^{-\delta/p}\chi_{F_k}$$

where

$$x_k = \tfrac{1}{2}\left[\tfrac{3}{2} \cdot 4^{k+1} + \tfrac{1}{2} \cdot 4^k\right] = 7/4k + 2$$

and

$$F_k = \left\{x \in R^n : |x - x_k| < \tfrac{1}{2} \cdot 4k + 1\right\}.$$

We seek conditions on r and δ so that $w_\gamma \notin W_{\alpha,p}$.
Now, $G_\alpha f = 4^{k[\delta/p - \bar{\alpha}]}$ on E_k, so

$$||G_\alpha f||^p_{p,w_\gamma} \geq \sum_{k=1}^{\infty} 4^{k[\delta - \alpha p + \gamma - n]} = \infty,$$

if $\delta - \alpha p + \gamma - n \geq 0$. By taking γ sufficiently close to β and δ sufficiently closed to n, this condition can be met. □

Theorem 4. *Suppose n, α, m and m_+ are as in Lemma 1. Fix $C > 1$ and set $k = [1 - C^{-2/\gamma}]^{1/2}$. Then, the weight-generating kernel for G_α corresponding to C is equivalent to*

$$|y|^{m_+}, |y| \leq 1,$$

and

$$\left[|y|^{-(m+1)/2} + |y|^{(1-n)/2}\right] e^{-k|y|},\ |y| \geq 1.$$

In particular, for $\alpha \in (0,2]$, the kernel is equivalent to $G_\alpha(ky) + G_2(ky)$.

Proof. In view of (12), the kernel is given by

$$(4\pi)^{-n/2} \int_0^\infty e^{-(r^2/4)t} e^{-1/t} t^{\frac{n}{2}-1} S(t)\, dt,$$

where $r = |y|$ and

$$S(t) = \sum_{j=1}^{\infty} \frac{[C^{-1}t^{-\alpha/2}]^j}{\Gamma(j\alpha/2)},$$

When $C^{-1}t^{-\alpha/2} \leq 1$, that is, $t \geq C^{-2/\alpha} \equiv c$, the sum $S(t)$ is, effectively, $t^{-\alpha/2}$, as is seen from the inequalities

$$\frac{C^{-1}t^{-\alpha/2}}{\Gamma(\alpha/2)} \leq S(t) \leq \frac{C^{-1}t^{-\alpha/2}}{\Gamma(\alpha/2)}\left[1 + \sum_{j=1}^{\infty} \frac{1}{\Gamma(j\alpha/2)}\right].$$

Here, we have used $\Gamma(x+y) \geq \Gamma(x)\Gamma(y)$ when $x, y > 0$.

For $t \leq c$, the asymptotic expression

$$\sum_{j=1}^{\infty} \frac{t^j}{\Gamma(\ell j)} = t^{1/l}e^{t^{1/l}}[1 + 0(t^{-1})], \text{ as } t \to \infty,$$

given in [8], yields

$$S(t) \approx t^{-1}e^{a/t},\ t \leq c.$$

Thus, the kernel is, essentially,

$$\int_0^c e^{-(r^2/4)t}e^{(c-1)/t}t^{(n/2)-2}\,dt + \int_c^{\infty} e^{-(r^2/4)t}e^{-1/t}t^{(n-\alpha)/2}\frac{dt}{t}. \tag{17}$$

Now, the first term in (17) is bounded on $0 \leq r \leq 1$, while the second term is equivalent to G_α for all $r \geq 0$. It only remains to show the first integral, I, satisfies $I \approx r^{(1-n)/2}e^{-kr}$ for $r \geq 1$. To this end set $s = rt/2$ in I to obtain

$$I \approx r^{(2-n)/2}e^{-kr}\int_0^{cr/2} e^{-r[\sqrt{s}-k/\sqrt{s}]^2/2} \cdot s^{\frac{n}{2}-2}\,ds$$

Next, let $y = \sqrt{s} - k/\sqrt{s}$ so that

$$I \approx r^{(2-n)/2}e^{-kr}\int_{-\infty}^{\beta(r)} e^{-ry^2/2}[y + f(y)]^{n-3}[1 + yf(y)^{-1}]\,dy,$$

where $\beta(r) = \sqrt{cr/2} - k\sqrt{2/cr}$ and $f(y) = \sqrt{y^2 + 4l} = \sqrt{s} + \frac{k}{\sqrt{s}}$.

Finally, take $z = \sqrt{r/2}y$ to get

$$I \approx r^{(1-n)/2}e^{-kr}\int_{-\infty}^{\gamma(r)} e^{-z^2}\left[\sqrt{2/r}z + f\left(\sqrt{2/r}z\right)\right]^{n-3} \left[1 + \sqrt{2/r}zf\left(\sqrt{2/r}z\right)^{-1}\right]dz,$$

with $\gamma(r) = \sqrt{c}r/2 - k/\sqrt{c}$. We have now just to observe that when $z \in \mathbb{R}$ and $r \geq 1$

$$0 \leq 1 + \sqrt{2/r}zf\left(\sqrt{2/r}z\right)^{-1} < 2$$

while $\sqrt{2/r}z + f\left(\sqrt{2/r}z\right)$ lies between $2k^{1/2}$ and $\sqrt{2z^2 + 4k}$. □

Typical of G_α weights are the exponential functions $e^{\beta x}$, $-1 < \beta < 1$.

Proposition 3. *Suppose $\alpha \in (0, 1/2)$ and $p \in (1, \infty)$. Set $w_\beta(f) = e^{\beta|x|}$, $x \in R^n$. Then, G_α is bounded on $L^p(w_\beta)$ if and only if $-1 < \beta < 1$. Moreover, on this range of β, one has*

$$\lim_{\alpha \to 0+} \|G_\alpha * f - f\|_{p,w_\beta} = 0$$

for all $f \in L^p(w_\beta)$.

Proof. Fix $\beta \in (-1,1)$. We show $C > 1$ exists, independent of $\alpha \in (0,1/2)$, such that

$$(G_\alpha w_\beta)(x) \le C w_\beta(x),\ x \in R^n.$$

The "if" part then follows by Remark 2.

Using the simple inequalities $|x+y| \le |x| + |y|$ when $\beta > 0$ and $|x-y| \ge |x| - |y|$ when $\beta < 0$ we obtain

$$(G_\alpha w_\beta)(x) \le w_\beta(x) \int_{R^n} e^{|\beta|\,|y|} G_\alpha(y)\,dy.$$

But, the proof of Lemma 1 shows

$$\int_{R^n} e^{|\beta|\,|y|} G_\alpha(y)\,dy \le \int_{|y|\le 1} e^{|\beta|\,|y|} |y|^{\alpha-n}\,dy + \int_{|y|>1} e^{[|\beta|-1]|y|} |y|^{-\frac{n}{2}-\frac{1}{4}}\,dy$$
$$\approx 1,$$

when $\alpha \in (0,1)$.

To prove the "only if" part, only the care $\beta = -1$ needs to be considered. We observed that $f(x) = \frac{e^{|x|}}{1+|x|^{n+1}}$ is in $L^p(w_{-1})$ and that G_α bounded on $L^p(w_{-1})$ implies the same of $G_{j\alpha}, j = 2,3,\ldots$. However, for $j \ge \dfrac{n+3}{\alpha}$, $G_{j\alpha} f \equiv \infty$. □

Example 2. *Consider the Bessel potential* $G_2(y)$ *so that the weight-generating kernels are equivalent to* $G_2(ky)$, $0 < k < 1$. *These are especially simple when the dimension,* n, *is 1 or 3. In the first case* $G_2(y)$ *is essentially equal to the Picard kernel,* $e^{-|y|}$, *and in the second case to* $|y|^{-1}e^{-|y|}$.

According to Corollary 1, then, T_{G_2} *is bounded on* $L^p(e^{k/p'|y|})$ *and* $L^p(e^{-k/p|y|})$ *when* $n = 1$; *on* $L^p\left(|y|^{1/p'}e^{k/p'|y|}\right)$ *and* $L^p\left(|y|^{1/p}e^{-k/p|y|}\right)$ *when* $n = 3$.

5. The Gauss–Weierstrass Operators

In this section, we briefly treat the Gauss–Weierstrass kernels, $\{W_t\}_{t>0}$, defined by

$$W_t(y) = (4\pi t)^{-n/2} \exp(-|y|^2/4t),\ y \in R^n.$$

The iterates of W_t satisfy $W_t^{(h)} = W_{ht}, h = 1,2,\ldots$.

Proposition 4. *Fix* $p \in (1,\infty)$ *and set* $w_\beta(x) = e^{\beta|x|}$. *Then,* W_t *is bounded on* $L^p(w_\beta)$ *for all* $\beta \in (-\infty,\infty)$. *Moreover, one has*

$$\lim_{t\to 0_+} \|W_t * f - f\|_{p,w_\beta} = 0, \tag{18}$$

for every $f \in L^p(w_\beta)$.

Proof. Only $\beta \ge 0$ need by considered, the result for $\beta < 0$ follows by duality.

It will suffice to show that for each $\beta \ge 0$,

$$(W_t * e^{\beta|\cdot|})(x) \le C e^{\beta|x|},$$

with $C > 1$ independent of $x \in R^n$ and $t \in (0,1)$.

Now,

$$\int_{R^n} W_t(y) e^{\beta|x-y|}\,dy \le \int_{R^n} W_t(y) e^{\beta[|x|+|y|]}\,dy = e^{\beta|x|} \int_{R^n} W_t(y) e^{\beta|y|}\,dy,$$

from which the boundedness assertion follows. Again $W_t(y)$ is an increasing function of t for fixed y with $|y| \geq \sqrt{2nt}$ so,

$$\begin{aligned}\int_{R^n} W_t(y)e^{\beta|y|}\,dy &= \left(\int_{|y|<\sqrt{2nt}} + \int_{|y|>\sqrt{2nt}}\right) W_t(y)e^{\beta|y|}\,dy \\ &\leq e^{\beta\sqrt{2nt}} \int_{|y|<\sqrt{2nt}} W_t(y)\,dy + \int_{|y|>\sqrt{2nt}} W_1(y)e^{\beta|y|}\,dy \\ &\leq e^{\beta\sqrt{2n}} + (4\pi)^{-n/2}\int_{R_n} \exp(-|y|^2/4)e^{\beta|y|}\,dy\end{aligned}$$

when $t \in (0,1)$, thereby yielding (18). □

Theorem 5. *Fix $C > 1$. Then, the weight-generating kernel for W_l corresponding to C is equivalent to*

$$t^{-\frac{n}{4}-\frac{1}{2}}|y|^{1-n/2}\exp(-t^{-1/2}k|y|),\ k = \sqrt{\log K},\ \textit{for some } K > 1,$$

with the constants of equivalence independent of $t \in (0,a)$, $|y| > 4ka^{1/2}$, where $0 < a < 1$.

Proof. The desired kernel is

$$\sum_{j=1}^{\infty} C^{-j}(4\pi tj)^{-n/2}\exp(-r^2/4jt) \tag{19}$$

where $r = |y|$.

Let $f(r,t,u) = C^{-u}(4\pi tu)^{-n/2}\exp(-r^2/4ut)$, $u > 0$, and let $\alpha = t^{-1/2}kr$. Denote by I_1, I_2 and I_3 the intervals $(0, \alpha/4k^2)$, $(\alpha/4k^2, 2\alpha/k^2)$ and $(2\alpha/k^2, \infty)$, respectively. It is easily shown that when $r > 1$ and $t \in (0,1)$, the function f, as a function of u, increases on I_1, decreases on I_3 and satisfies $K^{-1}f(r,t,u) \leq f(r,t,u+s) \leq Kf(r,t,u)$ for some $K > 1$ and all $u \in I_2$, $s \in (0,1)$. Thus, the study of the sum in (19) amounts to looking at the integrals

$$J_i = \int_{I_i} f(r,t,u)\,du,\ i = 1,2,3.$$

Indeed, $C^{-u} = e^{-k^2u}$, therefore,

$$\begin{aligned}C^{-1}(J_1+J_2+J_3) &= C^{-1}\left(\int_0^{[\alpha/4h^2]+1} + \int_{[\alpha/4k^2]+1}^{[2\alpha/k^2]} + \int_{[2\alpha/k^2]}^{\infty}\right) f(r,t,u)\,du \\ &\leq \sum_{j=1}^{\infty} f(r,t,j) \\ &= \left(\sum_{j=1}^{[\alpha/4k^2]} + \sum_{j=[\alpha/4k^2]+1}^{[2\alpha/k^2]} + \sum_{j=[2\alpha/k^2]+1}^{\infty}\right) f(r,t,u)\,du \\ &\leq C(J_1+J_2+J_3).\end{aligned}$$

We have

$$\begin{aligned}J_1 &\leq t^{-n/2}\left(\frac{\alpha}{4k^2}\right)^{-n/2}\exp\left(-k^2\frac{\alpha}{4k^2}\right)\exp\left(-|y|^2\Big/\frac{4\alpha t}{4k^2}\right)\frac{\alpha}{4k^2} \\ &\leq t^{-\frac{n}{4}-\frac{1}{2}}|y|^{1-\frac{n}{2}}\exp\left(-\frac{5}{4}t^{-1/2}k|y|\right)\end{aligned}$$

Again,

$$J_3 \leq t^{-n/2}\left(\frac{2\alpha}{k^2}\right)^{-n/2} \exp\left(-r^2\Big/\frac{4\alpha}{4k^2}t\right)\exp\left(-k^2\frac{\alpha}{4k^2}\right)$$
$$\leq t^{-n/4}|y|^{-n/2}\exp\left(-\frac{5}{4}t^{-1/2}k|y|\right) \leq J_1.$$

Finally, in J_2 take $u = \alpha v/2k^2$ to get

$$J_2 \leq t^{-n/4}|y|^{-n/2}\int_{1/2}^{4}\exp\left(-\frac{\alpha}{2}\left[v+\frac{1}{v}\right]\right)v^{-n/2}\,dv$$
$$\leq t^{-\frac{n}{4}-\frac{1}{2}}|y|^{1-\frac{n}{2}}\exp\left(-t^{-1/2}k|y|\right).$$

Altogether, then,

$$\int_0^\infty f(|y|,t,u)\,du \leq t^{-\frac{n}{4}-\frac{1}{2}}|y|^{1-\frac{n}{2}}\exp\left(-t^{-1/2}k|y|\right).$$

□

Remark 4. *The weight-generating kernels are similar to those of G_2 on R^1 and R^3 (see Example 2), whence the exponential weights of Proposition 4 are in some sense typical. This illustrates a general theorem of Lofstrom, [18], which asserts that no translation-invariant operator is bounded on $L^p(w)$, when w is a rapidly varying weight such as $w(\alpha) = \exp(|x|^\alpha), \alpha > 1$.*

6. The Hardy Averaging Operators

In this section we consider Lebesgue-measurable functions defined on the set

$$R^n_+ = \{y \in R^n : y_i > 0, i = 1,\ldots,n\},$$

where, as usual, we write $y = (y_1,\ldots,y_n)$. Given $x \in R^n_+$, we define the sets

$$E_n(x) = \{y \in R^n_+ : 0 < y_i < x_i, i = 1,\ldots,n\}$$

and

$$F_n(x) = \{y \in R^n_+ : 0 < x_i < y_i, i = 1,\ldots,n\}.$$

Finally, we denote the product $x_1^{-1}\ldots x_n^{-1}$ by x^{-1} or $\frac{1}{x}$ and the product $(\log\frac{x_1}{y_1})\ \ldots\ (\log\frac{x_n}{y_n})$ by $\log\frac{x}{y}$; here, $x = (x_1,\ldots,x_n)$ and $y = (y_1,\ldots,y_n)$ belong to R^n_+.

The Hardy averaging operators, P_n and Q_n, are defined at $f \in P(R^n_+)$, $x \in R^n_+$, by

$$(P_nf)(x) = x^{-1}\int_{E_n(x)} f(y)\,dy$$

and

$$(Q_nf)(x) = \int_{F_n(x)} f(y)\,\frac{dy}{y}.$$

These operators, which are the transposes of one another, are generalizations to n-dimensions of the well-known ones, considered in [5] for example. A simple induction argument leads to the following formulas for the iterates of P_n and Q_n :

$$\left(P_n^{(j)}f\right)(x) = \frac{x^{-1}}{\Gamma(j)^n}\int_{F_n(s)} f(y)[\log x/y]^{j-1}\frac{dy}{y},$$

and

$$\left(Q_n^{(j)} f\right)(x) = \frac{1}{\Gamma(j)^n} \int_{F_n(s)} f(y)[\log y/x]^{j-1} \frac{dy}{y},$$

in which $x \in R_+^n$ and $j = 0, 1, \ldots$.

From Theorem 1 of [19], we obtain the representations of the weight-generating kernels of P_n and Q_n described below.

Theorem 6. *For $C > 1$ and set $\alpha = nC^{-1/n}$. Then, the weight-generating kernels for P_n and Q_n corresponding to C are equivalent, respectively, to*

$$x^{-1}\left[1 + (\log x/y)^{1/2(n-1)} \exp[\alpha(\log x/y)^{1/n}]\right] \chi_{E_n(x)}(y) \tag{20}$$

and

$$y^{-1}\left[1 + (\log y/x)^{1/2(n-1)} \exp[\alpha(\log y/x)^{1/n}]\right] \chi_{F_n(x)}(y). \tag{21}$$

Proposition 5. *Let $w_\beta(x) = [1 + |x|]^\beta$, $\beta \in R$. Then P_n is bounded on $L^p(w_\beta)$ if and only if $\beta < 1/p'$; by duality, Q_n is bounded on $L^p(w_\beta)$ of and only if $\beta > -1/p$.*

Proof. For simplicity, we consider $n = 2$ only.

Take $\psi = w_\gamma$ and fix $\alpha \in (0, 2)$. Denote by g the weight-generating kernel (20) applied to ψ. The change of variable $y_1 = x_1 z_1$, $y_2 = x_2 z_2$ in the integral giving $g(x)$ yields

$$g(x) = \int_0^1 \int_0^1 \left[1 + \sqrt{x_1^2 z_1^2 + x_2^2 z_2^2}\right]^\gamma$$
$$\left[1 + (\log 1/z_1 \log 1/z_2)^{-1/4} \times \exp\left[\alpha(\log 1/z_1 \log 1/z_2)^{1/2}\right]\right] dz_1\, dz_2$$

Hence, when $r > -1$, we find

$$g(x) \approx \begin{cases} 1, & 0 < x_1,\ x_2 \le 1 \\ x_2^\gamma, & 0 < x_1 \le 1,\ x_2 > 1 \\ x_1^\gamma, & x_1 > 1,\ 0 < x_2 \le 1 \\ \max\left[x_1^\gamma, x_2^\gamma\right], & x_1,\ x_2 \ge 1; \end{cases}$$

that is, $g(x) \approx w_\gamma(x)$, provided $r > -1$. This proves the "if" part, since $\beta = -\gamma/p' < 1/p'$.

To see that we must have $\gamma < 1/p'$, note that $h = \chi_{E_2}(\dot{x})$, $\dot{x} = (1, 1)$, is in $L^p(w_\gamma)$ and

$$(P_2 h)(x) = \begin{cases} 1, & 0 < x_1,\ x_2 \le 1 \\ x_2^{-1}, & 0 < x_1 \le 1,\ x_2 > 1 \\ x_1^{-1}, & x_1 > 1,\ 0 < x_2 \le 1 \\ x_1^{-1} x_2^{-1}, & x_1,\ x_2 \ge 1 \end{cases}$$

so

$$\int_{R_+^2} [w_\beta P_2 h]^p = \infty, \text{ if } \beta \ge 1/p'.$$

□

Theorem 7. *Denote by G_1 and G_2 the positive integral operators on $P(R_+^n)$ with kernels (20) and (21), respectively. Suppose $\psi_i \in P(R_+^n)$ is such that $G_i\psi_i < \infty$ ae on R_+^n, $i = 1, 2$. Take $\phi_i = \psi_i + G_i\psi_i$, $i = 1, 2$ and set $w = \phi_1^{-\frac{1}{p'}} \phi_2^{\frac{1}{p}}$. Then,*

$$P_n \colon L^p(R_+^n) \to L^p(R_+^n). \tag{22}$$

Moreover, any weight w satisfying (22) is equivalent to one in the above form.

Proof. This result is a consequence of Corollary 1 and Theorem 2. □

Remark 5. *When $n = 1$, the functions x^β, $\beta > -1$, are eigenfunctions of the operator P corresponding to the eigenvalue $(\beta+1)^{-1}$. As a result, if $\phi(x) = \sum_{k=0}^{\infty} a_k x^k$ converges for all x and if $a_k > 0$, then there exists $\psi \in P(R_+)$ for which $\psi + \sum_{j=1}^{\infty} C^{-j} P^{(j)} \psi \approx \phi$, $C > 1$; namely. $\psi(x) = b_0 + \sum_{k=1}^{\infty} b_k x^k$, where $b_k = a_k \left(1 + \sum_{j=1}^{\infty} \frac{c^{-j}}{(k+1)^j} j\right)_k^{-1}$, $k = 0, 1, \ldots$.*

For example, $\phi_1(x) = e^{\beta p' e^x}$, $\beta > 0$, is an entire function with $\phi^{(k)}(0) > 0$, $k = 0, 1, \ldots$. Combining this $\phi_1(x)$ with $\phi_2(x) = x^{\gamma p}$ we obtain the P-weight $x^\gamma e^{-\beta e^x}$, $\gamma < 0 < \beta$. Interpolation with change of measure shows one can, in fact, take all $\gamma < 1/p'$.

Similar results are obtained when $\phi(x_1, \ldots, x_n)$ is everywhere on R^n the sum of a power series in $x_1, \ldots, x_n$ with nonnegative coefficients. To take a specific example, consider a power series in one variable, $\sum_{k=0}^{\infty} a_k x^k$, $a_k > 0$, which converges for all $x \in R$. Then, $\phi(x_1, \ldots, x_n) = \sum_{k=0}^{\infty} a_k (x_1 \ldots x_n)^k$ leads to the P_n-weights $w(x_1, \ldots, x_n) = x_1^{\gamma_1} \ldots x_n^{\gamma_n}\, \phi(x_1, \ldots, x_n)^{1/p'}$, where $\gamma_i < 1/p'$, $i = 1, \ldots, n$.

Criteria for the boundedness of Hardy operators between weighted Lebesgue spaces with possibly different weights are given in [5] for the case $n = 1$ and in [7] for the case $n = 2$.

Added in Proof: While this work was in press the author came across the paper [20]. In it Bloom proves our Theorem 1 using complex interpolation rather than interpolation with change of measure. A (typical) application of his result to the Hardy operators substitutes them in the necessary and sufficient conditions, thereby giving a criterion *for their two* weighted boundedness. This is in contrast to our Theorem 6, in which the explicit form of a *single weight* is given.

Acknowledgments: The author is grateful to son Ely and Vít Musil for technical aid.

References

1. Bardaro, C.; Karsli, H.; Vinti, G. On pointwise convergence of linear integral operators with homogeneous kernels. *Integral Transform. Spec. Funct.* **2008**, *16*, 429–439. [CrossRef]
2. Bloom, S.; Kerman, R. Weighted norm inequalities for operators of Hardy type. *Proc. Am. Math. Soc.* **1991**, *113*, 135–141. [CrossRef]
3. Gogatishvili, A.; Stepanov, V.D. Reduction theorems for weighted integral inequalities on the cone of monotone functions. *Russ. Math. Surv.* **2013**, *68*, 597–664. [CrossRef]
4. Luor, D.-H. Weighted estimates for integral transforms and a variant of Schur's Lemma. *Integral Transform. Spec. Funct.* **2014**, *25*, 571–587. [CrossRef]
5. Muckenhoupt, B. Hardy's inequality with weights. *Stud. Math.* **1972**, 44, 31–38. [CrossRef]
6. Muckenhoupt, B. Two weight norm inequalities for the Poisson integral. *Trans. Am. Math. Soc.* **1975**, *210*, 225–231. [CrossRef]
7. Sawyer, E.T. A characterization of a weighted norm inequality for the two-dimensionol Hardy operator. *Studia Math.* **1985**, *82*, 1–16. [CrossRef]
8. Sawyer, E.T. A characterization of two weight norm inequalities for fractional and Poisson integrals. *Trans. Am. Math. Soc.* **1988**, *308*, 533–545. [CrossRef]
9. Vinti, G.; Zampogni, L.; A unifying approach to convergence of linear sampling type operators in Orlicz spaces. *Adv. Diff. Equ.* **2011**, *16*, 573–600.

10. Weiss, G. Various Remarks Concerning Rubio de Francia's Proof of Peter Jones' Theorem and some Applications of Ideas in the Proof, Preprint.
11. Christ, M. Weighted norm inequalities and Schur's lemma. *Stud. Math.* **1984**, *78*, 309–319. [CrossRef]
12. Butzer, P.K.; Nessel, R.J. *Fourier Analysis and Approximation*; Birkhäuser Verlag: Basel, Switzerland, 1971; Volume I.
13. Stein, E.M. *Singular Integrals and Differentiability Properties of Functions*; Princeton University Press: Princeton, NJ, USA, 1970.
14. Muckenhoupt B. Weighted norm inequalities for the Hardy maximal function. *Trans. Am. Math. Soc.* **1972**, *165*, 207–226. [CrossRef]
15. Stein, E.M.; Weiss, G. Interpolation of operators with change of measures. *Trans. Am. Math. Soc.* **1958**, *87*, 159–172. [CrossRef]
16. Rosenblum, M. Summability of Fourier series in $L^p(d\mu)$. *Trans. Am. Math. Soc.* **1962**, *105*, 32–42.
17. Donoghue, W.F., Jr. *Distributions and Fourier Transforms*; Academic Press: Cambridge, MA, USA, 1969.
18. Lofstrom, J. A non-existence for translation-invariant operators on weighted L_p-spaces. *Math. Scand.* **1983**, *55*, 88–96. [CrossRef]
19. Wright, E.M. The asymptotic expansion of the generalized hypergeometric function. 1 *Lond. Math. Soc.* **1935**, *10*, 289–293. [CrossRef]
20. Bloom, S.; Solving weighted norm inequalities using the Rubio de Francia algorithm. *Proc. Am. Math. Soc.* **1987**, *101*, 306–312. [CrossRef]

7

Geometric Properties of Certain Classes of Analytic Functions Associated with a q-Integral Operator

Shahid Mahmood [1], Nusrat Raza [2,*], Eman S. A. Abujarad [3], Gautam Srivastava [4,5], H. M. Srivastava [6,7] and Sarfraz Nawaz Malik [8]

1. Department of Mechanical Engineering, Sarhad University of Science & Information Technology, Ring Road, Peshawar 25000, Pakistan; shahidmahmood757@gmail.com
2. Mathematics Section, Women's College, Aligarh Muslim University, Aligarh 202001, Uttar Pradesh, India
3. Department of Mathematics, Aligarh Muslim University, Aligarh 202001, Uttar Pradesh, India; emanjarad2@gmail.com
4. Department of Mathematics and Computer Science, Brandon University, 270 18th Street, Brandon, MB R7A 6A9, Canada; srivastavag@brandonu.ca
5. Research Center for Interneural Computing, China Medical University, Taichung 40402, Taiwan, Republic of China
6. Department of Mathematics and Statistics, University of Victoria, Victoria, BC V8W 3R4, Canada; harimsri@math.uvic.ca
7. Department of Medical Research, China Medical University Hospital, China Medical University, Taichung 40402, Taiwan, Republic of China
8. Department of Mathematics, COMSATS University Islamabad, Wah Campus 47040, Pakistan; snmalik110@yahoo.com

* Correspondence: nraza.maths@gmail.com

Abstract: This article presents certain families of analytic functions regarding q-starlikeness and q-convexity of complex order γ $(\gamma \in \mathbb{C}\setminus\{0\})$. This introduced a q-integral operator and certain subclasses of the newly introduced classes are defined by using this q-integral operator. Coefficient bounds for these subclasses are obtained. Furthermore, the (δ, q)-neighborhood of analytic functions are introduced and the inclusion relations between the (δ, q)-neighborhood and these subclasses of analytic functions are established. Moreover, the generalized hyper-Bessel function is defined, and application of main results are discussed.

Keywords: Geometric Function Theory; q-integral operator; q-starlike functions of complex order; q-convex functions of complex order; (δ, q)-neighborhood

MSC: 30C15; 30C45

---

1. Introduction

Recently, many researchers have focused on the study of q-calculus keeping in view its wide applications in many areas of mathematics, e.g., in the q-fractional calculus, q-integral calculus, q-transform analysis and others (see, for example, [1,2]). Jackson [3] was the first to introduce and develop the q-derivative and q-integral. Purohit [4] was the first one to introduce and analyze a class in open unit disk and he used a certain operator of fractional q-derivative. His remarkable contribution was to give q-extension of a number of results that were already known in analytic function theory. Later, the q-operator was studied by Mohammed and Darus regarding its geometric properties on certain analytic functions, see [5]. A very significant usage of the q-calculus in the context of Geometric Function Theory was basically furnished and the basic (or q-) hypergeometric functions were first used in Geometric Function Theory in a book chapter by Srivastava (see, for details, [6] pp. 347 et seq.;

see also [7]). Earlier, a class of q-starlike functions were introduced by Ismail et al. [8]. These are the generalized form of the known starlike functions by using the q-derivatives. Sahoo and Sharma [9] obtained many results of q-close-to-convex functions. Also, some recent results and investigations associated with the q-derivatives operator have been in [6,10–13].

It is worth mentioning here that the ordinary calculus is a limiting case of the quantum calculus. Now, we recall some basic concepts and definitions related to q-derivative, to be used in this work. For more details, see References [3,14–16].

The quantum derivative (named as q-derivative) of function f is defined as:

$$D_q f(z) = \frac{f(z) - f(qz)}{(1-q)z} \qquad (z \neq 0;\ 0 < q < 1).$$

We note that $D_q f(z) \longrightarrow f'(z)$ as $q \longrightarrow 1-$ and $D_q f(0) = f'(0)$, where f' is the ordinary derivative of f.

In particular, q-derivative of $h(z) = z^n$ is as follows :

$$D_q h(z) = [n]_q z^{n-1}, \tag{1}$$

where $[n]_q$ denotes q-number which is given as:

$$[n]_q = \frac{1-q^n}{1-q} \qquad (0 < q < 1). \tag{2}$$

Since we see that $[n]_q \longrightarrow n$ as $q \longrightarrow 1-$, therefore, in view of Equation (1), $D_q h(z) \longrightarrow h'(z)$ as $q \longrightarrow 1-$, where h' represents ordinary derivative of h.

The q-gamma function Γ_q is defined as:

$$\Gamma_q(t) = (1-q)^{1-t} \prod_{n=0}^{\infty} \frac{1-q^{n+1}}{1-q^{n+t}} \qquad (t > 0;\ 0 < q < 1), \tag{3}$$

which has the following properties:

$$\Gamma_q(t+1) = [t]_q \Gamma_q(t) \tag{4}$$

and

$$\Gamma_q(t+1) = [t]_q!\,, \tag{5}$$

where $t \in \mathbb{N}$ and $[.]_q!$ denotes the q-factorial and defined as:

$$[t]_q! = \begin{cases} [t]_q[t-1]_q \ldots [2]_q[1]_q, & t = 1,2,3,\ldots; \\ 1, & t = 0. \end{cases} \tag{6}$$

Also, the q-beta function B_q is defined as:

$$B_q(t,s) = \int_0^1 x^{t-1}(1-qx)_q^{s-1} d_q x \qquad (t,s > 0;\ 0 < q < 1), \tag{7}$$

which has the following property:

$$B_q(t,s) = \frac{\Gamma_q(s)\Gamma_q(t)}{\Gamma_q(s+t)}, \tag{8}$$

where Γ_q is given by Equation (3).

Furthermore, q-binomial coefficients are defined as [17]:

$$\binom{n}{k}_q = \frac{[n]_q!}{[k]_n![n-k]_q!}, \tag{9}$$

where $[.]_q!$ is given by Equation (6).

We consider the class $\mathcal{A}$ comprising the functions that are analytic in open unit disc $\mathbb{U} = \{z \in \mathbb{C} : |z| < 1\}$ and are of the form given as:

$$f(z) = z + \sum_{n=2}^{\infty} a_n z^n. \tag{10}$$

Using Equation (1), the q-derivative of f, defined by Equation (10) is as follows:

$$D_q f(z) = 1 + \sum_{n=2}^{\infty} [n]_q a_n z^{n-1} \qquad (z \in \mathbb{U};\ 0 < q < 1), \tag{11}$$

where $[n]_q$ is given by Equation (2).

The two important subsets of the class $\mathcal{A}$ are the families $\mathcal{S}^*$ consisting of those functions that are starlike with reference to origin and $\mathcal{C}$ which is the collection of convex functions. A function f is from S^*if for each point $x \in f(\mathbb{U})$ the linear segment between 0 and x is contained in $f(\mathbb{U})$. Also, a function $f \in \mathcal{C}$ if the image $f(\mathbb{U})$ is a convex subset of complex plane $\mathbb{C}$, i.e., $f(\mathbb{U})$ must have every line segment that joins its any two points.

Nasr and Aouf [18] defined the class of those functions which are starlike and are of complex order γ $(\gamma \in \mathbb{C}\backslash\{0\})$, denoted by $\mathcal{S}^*(\gamma)$ and Wiatrowski [19] gave the class of similar type convex functions i.e., of complex order γ $(\gamma \in \mathbb{C}\backslash\{0\})$, denoted by $\mathcal{C}(\gamma)$ as:

$$\mathcal{S}^*(\gamma) = \left\{ f \in \mathcal{A} : \Re\left(1 + \frac{1}{\gamma}\left(\frac{zf'(z)}{f(z)} - 1\right)\right) > 0 \ (z \in \mathbb{U}; \gamma \in \mathbb{C}\backslash\{0\}) \right\} \tag{12}$$

and

$$\mathcal{C}(\gamma) = \left\{ f \in \mathcal{A} : \Re\left(1 + \frac{1}{\gamma}\frac{zf''(z)}{f'(z)}\right) > 0 \ (z \in \mathbb{U}; \gamma \in \mathbb{C}\backslash\{0\}) \right\}, \tag{13}$$

respectively.

From Equations (12) and (13), it is clear that $\mathcal{S}^*(\gamma)$ and $\mathcal{C}(\gamma)$ are subclasses of the class $\mathcal{A}$.

The class denoted by $\mathcal{S}^*{}_q(\mu)$ of such q-starlike functions that are of order μ is defined as:

$$\mathcal{S}^*{}_q(\mu) = \left\{ f \in \mathcal{A} : \Re\left(\frac{zD_q f(z)}{f(z)}\right) > \mu \ \ (z \in \mathbb{U}; 0 \le \mu < 1) \right\}. \tag{14}$$

Also, the class $\mathcal{C}_q(\mu)$ of q-convex functions of order μ is defined as:

$$\mathcal{C}_q(\mu) = \left\{ f \in \mathcal{A} : \Re\left(\frac{D_q(zD_q f(z))}{D_q f(z)}\right) > \mu \ \ (z \in \mathbb{U}; 0 \le \mu < 1) \right\}. \tag{15}$$

For more detail, see [20]. From Equations (14) and (15), it is clear that $\mathcal{S}_q^*(\mu)$ and $\mathcal{C}_q(\mu)$ are subclasses of the class $\mathcal{A}$.

Next, we recall that the δ-neighborhood of the function $f(z) \in \mathcal{A}$ is defined as [21]:

$$\mathcal{N}_\delta(f) = \left\{ g(z) = z + \sum_{n=2}^{\infty} b_n z^n \,\middle|\, \sum_{n=2}^{\infty} n\,|a_n - b_n| \le \delta \right\} \qquad (\delta \ge 0). \tag{16}$$

In particular, the δ-neighborhood of the identity function $p(z) = z$ is defined as [21]:

$$\mathcal{N}_\delta(p) = \left\{ g(z) = z + \sum_{n=2}^{\infty} b_n z^n \middle| \sum_{n=2}^{\infty} n\,|b_n| \leq \delta \right\} \qquad (\delta \geq 0). \tag{17}$$

Finally, we recall that the Jung-Kim-Srivastava integral operator $\mathcal{Q}_\beta^\alpha : \mathcal{A} \to \mathcal{A}$ are defined as [22]:

$$\begin{aligned}\mathcal{Q}_\beta^\alpha f(z) &= \binom{\alpha+\beta}{\beta} \frac{\alpha}{z^\beta} \int_0^z t^{\beta-1} \left(1 - \frac{t}{z}\right)^{\alpha-1} f(t)dt \\ &= z + \frac{\Gamma(\alpha+\beta+1)}{\Gamma(\beta+1)} \sum_{n=2}^{\infty} \frac{\Gamma(\beta+n)}{\Gamma(\alpha+\beta+n)} a_n z^n \qquad (\beta > -1;\ \alpha > 0;\ f \in \mathcal{A}).\end{aligned} \tag{18}$$

The Bessel functions are associated with a wide range of problems in important areas of mathematical physics and Engineering. These functions appear in the solutions of heat transfer and other problems in cylindrical and spherical coordinates. Rainville [23] discussed the properties of the Bessel function.

The generalized Bessel functions $w_{\nu,b,d}(z)$ are defined as [24]:

$$w_{\nu,b,d}(z) = \sum_{n=0}^{\infty} \frac{(-d)^n}{n!\Gamma\left(\nu + n + \dfrac{b+1}{2}\right)} \left(\frac{z}{2}\right)^{2n+\nu}, \tag{19}$$

where $\nu, b, d, z \in \mathbb{C}$.

Orhan, Deniz and Srivastava [25] defined the function $\varphi_{\nu,b,d}(z) : \mathbb{U} \to \mathbb{C}$ as:

$$\varphi_{\nu,b,d}(z) = 2^\nu \Gamma\left(\nu + \frac{b+1}{2}\right) z^{-\frac{\nu}{2}} w_{\nu,b,d}(\sqrt{z}), \tag{20}$$

by using the Generalized Bessel function $w_{\nu,b,d}(z)$, given by Equation (12).

The power series representation for the function $\varphi_{\nu,b,d}(z)$ is as follows [25]:

$$\varphi_{\nu,b,d}(z) = \sum_{n=0}^{\infty} \frac{(-d/4)^n}{(c)_n n!} z^n, \tag{21}$$

where $c = \nu + \dfrac{b+1}{2} > 0$, $\nu, b, d \in \mathbb{R}$ and $z \in \mathbb{U} = \{z \in \mathbb{C} : |z| < 1\}$.

The hyper-Bessel function is defined as [26]:

$$J_{\alpha_d}(z) = \sum_{n=0}^{\infty} \frac{(z/d+1)^{\alpha_1 + \dots \alpha_d}}{\Gamma(\alpha_1+1)\dots\Gamma(\alpha_d+1)} {}_0F_d\left(-, (\alpha_d+1); -\left(\frac{z}{d+1}\right)^{d+1}\right), \tag{22}$$

where the hypergeometric function ${}_pF_q$ is defined by:

$${}_pF_q\left((\beta_p); (\eta_q); x\right) = \sum_{n=0}^{\infty} \frac{(\beta_1)_n(\beta_2)_n \dots (\beta_p)_n}{(\alpha_1)_n(\alpha_2)_n \dots (\alpha_q)_n} \frac{x^n}{n!}, \tag{23}$$

using above Equation (23) in Equation (22), then the function $J_{\alpha_d}(z)$ has the following power series:

$$J_{\alpha_d}(z) = \sum_{n=0}^{\infty} \frac{(-1)^n}{n!\Gamma(\alpha_1+n+1)\Gamma(\alpha_2+n+1)\dots\Gamma(\alpha_d+n+1)} \left(\frac{z}{d+1}\right)^{n(d+1)+\alpha_1+\dots\alpha_d}, \tag{24}$$

By choosing $d = 1$ and putting $\alpha_1 = \nu$, we get the classical Bessel function

$$J_\nu(z) = \sum_{n=0}^{\infty} \frac{(-1)^n}{n!\Gamma(\nu+n+1)} z^{2n+\nu}. \tag{25}$$

In the next section, we introduce the classes of q-starlike functions that are of complex order γ $(\gamma \in \mathbb{C}\setminus\{0\})$ and similarly, q-convex functions that are of complex order γ $(\gamma \in \mathbb{C}\setminus\{0\})$, which are denoted by $\mathcal{S}_q^*(\gamma)$ and $\mathcal{C}_q(\gamma)$, respectively. Also, we define a q-integral operator and define the subclasses $\mathcal{S}_q(\alpha,\beta,\gamma)$ and $\mathcal{C}_q(\alpha,\beta,\gamma)$ of the class $\mathcal{A}$ by using this q-integral operator. Then, we find the coefficient bounds for these subclasses.

First, we define the q-starlike function of complex order γ $(\gamma \in \mathbb{C}\setminus\{0\})$, denoted by $\mathcal{S}_q^*(\gamma)$ and the q-convex function of complex order γ $(\gamma \in \mathbb{C}\setminus\{0\})$, denoted by $\mathcal{C}_q(\gamma)$ by taking the q-derivative in place of ordinary derivatives in Equations (12) and (13), respectively.

The respective definitions of the classes $\mathcal{S}_q^*(\gamma)$ and $\mathcal{C}_q(\gamma)$ are as follows:

Definition 1. *The function $f \in \mathcal{A}$ will belong to the class $\mathcal{S}_q^*(\gamma)$ if it satisfies the following inequality:*

$$\Re\left(1+\frac{1}{\gamma}\left(\frac{zD_qf(z)}{f(z)}-1\right)\right) > 0 \qquad (\gamma \in \mathbb{C}\setminus\{0\},\ 0<q<1). \tag{26}$$

Definition 2. *The function $f \in \mathcal{A}$ will belong to the class $\mathcal{C}_q(\gamma)$ if it satisfies the following inequality:*

$$\Re\left(1+\frac{1}{\gamma}\left(\frac{D_q(zD_qf(z))}{D_qf(z)}\right)\right) > 0 \qquad (\gamma \in \mathbb{C}\setminus\{0\},0<q<1). \tag{27}$$

Remark 1. *(i) If $\gamma \in \mathbb{R}$ and $\gamma = 1-\mu$ $(0 \le \mu < 1)$, then the subclasses $\mathcal{S}_q^*(\gamma)$ and $\mathcal{C}_q(\gamma)$ give the sub classes $\mathcal{S}_q^*(\mu)$ and $\mathcal{C}_q(\mu)$, respectively.*

(ii) Using the fact that $\lim_{q\to 1-} D_qf(z) = f'(z)$, we get that $\lim_{q\to 1-} \mathcal{S}_q^*(\gamma) = \mathcal{S}^*(\gamma)$ and $\lim_{q\to 1-} \mathcal{C}_q(\gamma) = \mathcal{C}(\gamma)$.

Now, we introduce the q-integral operator $\chi_{\beta,q}^{\alpha}$ as:

$$\chi_{\beta,q}^{\alpha}f(z) = \binom{\alpha+\beta}{\beta}_q \frac{[\alpha]_q}{z^\beta}\int_0^z t^{\beta-1}\left(1-\frac{qt}{z}\right)_q^{\alpha-1} f(t)d_qt$$
$$(\alpha>0;\ \beta>-1;\ 0<q<1;\ |z|<1;\ f\in\mathcal{A}). \tag{28}$$

It is clear that $\chi_{\beta,q}^{\alpha}f(z)$ is analytic in open disc $\mathbb{U}$.

Using Equations (4), (5) and (7)–(9), we get the following power series for the function $\chi_{\beta,q}^{\alpha}f$ in $\mathbb{U}$:

$$\chi_{\beta,q}^{\alpha}f(z) = z + \sum_{n=2}^{\infty}\frac{\Gamma_q(\beta+n)\Gamma_q(\alpha+\beta+1)}{\Gamma_q(\alpha+\beta+n)\Gamma_q(\beta+1)}a_nz^n \qquad (\alpha>0;\ \beta>-1;\ 0<q<1;\ f\in\mathcal{A}). \tag{29}$$

Remark 2. *For $q \longrightarrow 1-$, Equation (29), gives the Jung-Kim-Srivastava integral operator $\mathcal{Q}_\beta^\alpha$, given by Equation (18).*

Remark 3. *Taking $\alpha = 1$ in Equation (28) and using Equations (4), (5) and (9), we get the q-Bernardi integral operator, defined as [27]:*

$$\mathcal{F}(z) = \frac{[1+\beta]_q}{z^\beta}\int_0^z t^{\beta-1}f(t)d_qt \qquad \beta = 1,2,3,\ldots$$

Next, in view of the Definitions 1 and 2 and the fact that $\Re(z) < |z|$, we introduce the subclasses $\mathcal{S}_q(\alpha,\beta,\gamma)$ and $\mathcal{C}_q(\alpha,\beta,\gamma)$ of the classes $\mathcal{S}_q^*(\gamma)$ and $\mathcal{C}_q(\gamma)$, respectively, by using the operator $\chi^{\alpha}_{\beta,q}$, as:

Definition 3. *The function $f \in \mathcal{A}$ will belong to $\mathcal{S}_q(\alpha,\beta,\gamma)$ if it satisfies the following inequality:*

$$\left|\frac{zD_q(\chi^{\alpha}_{\beta,q}f(z))}{\chi^{\alpha}_{\beta,q}f(z)} - 1\right| < |\gamma|, \tag{30}$$

where $\alpha > 0;\ \beta > -1;\ 0 < q < 1;\ \gamma \in \mathbb{C}\setminus\{0\}$.

Definition 4. *The function $f \in \mathcal{A}$ will belong to $\mathcal{C}_q(\alpha,\beta,\gamma)$ if it satisfies the following inequality:*

$$\left|\frac{D_q\left(zD_q\chi^{\alpha}_{\beta,q}f(z)\right)}{D_q\chi^{\alpha}_{\beta,q}f(z)}\right| < |\gamma|, \tag{31}$$

where $\alpha > 0;\ \beta > -1;\ 0 < q < 1; \gamma \in \mathbb{C}\setminus\{0\}$.

Now, we establish the following result, which gives the coefficient bound for the subclass $\mathcal{S}_q(\alpha,\beta,\gamma)$:

Lemma 1. *If f is an analytic function such that it belongs to the class $\mathcal{S}_q(\alpha,\beta,\gamma)$, then*

$$\sum_{n=2}^{\infty}\frac{\Gamma_q(\beta+n)\Gamma_q(\alpha+\beta+1)}{\Gamma_q(\alpha+\beta+n)\Gamma_q(\beta+1)}\left([n]_q - |\gamma| - 1\right)a_n < |\gamma| \quad (\alpha > 0;\ \beta > -1;\ 0 < q < 1; \gamma \in \mathbb{C}\setminus\{0\}), \tag{32}$$

where Γ_q and $[n]_q$ are given by Equations (3) and (2), respectively.

Proof. Let $f \in \mathcal{A}$, then using Equations (11) and (29), we have

$$\left|\frac{zD_q(\chi^{\alpha}_{\beta,q}f(z))}{\chi^{\alpha}_{\beta,q}f(z)} - 1\right| = \left|\frac{z + \sum_{n=2}^{\infty}\frac{\Gamma_q(\beta+n)\Gamma_q(\alpha+\beta+1)}{\Gamma_q(\alpha+\beta+n)\Gamma_q(\beta+1)}[n]_q a_n z^n}{z + \sum_{n=2}^{\infty}\frac{\Gamma_q(\beta+n)\Gamma_q(\alpha+\beta+1)}{\Gamma_q(\alpha+\beta+n)\Gamma_q(\beta+1)}a_n z^n} - 1\right|. \tag{33}$$

If $f \in \mathcal{S}_q(\alpha,\beta,\gamma)$, then in view of Definition 3 and Equation (33), we have

$$\left|\frac{z + \sum_{n=2}^{\infty}\frac{\Gamma_q(\beta+n)\Gamma_q(\alpha+\beta+1)}{\Gamma_q(\alpha+\beta+n)\Gamma_q(\beta+1)}[n]_q a_n z^n}{z + \sum_{n=2}^{\infty}\frac{\Gamma_q(\beta+n)\Gamma_q(\alpha+\beta+1)}{\Gamma_q(\alpha+\beta+n)\Gamma_q(\beta+1)}a_n z^n} - 1\right| < |\gamma|,$$

which, on simplifying, gives

$$\left|\frac{\sum_{n=2}^{\infty}\frac{\Gamma_q(\beta+n)\Gamma_q(\alpha+\beta+1)}{\Gamma_q(\alpha+\beta+n)\Gamma_q(\beta+1)}\left([n]_q - 1\right)a_n z^{n-1}}{1 + \sum_{n=2}^{\infty}\frac{\Gamma_q(\beta+n)\Gamma_q(\alpha+\beta+1)}{\Gamma_q(\alpha+\beta+n)\Gamma_q(\beta+1)}a_n z^{n-1}}\right| < |\gamma|. \tag{34}$$

Now, using the fact that $\Re(z) < |z|$ in the Inequality (34), we get

$$\Re\left(\frac{\sum_{n=2}^{\infty}\frac{\Gamma_q(\beta+n)\Gamma_q(\alpha+\beta+1)}{\Gamma_q(\alpha+\beta+n)\Gamma_q(\beta+1)}\left([n]_q-1\right)a_n z^{n-1}}{1+\sum_{n=2}^{\infty}\frac{\Gamma_q(\beta+n)\Gamma_q(\alpha+\beta+1)}{\Gamma_q(\alpha+\beta+n)\Gamma_q(\beta+1)}a_n z^{n-1}}\right) < |\gamma|. \tag{35}$$

Since $\chi^{\alpha}_{\beta,q}f(z)$ is analytic in $\mathbb{U}$, therefore taking limit $z \to 1-$through real axis, Inequality (35), gives the Assertion (32). □

Also, we establish the following result, which gives the coefficient bound for the subclass $\mathcal{C}_q(\alpha,\beta,\gamma)$:

Lemma 2. *If f is an analytic function such that it belongs to the class $\mathcal{C}_q(\alpha,\beta,\gamma)$ and $|\gamma| \geq 1$ then*

$$\sum_{n=2}^{\infty}\frac{\Gamma_q(\beta+n)\Gamma_q(\alpha+\beta+1)}{\Gamma_q(\alpha+\beta+n)\Gamma_q(\beta+1)}([n]_q\left([n]_q-|\gamma|\right))a_n < |\gamma|-1 \quad (\alpha>0;\ \beta>-1;\ 0<q<1; \gamma\in\mathbb{C}\setminus\{0\}), \tag{36}$$

where Γ_q and $[n]_q$ are given by Equations (3) and (2), respectively.

Proof. Let $f \in \mathcal{A}$, then using Equations (11) and (29), we get

$$\left|\frac{D_q\left(zD_q\chi^{\alpha}_{\beta,q}f(z)\right)}{D_q\chi^{\alpha}_{\beta,q}f(z)}\right| = \left|\frac{1+\sum_{n=2}^{\infty}\frac{\Gamma_q(\beta+n)\Gamma_q(\alpha+\beta+1)}{\Gamma_q(\alpha+\beta+n)\Gamma_q(\beta+1)}([n]_q)^2 a_n z^{n-1}}{1+\sum_{n=2}^{\infty}\frac{\Gamma_q(\beta+n)\Gamma_q(\alpha+\beta+1)}{\Gamma_q(\alpha+\beta+n)\Gamma_q(\beta+1)}[n]_q a_n z^{n-1}}\right|. \tag{37}$$

If $f \in \mathcal{C}_q(\alpha,\beta,\gamma)$, then in view of Definition 4 and Equation (37), we have

$$\left|\frac{1+\sum_{n=2}^{\infty}\frac{\Gamma_q(\beta+n)\Gamma_q(\alpha+\beta+1)}{\Gamma_q(\alpha+\beta+n)\Gamma_q(\beta+1)}([n]_q)^2 a_n z^{n-1}}{1+\sum_{n=2}^{\infty}\frac{\Gamma_q(\beta+n)\Gamma_q(\alpha+\beta+1)}{\Gamma_q(\alpha+\beta+n)\Gamma_q(\beta+1)}[n]_q a_n z^{n-1}}\right| < |\gamma| \tag{38}$$

Now, using the fact that $\Re(z) < |z|$ in Inequality (38), we get

$$\Re\left(\frac{1+\sum_{n=2}^{\infty}\frac{\Gamma_q(\beta+n)\Gamma_q(\alpha+\beta+1)}{\Gamma_q(\alpha+\beta+n)\Gamma_q(\beta+1)}([n]_q)^2 a_n z^{n-1}}{1+\sum_{n=2}^{\infty}\frac{\Gamma_q(\beta+n)\Gamma_q(\alpha+\beta+1)}{\Gamma_q(\alpha+\beta+n)\Gamma_q(\beta+1)}[n]_q a_n z^{n-1}}\right) < |\gamma| \tag{39}$$

Since $\chi^{\alpha}_{\beta,q}f(z)$ is analytic in $\mathbb{U}$, therefore taking limit $z \to 1-$ through real axis, Inequality (39) gives the Assertion (36). □

In the next section, we define (δ,q)-neighborhood of the function $f \in \mathcal{A}$ and establish the inclusion relations of the subclasses $\mathcal{S}_q(\alpha,\beta,\gamma)$ and $\mathcal{C}_q(\alpha,\beta,\gamma)$ with the (δ,q)-neighborhood of the identity function $p(z) = z$.

2. The Classes $\mathcal{N}_{\delta,q}(f)$ and $\mathcal{N}_{\delta,q}(p)$

In view of Equation (16), we define the (δ,q)-neighborhood of the function $f \in \mathcal{A}$ as:

$$\mathcal{N}_{\delta,q}(f) = \left\{g(z) = z + \sum_{n=2}^{\infty} b_n z^n \,\middle|\, \sum_{n=2}^{\infty}[n]_q\,|a_n - b_n| \leq \delta\right\} \qquad (\delta \geq 0,\ 0<q<1), \tag{40}$$

where $[n]_q$ is given by Equation (2).

In particular, the (δ, q)-neighborhood of the identity function $p(z) = z$, defined as:

$$\mathcal{N}_{\delta,q}(p) = \left\{ g(z) = z + \sum_{n=2}^{\infty} b_n z^n \middle| \sum_{n=2}^{\infty} [n]_q |b_n| \leq \delta \right\} \qquad (\delta \geq 0,\ 0 < q < 1). \tag{41}$$

Since $[n]_q$ approaches n as q approaches $1-$, therefore, from Equations (16) and (40), we note that $\lim_{q\to 1-} \mathcal{N}_{\delta,q}(f) = \mathcal{N}_{\delta}(f)$, where $\mathcal{N}_{\delta}(f)$ is defined by Equation (16). In particular, $\lim_{q\to 1-} \mathcal{N}_{\delta,q}(p) = \mathcal{N}_{\delta}(p)$.

Now, we establish the following inclusion relation between the class $\mathcal{S}_q(\alpha, \beta, \gamma)$ and (δ, q)-neighborhood $\mathcal{N}_{\delta,q}(p)$ of identity function p for the specified range of values of δ:

Theorem 1. *If* $-1 < \beta \leq 0$, $|\gamma| \leq [n]_q - 1 \ \ (n = 2, 3, \dots)$ *and*

$$\delta \geq \frac{|\gamma| [2]_q \Gamma_q(\alpha + \beta + 2) \Gamma_q(\beta + 1)}{([2]_q - |\gamma| - 1) \Gamma_q(\beta + 2) \Gamma_q(\alpha + \beta + 1)}, \tag{42}$$

then

$$\mathcal{S}_q(\alpha, \beta, \gamma) \subset \mathcal{N}_{\delta,q}(p) \qquad (\gamma \in \mathbb{C} \setminus \{0\};\ \alpha > 0;\ 0 < q < 1). \tag{43}$$

Proof. Let $f \in \mathcal{S}_q(\alpha, \beta, \gamma)$, then, in view of Lemma 1, Inequality (32) holds. Since for $\alpha > 0$, $-1 < \beta \leq 0$, the sequence $\left\{ \frac{\Gamma_q(\beta + n)}{\Gamma_q(\alpha + \beta + n)} \right\}_{n=2}^{\infty}$ is non-decreasing, therefore, we have

$$\frac{\Gamma_q(\beta + 2)\Gamma_q(\alpha + \beta + 1)}{\Gamma_q(\alpha + \beta + 2)\Gamma_q(\beta + 1)} ([2]_q - |\gamma| - 1) \sum_{n=2}^{\infty} a_n \leq \sum_{n=2}^{\infty} \frac{\Gamma_q(\beta + n)\Gamma_q(\alpha + \beta + 1)}{\Gamma_q(\alpha + \beta + n)\Gamma_q(\beta + 1)} ([n]_q - |\gamma| - 1)\, a_n,$$

which in view of Inequality (32), gives

$$\frac{\Gamma_q(\beta + 2)\Gamma_q(\alpha + \beta + 1)}{\Gamma_q(\alpha + \beta + 2)\Gamma_q(\beta + 1)} ([2]_q - |\gamma| - 1) \sum_{n=2}^{\infty} a_n < |\gamma|, \tag{44}$$

or, equivalently,

$$\sum_{n=2}^{\infty} a_n < \frac{|\gamma| \Gamma_q(\alpha + \beta + 2)\Gamma_q(\beta + 1)}{\Gamma_q(\beta + 2)\Gamma_q(\alpha + \beta + 1) \left([2]_q - |\gamma| - 1 \right)}. \tag{45}$$

Again, using the fact that the sequence $\left\{ \frac{\Gamma_q(\beta + n)}{\Gamma_q(\alpha + \beta + n)} \right\}_{n=2}^{\infty}$ is non-decreasing for $\alpha > 0$ and $-1 < \beta \leq 0$, Inequality (32), gives

$$\frac{\Gamma_q(\beta + 2)\Gamma_q(\alpha + \beta + 1)}{\Gamma_q(\alpha + \beta + 2)\Gamma_q(\beta + 1)} \sum_{n=2}^{\infty} ([n]_q - |\gamma| - 1) a_n < |\gamma|,$$

or, equivalently,

$$\frac{\Gamma_q(\beta + 2)\Gamma_q(\alpha + \beta + 1)}{\Gamma_q(\alpha + \beta + 2)\Gamma_q(\beta + 1)} \sum_{n=2}^{\infty} [n]_q a_n < |\gamma| + \frac{(1 + |\gamma|)\Gamma_q(\beta + 2)\Gamma_q(\alpha + \beta + 1)}{\Gamma_q(\alpha + \beta + 2)\Gamma_q(\beta + 1)} \sum_{n=2}^{\infty} a_n, \tag{46}$$

which on using the Inequality (45), gives

$$\frac{\Gamma_q(\beta + 2)\Gamma_q(\alpha + \beta + 1)}{\Gamma_q(\alpha + \beta + 2)\Gamma_q(\beta + 1)} \sum_{n=2}^{\infty} [n]_q a_n < |\gamma| + \frac{(1 + |\gamma|)|\gamma|}{[2]_q - |\gamma| - 1},$$

or, equivalently,

$$\sum_{n=2}^{\infty}[n]_q\, a_n < \frac{|\gamma|[2]_q\Gamma_q(\alpha+\beta+2)\Gamma_q(\beta+1)}{([2]_q-|\gamma|-1)\Gamma_q(\beta+2)\Gamma_q(\alpha+\beta+1)}. \tag{47}$$

Now, if we take $\delta \geq \frac{|\gamma|[2]_q\Gamma_q(\alpha+\beta+2)\Gamma_q(\beta+1)}{([2]_q-|\gamma|-1)\Gamma_q(\beta+2)\Gamma_q(\alpha+\beta+1)}$, then in view of Equation (41) and Inequality (47), we obtain that $f(z) \in \mathcal{N}_{\delta,q}(p)$, which proves the inclusion Relation (43). □

Next, we establish the following inclusion relation between the class $\mathcal{C}_q(\alpha,\beta,\gamma)$ and (δ,q)-neighborhood $\mathcal{N}_{\delta,q}(p)$ of identity function p for the specified range of values of δ:

Theorem 2. *If* $-1 < \beta \leq 0$, $|\gamma| \geq 1$ *and*

$$\delta \geq \frac{(|\gamma|-1)\Gamma_q(\alpha+\beta+2)\Gamma_q(\beta+1)}{([2]_q-|\gamma|)\,\Gamma_q(\beta+2)\Gamma_q(\alpha+\beta+1)}, \tag{48}$$

then

$$\mathcal{C}_q(\alpha,\beta,\gamma) \subset \mathcal{N}_{\delta,q}(p) \qquad (\alpha>0;\ \gamma\in\mathbb{C}\setminus\{0\};\ 0<q<1). \tag{49}$$

Proof. Let $f \in \mathcal{C}_q(\alpha,\beta,\gamma)$, then, in view of Lemma 2, Inequality (36) holds. Since for $\alpha > 0$, $-1 < \beta \leq 0$, the sequence $\left\{\frac{\Gamma_q(\beta+n)}{\Gamma_q(\alpha+\beta+n)}\right\}_{n=2}^{\infty}$ is non-decreasing, therefore we have

$$\frac{\Gamma_q(\beta+2)\Gamma_q(\alpha+\beta+1)}{\Gamma_q(\alpha+\beta+2)\Gamma_q(\beta+1)}\left([2]_q-|\gamma|\right)\sum_{n=2}^{\infty}[n]_q a_n \leq \sum_{n=2}^{\infty}\frac{\Gamma_q(\beta+n)\Gamma_q(\alpha+\beta+1)}{\Gamma_q(\alpha+\beta+n)\Gamma_q(\beta+1)}\left([n]_q\left([n]_q-|\gamma|\right)\right)a_n,$$

which, in view of Inequality (36), gives

$$\frac{\Gamma_q(\beta+2)\Gamma_q(\alpha+\beta+1)}{\Gamma_q(\alpha+\beta+2)\Gamma_q(\beta+1)}\left([2]_q-|\gamma|\right)\sum_{n=2}^{\infty}[n]_q a_n < |\gamma|-1, \tag{50}$$

or, equivalently

$$\sum_{n=2}^{\infty}[n]_q a_n < \frac{(|\gamma|-1)\Gamma_q(\alpha+\beta+2)\Gamma_q(\beta+1)}{([2]_q-|\gamma|)\,\Gamma_q(\beta+2)\Gamma_q(\alpha+\beta+1)}. \tag{51}$$

Now, if we take $\delta \geq \frac{(|\gamma|-1)\Gamma_q(\alpha+\beta+2)\Gamma_q(\beta+1)}{([2]_q-|\gamma|)\,\Gamma_q(\beta+2)\Gamma_q(\alpha+\beta+1)}$, then in view of Equation (41) and Inequality (51), we obtain that $f(z) \in \mathcal{N}_{\delta,q}(p)$, which proves the inclusion Relation (49). □

3. The Classes $\mathcal{S}_q^{(\eta)}(\alpha,\beta,\gamma)$ and $\mathcal{C}_q^{(\eta)}(\alpha,\beta,\gamma)$

In this section, the classes $\mathcal{S}_q^{(\eta)}(\alpha,\beta,\gamma)$ and $\mathcal{C}_q^{(\eta)}(\alpha,\beta,\gamma)$ are defined. Then, we establish the inclusion relations between the neighborhood of a function belonging to $\mathcal{S}_q(\alpha,\beta,\gamma)$ and $\mathcal{C}_q(\alpha,\beta,\gamma)$ with $\mathcal{S}_q^{(\eta)}(\alpha,\beta,\gamma)$ and $\mathcal{C}_q^{(\eta)}(\alpha,\beta,\gamma)$, respectively. First, we define the class $\mathcal{S}_q^{(\eta)}(\alpha,\beta,\gamma)$ as follows.

Definition 5. *The function* $f \in \mathcal{A}$, *belongs to* $\mathcal{S}_q^{(\eta)}(\alpha,\beta,\gamma)$ $(\alpha>0;\ -1<\beta;\ \gamma\in\mathbb{C}\setminus\{0\};\ 0<q<1;\ 0\leq\eta<1)$ *if there exists a function* $g(z) \in \mathcal{S}_q(\alpha,\beta,\gamma)$ *that satisfies*

$$\left|\frac{f(z)}{g(z)}-1\right| < 1-\eta, \tag{52}$$

where

$$g(z) = z + \sum_{n=2}^{\infty} b_n z^n. \tag{53}$$

Similarly, we define the class $\mathcal{S}_q^{(\eta)}(\alpha,\beta,\gamma)$ as:

Definition 6. *The function $f \in \mathcal{A}$, belongs to $\mathcal{C}_q^{(\eta)}(\alpha,\beta,\gamma)$ $(\alpha > 0;\ -1 < \beta;\ \gamma \in \mathbb{C}\setminus\{0\};\ 0 < q < 1;\ 0 \leq \eta < 1)$ if there exists a function g, given by Equation (53), in the class $\mathcal{C}_q(\alpha,\beta,\gamma)$, satisfying the Inequality (52).*

Now, we establish the following inclusion relation between a neighborhood $\mathcal{N}_{\delta,q}(g)$ of any function $g \in \mathcal{S}_q(\alpha,\beta,\gamma)$ and the class $\mathcal{S}_q^{(\eta)}(\alpha,\beta,\gamma)$ for the specified range of values of η:

Theorem 3. *Let the function g, given by Equation (53), belongs to the class $\mathcal{S}_q(\alpha,\beta,\gamma)$ and*

$$\eta < 1 - \frac{\delta\Gamma_q(\beta+2)\Gamma_q(\alpha+\beta+1)\left([2]_q - |\gamma| - 1\right)}{[2]_q\left(\left([2]_q - |\gamma| - 1\right)\Gamma_q(\beta+2)\Gamma_q(\alpha+\beta+1) - |\gamma|\Gamma_q(\alpha+\beta+2)\Gamma_q(\beta+1)\right)}, \tag{54}$$

then

$$\mathcal{N}_{\delta,q}(g) \subset \mathcal{S}_q^{(\eta)}(\alpha,\beta,\gamma), \tag{55}$$

where $\alpha > 0;\ -1 < \beta \leq 0;\ \gamma \in \mathbb{C}\setminus\{0\};\ \delta \geq 0;\ 0 < q < 1;\ 0 \leq \eta < 1.$

Proof. We assume that $f \in \mathcal{N}_{\delta,q}(g)$, then in view of Relation (40), we have

$$\sum_{n=2}^{\infty} [n]_q\,|a_n - b_n| \leq \delta. \tag{56}$$

Since $\{[n]_q\}_{n=2}^{\infty}$ is non-decreasing sequence, therefore

$$\sum_{n=2}^{\infty} [2]_q\,|a_n - b_n| \leq \sum_{n=2}^{\infty} [n]_q\,|a_n - b_n|,$$

This implies that

$$[2]_q \sum_{n=2}^{\infty} |a_n - b_n| \leq \sum_{n=2}^{\infty} [n]_q\,|a_n - b_n|,$$

which in view of Inequality (56) gives

$$[2]_q \sum_{n=2}^{\infty} |a_n - b_n| \leq \delta,$$

or, equivalently

$$\sum_{n=2}^{\infty} |a_n - b_n| \leq \frac{\delta}{[2]_q} \qquad (0 < q < 1;\ \delta \geq 0). \tag{57}$$

Since $-1 < \beta \leq 0$, therefore, for the function g, given by Equation (53), in the class $\mathcal{S}_q(\alpha,\beta,\gamma)$, using Inequality (45), we get

$$\sum_{n=2}^{\infty} b_n \leq \frac{|\gamma|\Gamma_q(\alpha+\beta+2)\Gamma_q(\beta+1)}{\Gamma_q(\beta+2)\Gamma_q(\alpha+\beta+1)\left([2]_q + |\gamma| - 1\right)}. \tag{58}$$

Using Equations (10), (53) and the fact that $|z| < 1$, we get

$$\left|\frac{f(z)}{g(z)} - 1\right| = \left|\frac{\sum_{n=2}^{\infty}(a_n - b_n)z^{n-1}}{1 + \sum_{n=2}^{\infty} b_n z^{n-1}}\right| \leq \frac{\sum_{n=2}^{\infty}|a_n - b_n|}{1 - \sum_{n=2}^{\infty}|b_n|} \leq \frac{\sum_{n=2}^{\infty}|a_n - b_n|}{1 - \sum_{n=2}^{\infty} b_n}, \tag{59}$$

Now, using Inequalities (57) and (58) in Inequality (59), we get

$$\left|\frac{f(z)}{g(z)}-1\right| \le \frac{\delta\Gamma_q(\beta+2)\Gamma_q(\alpha+\beta+1)\left([2]_q-|\gamma|-1\right)}{[2]_q\left(\left([2]_q-|\gamma|-1\right)\Gamma_q(\beta+2)\Gamma_q(\alpha+\beta+1)-|\gamma|\Gamma_q(\alpha+\beta+2)\Gamma_q(\beta+1)\right)}. \tag{60}$$

If we take $\eta < 1 - \dfrac{\delta\Gamma_q(\beta+2)\Gamma_q(\alpha+\beta+1)\left([2]_q-|\gamma|-1\right)}{[2]_q\left(\left([2]_q-|\gamma|-1\right)\Gamma_q(\beta+2)\Gamma_q(\alpha+\beta+1)-|\gamma|\Gamma_q(\alpha+\beta+2)\Gamma_q(\beta+1)\right)}$, then in view of Definition 5 and Inequality (60), we obtain that $f \in \mathcal{S}_q^{(\eta)}(\alpha,\beta,\gamma)$, which proves the inclusion Relation (55). □

Next, we establish the following inclusion relation between a neighborhood $\mathcal{N}_{\delta,q}(g)$ of any function $g \in \mathcal{C}_q(\alpha,\beta,\gamma)$ and the class $\mathcal{C}_q^{(\eta)}(\alpha,\beta,\gamma)$ for the specified range of values of η:

Theorem 4. *Let the function g, given by Equation (53), belongs to the class $\mathcal{C}_q(\alpha,\beta,\gamma)$ and*

$$\eta < 1 - \frac{\delta[2]_q\left([2]_q-|\gamma|\right)\Gamma_q(\beta+2)\Gamma_q(\alpha+\beta+1)}{[2]_q\left([2]_q\left([2]_q-|\gamma|\right)\Gamma_q(\beta+2)\Gamma_q(\alpha+\beta+1)-(|\gamma|-1)\Gamma_q(\alpha+\beta+2)\Gamma_q(\beta+1)\right)}, \tag{61}$$

then

$$\mathcal{N}_{\delta,q}(g) \subset \mathcal{C}_q^{(\eta)}(\alpha,\beta,\gamma), \tag{62}$$

where $|\gamma|>1$, $\alpha>0$; $-1<\beta\le 0$; $\gamma \in \mathbb{C}\setminus\{0\}$; $0<q<1$; $\delta \ge 0$; $0 \le \eta < 1$.

Proof. If we take any $f \in \mathcal{N}_{\delta,q}(g)$, then Inequality (57) holds.

Now, since $-1 < \beta \le 0$, therefore, for any function g, given by Equation (53), in the class $\mathcal{C}_q(\alpha,\beta,\gamma)$, using Inequality (51) and the fact that the sequence $\{[n]_q\}_{n=2}^{\infty}$ is non-decreasing, we get

$$\sum_{n=2}^{\infty} b_n < \frac{(|\gamma|-1)\Gamma_q(\alpha+\beta+2)\Gamma_q(\beta+1)}{[2]_q\left([2]_q-|\gamma|\right)\Gamma_q(\beta+2)\Gamma_q(\alpha+\beta+1)}. \tag{63}$$

Using Inequalities (57) and (63) in Inequality (59), we get

$$\left|\frac{f(z)}{g(z)}-1\right| \le \frac{\delta[2]_q\left([2]_q-|\gamma|\right)\Gamma_q(\beta+2)\Gamma_q(\alpha+\beta+1)}{[2]_q\left([2]_q\left([2]_q-|\gamma|\right)\Gamma_q(\beta+2)\Gamma_q(\alpha+\beta+1)-(|\gamma|-1)\Gamma_q(\alpha+\beta+2)\Gamma_q(\beta+1)\right)}. \tag{64}$$

If we take $\eta < 1 - \dfrac{\delta[2]_q\left([2]_q-|\gamma|\right)\Gamma_q(\beta+2)\Gamma_q(\alpha+\beta+1)}{[2]_q\left([2]_q\left([2]_q-|\gamma|\right)\Gamma_q(\beta+2)\Gamma_q(\alpha+\beta+1)-(|\gamma|-1)\Gamma_q(\alpha+\beta+2)\Gamma_q(\beta+1)\right)}$, then in view of Definition 6 and Inequality (64), we obtain that $f \in \mathcal{C}_q^{(\eta)}(\alpha,\beta,\gamma)$, which proves the Assertion (61). □

4. Application

First, we define the generalized hyper-Bessel function $w_{c,b,\alpha_d}(z)$ as :

$$w_{c,b,\alpha_d}(z) = \sum_{n=0}^{\infty} \frac{(-c)^n}{n!\prod_{i=1}^{d}\Gamma\left(\alpha_i+n+\dfrac{b+1}{2}\right)}\left(\frac{z}{d+1}\right)^{n(d+1)+\sum_{i=1}^{d}\alpha_i} \tag{65}$$

where $\nu, b, d, z \in \mathbb{C}$.

Second, we define the function $\varphi_{\alpha_d,b,c}(z):\mathbb{U}\to\mathbb{C}$ as:

$$\varphi_{\alpha_d,b,c}(z)=(d+1)^{\sum_{i=1}^{d}\alpha_i}\prod_{i=1}^{d}\Gamma\left(\alpha_i+\frac{b+1}{2}\right)z^{1-\frac{\sum_{i=1}^{d}\alpha_i}{d+1}}w_{\alpha_d,b,c}(z^{1/d+1}), \tag{66}$$

by using Equation (65) in Equation (66), we get

$$\begin{aligned}\varphi_{c,b,\alpha_d}(z)&=\sum_{n=0}^{\infty}\frac{(-c)^n}{n!\prod_{i=1}^{d}\left(\alpha_i+\frac{b+1}{2}\right)_n(d+1)^{n(d+1)}}z^{n+1}\\&=z+\sum_{n=2}^{\infty}\frac{(-c)^{n-1}}{(n-1)!\prod_{i=1}^{d}\left(\alpha_i+\frac{b+1}{2}\right)_{n-1}(d+1)^{(n-1)(d+1)}}z^n\end{aligned} \tag{67}$$

by choosing $d=1$ and $\alpha_1=\nu$, then the functions $w_{c,b,\alpha_d}(z)$ and $\varphi_{\alpha_d,b,c}(z)$ are reduce to $w_{\nu,b,d}(z)$ and $\phi_{\nu,b,d}(z)$, respectively.

Third, we applying the introduced function $\varphi_{c,b,\alpha_d}(z)$, given by Equation (67) in the results of Lemma 1 and Lemma 2, we get the conditions for that function $\varphi_{c,b,\alpha_d}(z)$ to be in the classes $\mathcal{S}_q(\alpha,\beta,\gamma)$ and $\mathcal{C}_q(\alpha,\beta,\gamma)$ in the following corollaries, respectively:

Corollary 1. *If $\varphi_{c,b,\alpha_d}(z)$ is an analytic function such that it belongs to the class $\mathcal{S}_q(\alpha,\beta,\gamma)$, then*

$$\sum_{n=2}^{\infty}\frac{(-c)^{n-1}\Gamma_q(\beta+n)\Gamma_q(\alpha+\beta+1)}{(n-1)!\prod_{i=1}^{d}\left(\alpha_i+\frac{b+1}{2}\right)_{n-1}(d+1)^{(n-1)(d+1)}\Gamma_q(\alpha+\beta+n)\Gamma_q(\beta+1)}\times([n]_q-|\gamma|-1)<|\gamma|\quad(\alpha>0;\ \beta>-1;\ 0<q<1;\gamma\in\mathbb{C}\backslash\{0\}),$$

where Γ_q and $[n]_q$ are given by Equations (2) and (3), respectively.

Corollary 2. *If $\varphi_{c,b,\alpha_d}(z)$ is an analytic function such that it belongs to the class $\mathcal{C}_q(\alpha,\beta,\gamma)$ and $|\gamma|\geq 1$ then*

$$\sum_{n=2}^{\infty}\frac{(-c)^{n-1}\Gamma_q(\beta+n)\Gamma_q(\alpha+\beta+1)}{(n-1)!\prod_{i=1}^{d}\left(\alpha_i+\frac{b+1}{2}\right)_{n-1}(d+1)^{(n-1)(d+1)}\Gamma_q(\alpha+\beta+n)\Gamma_q(\beta+1)}\times([n]_q([n]_q-|\gamma|))a_n<|\gamma|-1\quad(\alpha>0;\ \beta>-1;\ 0<q<1;\gamma\in\mathbb{C}\backslash\{0\}),$$

where Γ_q and $[n]_q$ are given by Equations (2) and (3), respectively.

5. Discussion of Results and Future Work

The concept of q-derivatives has so far been applied in many areas of not only mathematics but also physics, including fractional calculus and quantum physics. However, research on q-calculus is in connection with function theory and especially geometric properties of analytic functions such as starlikeness and convexity, which is fairly familiar on this topic. Finding sharp coefficient bounds for analytic functions belonging to Classes of starlikeness and convexity defined by q-calculus operators is of particular importance since any information can shed light on the study of the geometric properties of such functions. Our results are applicable by using any analytic functions.

6. Conclusions

In this paper, we have used q-calculus to introduce a new q-integral operator which is a generalization of the known Jung-Kim-Srivastava integral operator. Also, a new subclass involving the q-integral operator introduced has been defined. Some interesting coefficient bounds for these subclasses of analytic functions have been studied. Furthermore, the (δ, q)-neighborhood of analytic functions and the inclusion relation between the (δ, q)-neighborhood and the subclasses involving the q-integral operator have been derived. The ideas of this paper may stimulate further research in this field.

Author Contributions: Conceptualization, H.M.S. and E.S.A.A.; Formal analysis, H.M.S. and S.M.; Funding acquisition, S.M. and G.S.; Investigation, E.S.A.A. and S.M..; Methodology, E.S.A.A. and S.M.; Supervision, H.M.S. and N.R.; Validation, S.N.M. and S.M.; Visualization, G.S.;Writing—original draft, E.S.A.A. and G.S.; Writing—review & editing, E.S.A.A., G.S. and S.N.M.

References

1. Gauchman, H. Integral inequalities in q-calculus. *Comput. Math. Appl.* **2004**, *47*, 281–300. [CrossRef]
2. Tang, Y.; Tie, Z. A remark on the q-fractional order differential equations. *Appl. Math. Comput.* **2019**, *350*, 198–208. [CrossRef]
3. Jackson, F.H. On q-definite integrals. *Quart. J. Pure Appl. Math.* **1910**, *41*, 193–203.
4. Purohit, S.D. A new class of multivalently analytic functions associated with fractional q-calculus operators. *Fract. Differ. Calc.* **2012**, *2*, 129–138. [CrossRef]
5. Mohammed, A.; Darus, M. A generalized operator involving the q-hypergeometric function. *Mat. Vesnik* **2013**, *65*, 454–465.
6. Srivastava, H.M. Univalent functions, fractional calculus and associated generalizes hypergeometric functions. In *Univalent Functions, Fractional Calculus, and Their Applications*; Srivastava, H.M., Owa, S., Eds.; John Wiley and Sons: New York, NY, USA; Chichester, UK; Brisbane, Australia; Toronto, ON, Canada, 1989; pp. 329–354.
7. Srivastava, H.M.; Bansal, D. Close-to-convexity of a certain family of q-Mittag-Leffler functions. *J. Nonlinear Var. Anal.* **2017**, *1*, 61–69.
8. Ismail, M.E.H. A generalization of starlike functions. *Complex Var. Theory Appl. Int. J.* **1990**, *14*, 77–84. [CrossRef]
9. Sahoo, S.K.; Sharma, N.L. On a generalization of close-to-convex functions. *arXiv* **2014**, arXiv:1404.3268.
10. Mahmood, S.; Jabeen, M.; Malik, S.N.; Srivastava, H.M.; Manzoor, R.; Riaz, S.M.J. Some coefficient inequalities of q-starlike functions associated with conic domain defined by q-derivative. *J. Funct. Spaces* **2018**, *2018*, 8492072. [CrossRef]
11. Mahmood, S.; Srivastava, H.M.; Khan, N.; Ahmed, Q.Z.; Khan, B.; Ali, I. Upper Bound of the Third Hankel Determinant for a Subclass of q-starlike Functions. *Symmetry* **2019**, *11*, 347. [CrossRef]
12. Srivastava, H.M.; Tahir, M; Khan, B.; Ahmed, Q.Z.; Khan, N. Some general classes of q-starlike functions associated with the Janowski functions. *Symmetry* **2019**, *11*, 292. [CrossRef]
13. Uçar, H.E.Ö. Coefficient inequality for q-starlike functions. *Appl. Math. Comput.* **2016**, *276*, 122–126.
14. Ezeafulukwe, U.A.; Darus, M. A Note on q-Calculus. *Fasc. Math.* **2015**, *55*, 53–63. [CrossRef]
15. Ezeafulukwe, U.A.; Darus, M. Certain properties of q-hypergeometric functions. *Int. J. Math. Math. Sci.* **2015**, *2015*, 489218. [CrossRef]
16. Jackson, F.H. q-difference equations. *Am. J. Math.* **1910**, *32*, 305–314. [CrossRef]
17. Corcino, R.B. On P, Q-Binomial Coefficients. *Integers* **2008**, *8*, A29.
18. Nasr, M.A.; Aouf, M.K. Starlike function of complex order. *J. Natur. Sci. Math.* **1985**, *25*, 1–12.
19. Wiatrowski, P. The coefficients of a certain family of hololorphic functions. *Zeszyt Nauk. Univ. Lodz. Nauki Mat. Przyrod. Ser. II Zeszyt* **1971**, *39*, 75–85.
20. Seoudy, T.M.; Aouf, M.K. Coefficient estimates of new classes of q-starlike and q-convex functions of complex order. *J. Math. Inequal* **2016**, *10*, 135–145. [CrossRef]
21. Ruscheweyh, S. Neighborhoods of univalent functions. *Proc. Am. Math. Soc.* **1981**, *81*, 521–527. [CrossRef]
22. Jung, I.B.; Kim, Y.C.; Srivastava, H.M. The Hardy space of analytic functions associated with certain one-parameter families of integral operators. *J. Math. Anal. Appl.* **1993**, *176*, 138–147. [CrossRef]
23. Rainville, E.D. *Special Functions*; Chelsea: New York, NY, USA, 1971.
24. Baricz, Á. *Generalized Bessel Functions of the Frst Kind*; Springer: Berlin, Germany, 2010.

25. Deniz, E.; Orhan, H.; Srivastava, H. Some sufficient conditions for univalence of certain families of integral operators involving generalized Bessel functions. *Taiwan. J. Math.* **2011**, *15*, 883–917.
26. Chaggara, H.; Romdhane, N.B. On the zeros of the hyper-Bessel function. *Integral Transforms Spec. Funct.* **2015**, *2*, 26. [CrossRef]
27. Noor, K.I.; Riaz, S.; Noor, M.A. On q-Bernardi integral Operator. *TWMS J. Pure Appl. Math.* **2017**, *8*, 3–11.

Construction of Stancu-Type Bernstein Operators based on Bézier Bases with Shape Parameter λ

Hari M. Srivastava [1,2,*], Faruk Özger [3] and S. A. Mohiuddine [4]

1 Department of Mathematics and Statistics, University of Victoria, Victoria, BC V8W 3R4, Canada
2 Department of Medical Research, China Medical University Hospital, China Medical University, Taichung 40402, Taiwan
3 Department of Engineering Sciences, İzmir Katip Çelebi University, İzmir 35620, Turkey; farukozger@gmail.com
4 Operator Theory and Applications Research Group, Department of Mathematics, Faculty of Science, King Abdulaziz University, P.O. Box 80203, Jeddah 21589, Saudi Arabia; mohiuddine@gmail.com
* Correspondence: harimsri@math.uvic.ca

Abstract: We construct Stancu-type Bernstein operators based on Bézier bases with shape parameter $\lambda \in [-1,1]$ and calculate their moments. The uniform convergence of the operator and global approximation result by means of Ditzian-Totik modulus of smoothness are established. Also, we establish the direct approximation theorem with the help of second order modulus of smoothness, calculate the rate of convergence via Lipschitz-type function, and discuss the Voronovskaja-type approximation theorems. Finally, in the last section, we construct the bivariate case of Stancu-type λ-Bernstein operators and study their approximation behaviors.

Keywords: Stancu-type Bernstein operators; Bézier bases; Voronovskaja-type theorems; modulus of continuity; rate of convergence; bivariate operators; approximation properties

MSC: 41A25; 41A35

---

1. Introduction

A famous mathematician Bernstein [1] constructed polynomials nowadays called Bernstein polynomials, which are familiar and widely investigated polynomials in theory of approximation. Bernstein gave a simple and very elegant way to obtain Weierstrass approximation theorem with the help of his newly constructed polynomials. For any continuous function $f(x)$ defined on $C[0,1]$, Bernstein polynomials of order n are given by

$$B_n(f;x) = \sum_{i=0}^{n} f\left(\frac{i}{n}\right) b_{n,i}(x) \qquad (x \in [0,1]), \tag{1}$$

where the Bernstein basis functions $b_{n,i}(x)$ are defined by

$$b_{n,i}(x) = \binom{n}{i} x^i (1-x)^{n-i} \qquad (i = 0, \ldots, n).$$

Stancu [2] presented a generalization of Bernstein polynomials with the help of two parameters α and β such that $0 \leq \alpha \leq \beta$, as follows:

$$S_{n,\alpha,\beta}(f;x) = \sum_{i=0}^{n} f\left(\frac{i+\alpha}{n+\beta}\right) \binom{n}{i} x^i (1-x)^{n-i} \qquad (x \in [0,1]). \tag{2}$$

If we take both the parameters $\alpha = \beta = 0$, then we get the classical Bernstein polynomials. The operators defined by (2) are called Bernstein–Stancu operators. For some recent work, we refer to [3–6].

In the recent past, Cai et al. [7] presented a new construction of Bernstein operators with the help of Bézier bases with shape parameter λ and called it λ-Bernstein operators, which are defined by

$$B_n^{\lambda}(f;x) = \sum_{i=0}^{n} f\left(\frac{i}{n}\right) \tilde{b}_{n,i}(\lambda;x) \qquad (n \in \mathbb{N}) \tag{3}$$

where $\tilde{b}_{n,i}(\lambda;x)$ are Bézier bases with shape parameter λ (see [8]), defined by

$$\begin{aligned} \tilde{b}_{n,0}(\lambda;x) &= b_{n,0}(x) - \frac{\lambda}{n+1} b_{n+1,1}(x), \\ \tilde{b}_{n,i}(\lambda;x) &= b_{n,i}(x) + \frac{n-2i+1}{n^2-1}\lambda b_{n+1,i}(x) - \frac{n-2i-1}{n^2-1}\lambda b_{n+1,i+1}(x), \quad i = 1,2\dots,n-1, \\ \tilde{b}_{n,n}(\lambda;x) &= b_{n,n}(x) - \frac{\lambda}{n+1} b_{n+1,n}(x), \end{aligned} \tag{4}$$

in this case $\lambda \in [-1,1]$ and $b_{n,i}(x)$ are the Bernstein basis functions. By taking the above operators into account, they established various approximation results, namely, Korovkin- and Voronovskaja-type theorems, rate of convergence via Lipschitz continuous functions, local approximation and other related results. In the same year, Cai [9] generalized λ-Bernstein operators by constructing the Kantorovich-type λ-Bernstein operators, as well as its Bézier variant, and studied several approximation results. Later, various approximation properties and asymptotic type results of the Kantorovich-type λ-Bernstein operators have been studied by Acu et al. [10]. Very recently, Özger [11] obtained statistical approximation for λ-Bernstein operators including a Voronovskaja-type theorem in statistical sense. In the same article, he also constructed bivariate λ-Bernstein operators and studied their approximation properties.

The Bernstein operators are some of the most studied positive linear operators which were modified by many authors, and we are mentioning some of them and other related work [12–23].

We are now ready to construct our new operators as follows: Suppose that α and β are two non-negative parameters such that $0 \leq \alpha \leq \beta$. Then, the Stancu-type modification of λ-Bernstein operators $B_{n,\alpha,\beta}^{\lambda}(f;x) : C[0,1] \longrightarrow C[0,1]$ is defined by

$$B_{n,\alpha,\beta}^{\lambda}(f;x) = \sum_{i=0}^{n} f\left(\frac{i+\alpha}{n+\beta}\right) \tilde{b}_{n,i}(\lambda;x) \tag{5}$$

for any $n \in \mathbb{N}$ and we call it Stancu-type λ-Bernstein operators or λ-Bernstein–Stancu operators, where Bézier bases $\tilde{b}_{n,i}(\lambda;x)$ are defined in (4).

Remark 1. *We have the following results for Stancu-type λ-Bernstein operators:*

(i) If we take $\lambda = 0$ in (5), then Stancu-type λ-Bernstein Stancu operators reduce to the classical Bernstein–Stancu operators defined in [2].
(ii) The choice of $\alpha = \beta = 0$ in (5) gives λ-Bernstein operators defined by Cai et al. [7].
(iii) If we choose $\alpha = \beta = \lambda = 0$, then (5) reduces to the classical Bernstein operators defined in [1].

The rest of the paper is organized as follows: In Section 2, we calculate the moments of (5) and prove global approximation formula in terms of Ditzian–Totik uniform modulus of smoothness of first and second order. The local direct estimate of the rate of convergence by Lipschitz-type function involving two parameters for λ-Bernstein–Stancu operators is investigated. In Section 3, we establish quantitative Voronovskaja-type theorem for our operators. The final section of the paper is devoted to study the bivariate case of λ-Bernstein–Stancu operators .

2. Some Auxiliary Lemmas and Approximation by Stancu-Type λ-Bernstein Operators

In this section, we first prove some lemma which will be used to study the approximation results of (5).

Lemma 1. *For $x \in [0,1]$, the moments of Stancu-type λ-Bernstein operators are given as:*

$$B^{\lambda}_{n,\alpha,\beta}(1;x) = 1;$$

$$B^{\lambda}_{n,\alpha,\beta}(t;x) = \frac{\alpha+nx}{n+\beta} + \lambda\left[\frac{1-2x+x^{n+1}+(\alpha-1)(1-x)^{n+1}}{(n+\beta)(n-1)} + \frac{\alpha x(1-x)^n}{n+\beta}\right];$$

$$\begin{aligned} B^{\lambda}_{n,\alpha,\beta}(t^2;x) &= \frac{1}{(n+\beta)^2}\left\{n(n-1)x^2+(1+2\alpha)nx+\alpha^2\right\} \\ &+\lambda\left[\frac{2nx-1-4nx^2+(2n+1)x^{n+1}+(1-x)^{n+1}}{(n+\beta)^2(n-1)} + \frac{\alpha^2-4\alpha x}{(n+\beta)^2(n-1)}\right. \\ &+\left.\frac{2\alpha n-2\alpha(\alpha+n)(x^{n+1}+(1-x)^n)+\alpha^2 x(n^2+1)(1-x)^n}{(n+\beta)^2(n^2-1)}\right]. \end{aligned}$$

Proof. Using the definition of operators (5) and Bézier–Bernstein bases $\tilde{b}_{n,i}(\lambda;x)$ (4), we write

$$\begin{aligned} B^{\lambda}_{n,\alpha,\beta}(t;x) &= \sum_{i=0}^{n}\frac{i+\alpha}{n+\beta}\tilde{b}_{n,i}(\lambda;x) = \frac{\alpha}{n+\beta}b_{n,0}(x) - \frac{\alpha}{n+\beta}\frac{\lambda}{n+1}b_{n+1,1}(x) \\ &+\sum_{i=1}^{n-1}\frac{i+\alpha}{n+\beta}\left[b_{n,i}(x)+\lambda\left(\frac{n-2i+n+1}{n^2-1}b_{n+1,i}(x) - \frac{n-2i-1}{n^2-1}b_{n+1,i+1}(x)\right)\right] \\ &+\frac{n+\alpha}{n+\beta}b_{n,n}(x) - \frac{n+\alpha}{n+\beta}\frac{\lambda}{n+1}b_{n+1,n}(x) \\ &= \sum_{i=0}^{n}\frac{i+\alpha}{n+\beta}b_{n,i}(x) + \lambda\left(\theta_1(n,\alpha,\beta,x) - \theta_2(n,\alpha,\beta,x)\right), \end{aligned}$$

where

$$\begin{aligned} \theta_1(n,\alpha,\beta,x) &= \sum_{i=0}^{n}\frac{i+\alpha}{n+\beta}\frac{n-2i+1}{n^2-1}b_{n+1,i}(x); \\ \theta_2(n,\alpha,\beta,x) &= \sum_{i=1}^{n-1}\frac{i+\alpha}{n+\beta}\frac{n-2i-1}{n^2-1}b_{n+1,i+1}(x). \end{aligned}$$

Now, we compute the expressions $\theta_1(n,\alpha,\beta,x)$ and $\theta_2(n,\alpha,\beta,x)$. Since the Bernstein–Stancu operators are linear, and Bernstein–Stancu operators and fundamental Bernstein bases satisfy the following equality:

$$\sum_{i=1}^{n}\frac{i+\alpha}{n+\beta}b_{n,i}(x) = \frac{nx}{n+\beta} + \frac{\alpha}{n+\beta},$$

one writes

$$\begin{aligned} \theta_1(n,\alpha,\beta,x) &= \frac{1}{n-1}\sum_{i=0}^{n}\frac{i+\alpha}{n+\beta}b_{n+1,i}(x) - \frac{2}{n^2-1}\sum_{i=0}^{n}\frac{i^2+\alpha i}{n+\beta}b_{n+1,i}(x) \\ &= \frac{1}{n-1}\sum_{i=0}^{n}\frac{i}{n+\beta}b_{n+1,i}(x) + \frac{1}{n-1}\sum_{i=0}^{n}\frac{\alpha}{n+\beta}b_{n+1,i}(x) \\ &- \frac{2}{n^2-1}\sum_{i=1}^{n}\frac{i^2}{n+\beta}b_{n+1,i}(x) - \frac{2}{n^2-1}\sum_{i=1}^{n}\frac{\alpha i}{n+\beta}b_{n+1,i}(x) \end{aligned}$$

$$
\begin{aligned}
&= \frac{(n+1)x-2x}{(n+\beta)(n-1)}\sum_{i=0}^{n-1} b_{n,i}(x) + \frac{\alpha}{(n+\beta)(n-1)}\sum_{i=0}^{n} b_{n+1,i}(x) \\
&\quad - \frac{2\alpha x}{(n+\beta)(n-1)}\sum_{i=0}^{n-1} b_{n,i}(x) - \frac{2nx^2}{(n+\beta)(n-1)}\sum_{i=0}^{n-2} b_{n-1,i}(x) \\
&= \frac{x-x^{n+d+1}}{n+\beta} - \frac{2nx^2-2nx^{n+1}}{(n+\beta)(n-1)} + \frac{x-x^{n+d+1}}{n+\beta} - \frac{\alpha-2\alpha x+\alpha x^{n+1}}{(n+\beta)(n-1)}
\end{aligned}
$$

$$
\begin{aligned}
\theta_2(n,\alpha,\beta,x) &= \frac{1}{n+1}\sum_{i=1}^{n-1}\frac{i+\alpha}{n+\beta}b_{n+1,i+1}(x) - \frac{2}{n^2-1}\sum_{i=1}^{n-1}\frac{i^2+\alpha i}{n+\beta}b_{n+1,i+1}(x) \\
&= \frac{1}{n+1}\sum_{i=1}^{n-1}\frac{i}{n+\beta}b_{n+1,i+1}(x) + \frac{1}{n+1}\sum_{i=1}^{n-1}\frac{\alpha}{n+\beta}b_{n+1,i+1}(x) \\
&\quad - \frac{2}{n^2-1}\sum_{i=1}^{n-1}\frac{i^2}{n+\beta}b_{n+1,i+1}(x) - \frac{2}{n^2-1}\sum_{i=1}^{n-1}\frac{\alpha i}{n+\beta}b_{n+1,i+1}(x) \\
&= \frac{x}{n+\beta}\sum_{i=1}^{n-1} b_{n,i}(x) - \frac{1}{(n+\beta)(n+1)}\sum_{i=1}^{n-1} b_{n+1,i+1}(x) \\
&\quad - \frac{2nx^2}{(n+\beta)(n+1)}\sum_{i=0}^{n-2} b_{n-1,i}(x) + \frac{2x}{(n+\beta)(n+1)}\sum_{i=1}^{n-1} b_{n,i}(x) \\
&\quad - \frac{2}{(n+\beta)(n^2-1)}\sum_{i=1}^{n-1} b_{n+1,i+1}(x) + \frac{\alpha}{(n+\beta)(n+1)}\sum_{i=1}^{n-1} b_{n+1,i+1}(x) \\
&\quad - \frac{2\alpha x}{(n+\beta)(n-1)}\sum_{i=1}^{n-1} b_{n,i}(x) + \frac{2\alpha}{(n+\beta)(n^2-1)}\sum_{i=1}^{n-1} b_{n+1,i+1}(x) \\
&= \frac{x-x^{n+1}}{n+\beta} - \frac{x(1-x)^n}{n+\beta} - \frac{1-(1-x)^{n+1}-x(n+1)(1-x)^n-x^{n+1}}{(n+\beta)(n+1)} \\
&\quad - \frac{2-(1-x)^{n+1}-2x(n+1)(1-x)^n-2x^{n+1}}{(n+\beta)(n^2-1)} + \frac{\alpha-\alpha(1-x)^{n+1}-\alpha x^{n+1}}{(n+\beta)(n+1)} \\
&\quad + \frac{2x-2x(1-x)^n-2x^{n+1}}{(n+\beta)(n-1)} - \frac{2nx^2-2nx^{n+1}}{(n+\beta)(n-1)} - \frac{\alpha x(1-x)^n}{n+\beta} \\
&\quad - \frac{2\alpha x-2\alpha x^{n+1}}{(n+\beta)(n+1)} + \frac{2\alpha-2\alpha(1-x)^{n+1}-2\alpha x^{n+1}}{(n+\beta)(n^2-1)}.
\end{aligned}
$$

We get the desired result for $B^{\lambda}_{n,\alpha,\beta}(t;x)$ by combining the results obtained for $\theta_1(n,\alpha,\beta,x)$ and $\theta_2(n,\alpha,\beta,x)$.

Again, by using the following identity;

$$
\sum_{i=1}^{n}\frac{(i+\alpha)^2}{(n+\beta)^2}b_{n,i}(x) = \frac{1}{(n+\beta)^2}\left\{n(n-1)x^2+(1+2\alpha)nx+\alpha^2\right\}
$$

together with (4) and (5), we can write

$$
\begin{aligned}
B^{\lambda}_{n,\alpha,\beta}(t^2;x) &= \sum_{i=0}^{n}\frac{(i+\alpha)^2}{(n+\beta)^2}\tilde{b}_{n,i}(\lambda;x) = \frac{\alpha^2}{(n+\beta)^2}b_{n,0}(x) - \frac{\alpha^2}{(n+\beta)^2}\frac{\lambda}{n+1}b_{n+1,1}(x) \\
&\quad + \sum_{i=1}^{n-1}\frac{(i+\alpha)^2}{(n+\beta)^2}\left[b_{n,i}(x)+\lambda\left(\frac{n-2i+1}{n^2-1}b_{n+1,i}(x)-\frac{n-2i-1}{n^2-1}b_{n+1,i+1}(x)\right)\right] \\
&\quad + \frac{(n+\alpha)^2}{(n+\beta)^2}b_{n,n}(x) - \frac{(n+\alpha)^2}{(n+\beta)^2}\frac{\lambda}{n+1}b_{n+1,n}(x)
\end{aligned}
$$

$$= \sum_{i=0}^{n} \frac{(i+\alpha)^2}{(n+\beta)^2} b_{n,i}(x) + \lambda \left(\theta_3(n,\alpha,\beta,x) - \theta_4(n,\alpha,\beta,x)\right),$$

where

$$\theta_3(n,\alpha,\beta,x) = \sum_{i=0}^{n} \frac{(i+\alpha)^2}{(n+\beta)^2} \frac{n-2i+1}{n^2-1} b_{n+1,i}(x);$$
$$\theta_4(n,\alpha,\beta,x) = \sum_{i=1}^{n-1} \frac{(i+\alpha)^2}{(n+\beta)^2} \frac{n-2i-1}{n^2-1} b_{n+1,i+1}(x).$$

We now compute the expressions $\theta_3(n,\alpha,\beta,x)$ and $\theta_4(n,\alpha,\beta,x)$ as follows:

$$\begin{aligned}
\theta_3(n,\alpha,\beta,x) &= \frac{1}{n-1}\sum_{i=0}^{n} \frac{(i+\alpha)^2}{(n+\beta)^2} b_{n+1,i}(x) - \frac{2}{n^2-1}\sum_{i=0}^{n} \frac{(i+\alpha)^2 i}{(n+\beta)^2} b_{n+1,i}(x) \\
&= \frac{1}{n-1}\sum_{i=0}^{n} \frac{i^2}{(n+\beta)^2} b_{n+1,i}(x) + \frac{2\alpha}{n-1}\sum_{i=0}^{n} \frac{i}{(n+\beta)^2} b_{n+1,i}(x) \\
&+ \frac{\alpha^2}{n-1}\sum_{i=0}^{n} b_{n+1,i}(x) - \frac{2}{n^2-1}\sum_{i=0}^{n} \frac{i^3}{(n+\beta)^2} b_{n+1,i}(x) \\
&- \frac{4\alpha}{n^2-1}\sum_{i=0}^{n} \frac{i^2}{(n+\beta)^2} b_{n+1,i}(x) - \frac{2\alpha^2}{n^2-1}\sum_{i=0}^{n} \frac{i}{(n+\beta)^2} b_{n+1,i}(x) \\
&= \frac{n(n+1)x^2}{(n+\beta)^2(n-1)}\sum_{i=0}^{n-2} b_{n-1,i}(x) + \frac{(n+1)x}{(n+\beta)^2(n-1)}\sum_{i=0}^{n-1} b_{n,i}(x) \\
&- \frac{2nx^3}{(n+\beta)^2}\sum_{i=0}^{n-3} b_{n-2,i}(x) - \frac{6nx^2}{(n+\beta)^2(n-1)}\sum_{i=0}^{n-2} b_{n-1,i}(x) \\
&- \frac{x}{(n+\beta)^2(n-1)}\sum_{i=0}^{n-1} b_{n,i}(x) + \frac{2\alpha x(n+1)}{(n+\beta)^2(n-1)}\sum_{i=0}^{n-1} b_{n,i}(x) \\
&+ \frac{\alpha^2}{(n+\beta)^2(n-1)}\sum_{i=0}^{n} b_{n+1,i}(x) - \frac{4\alpha n x^2}{(n+\beta)^2(n-1)}\sum_{i=0}^{n-2} b_{n-1,i}(x) \\
&- \frac{4\alpha x}{(n+\beta)^2(n-1)}\sum_{i=0}^{n-1} b_{n,i}(x) - \frac{2\alpha^2 x}{(n+\beta)^2(n-1)}\sum_{i=0}^{n-1} b_{n,i}(x) \\
&= \frac{2n(x^{n+1}-x^3)}{(n+\beta)^2} + \frac{x-x^{n+1}}{(n+\beta)^2} + \frac{(n^2-5n)(x^2-x^{n+1})}{(n+\beta)^2(n-1)} \\
&+ \frac{2\alpha(n+1)x^{n+1} + \alpha^2(1-x+x^{n+1}) - 4\alpha n x^2}{(n+\beta)^2(n-1)} + \frac{2\alpha x}{(n+\beta)^2}.
\end{aligned}$$

$$\begin{aligned}
\theta_4(n,\alpha,\beta,x) &= \frac{1}{n+1}\sum_{i=1}^{n-1} \frac{(i+\alpha)^2}{(n+\beta)^2} b_{n+1,i+1}(x) - \frac{2}{n^2-1}\sum_{i=1}^{n-1} \frac{(i+\alpha)^2 i}{(n+\beta)^2} b_{n+1,i+1}(x) \\
&= \frac{1}{n+1}\sum_{i=1}^{n-1} \frac{i^2}{(n+\beta)^2} b_{n+1,i+1}(x) + \frac{2\alpha}{n+1}\sum_{i=1}^{n-1} \frac{i}{(n+\beta)^2} b_{n+1,i+1}(x) \\
&+ \frac{\alpha^2}{n+1}\sum_{i=1}^{n-1} \frac{1}{(n+\beta)^2} b_{n+1,i+1}(x) - \frac{2}{n^2-1}\sum_{i=1}^{n-1} \frac{i^3}{(n+\beta)^2} b_{n+1,i+1}(x) \\
&- \frac{4\alpha}{n^2-1}\sum_{i=1}^{n-1} \frac{i^2}{(n+\beta)^2} b_{n+1,i+1}(x) - \frac{2\alpha^2}{n^2-1}\sum_{i=1}^{n-1} \frac{i}{(n+\beta)^2} b_{n+1,i+1}(x)
\end{aligned}$$

$$
\begin{aligned}
&= \frac{nx^2}{(n+\beta)^2}\sum_{i=0}^{n-2} b_{n-1,i}(x) + \frac{1}{(n+\beta)^2(n-1)}\sum_{i=1}^{n-1} b_{n+1,i+1}(x) \\
&\quad - \frac{2nx^3}{(n+\beta)^2}\sum_{i=0}^{n-3} b_{n-2,i}(x) - \frac{-2x}{(n+\beta)^2(n-1)}\sum_{i=1}^{n-1} b_{n,i}(x) \\
&\quad + \frac{2}{(n+\beta)^2(n-1)}\sum_{i=1}^{n-1} b_{n+1,i+1}(x) - \frac{x}{(n+\beta)^2}\sum_{i=1}^{n-1} b_{n,i}(x) \\
&\quad + \frac{2\alpha x}{(n+\beta)^2}\sum_{i=1}^{n-1} b_{n,i}(x) - \frac{2\alpha}{(n+\beta)^2(n+1)}\sum_{i=1}^{n-1} b_{n+1,i+1}(x) \\
&\quad + \frac{\alpha^2}{(n+\beta)^2(n+1)}\sum_{i=1}^{n-1} b_{n+1,i+1}(x) - \frac{4\alpha n x^2}{(n+\beta)^2(n-1)}\sum_{i=0}^{n-2} b_{n-1,i}(x) \\
&\quad + \frac{4\alpha x}{(n+\beta)^2(n-1)}\sum_{i=1}^{n-1} b_{n,i}(x) - \frac{4\alpha}{(n+\beta)^2(n^2-1)}\sum_{i=1}^{n-1} b_{n+1,i+1}(x) \\
&\quad - \frac{2\alpha^2 x}{(n+\beta)^2(n-1)}\sum_{i=1}^{n-1} b_{n,i}(x) + \frac{2\alpha}{(n+\beta)^2(n^2-1)}\sum_{i=1}^{n-1} b_{n+1,i+1}(x) \\
&= \frac{nx^2+(n+1)x^{n+1}-x-2nx^3}{(n+\beta)^2} + \frac{1-(1-x)^{n+1}-x^{n+1}}{(n+\beta)^2(n+1)} \\
&\quad + \frac{2x^{n+1}-2x}{(n+\beta)^2(n-1)} + \frac{2-2(1-x)^{n+1}-2x^{n+1}}{(n+\beta)^2(n^2-1)} \\
&\quad + \frac{2\alpha x+2\alpha x^{n+1}-\alpha^2 x(1-x)^n}{(n+\beta)^2} + \frac{\alpha(\alpha-2)(1-x^{n+1}-(1-x)^{n+1})}{(n+\beta)^2(n+1)} \\
&\quad + \frac{2\alpha(\alpha-2)x((1-x)^{n+1}-1)+2\alpha^2x^{n+1}}{(n+\beta)^2(n-1)} + \frac{2\alpha(x^{n+1}+(1-x)^{n+1}-1)}{(n+\beta)^2(n^2-1)},
\end{aligned}
$$

which completes the result for $B^{\lambda}_{n,\alpha,\beta}(t^2;x)$ by combining the results obtained for $\theta_3(n,\alpha,\beta,x)$ and $\theta_4(n,\alpha,\beta,x)$. □

Corollary 1. *The following relations hold:*

$$
\begin{aligned}
B^{\lambda}_{n,\alpha,\beta}(t-x;x) &= \sum_{i=0}^{n}\frac{i+\alpha}{n+\beta}\tilde{b}_{n,i}(\lambda;x) - x\sum_{i=0}^{n}\tilde{b}_{n,i}(\lambda;x) \\
&= \frac{\alpha-\beta x}{n+\beta} + \lambda\frac{1-2x+x^{n+1}-(1-x)^{n+1}}{(n+\beta)(n-1)} \\
&\quad + \lambda\frac{\alpha x(1-x)^n}{n+\beta} + \lambda\frac{\alpha(1-x)^{n+1}}{(n+\beta)(n-1)}; \\
B^{\lambda}_{n,\alpha,\beta}((t-x)^2;x) &= \sum_{i=0}^{n}\left(\frac{i+\alpha}{n+\beta}\right)^2\tilde{b}_{n,i}(\lambda;x) - 2x\sum_{i=0}^{n}\frac{i+\alpha}{n+\beta}\tilde{b}_{n,i}(\lambda;x) + x^2\sum_{i=0}^{n}\tilde{b}_{n,i}(\lambda;x) \\
&= \frac{nx(1-x)+(\beta x-\alpha)^2}{(n+\beta)^2} \\
&\quad + \lambda\left[\frac{4x^2-2x-2x^{n+2}-2(\alpha-1)x(1-x)^{n+1}}{(n+\beta)(n-1)} - \frac{2\alpha x^2(1-x)^n}{n+\beta}\right] \\
&\quad + \lambda\frac{2nx-1-4nx^2+(2n+1)x^{n+1}+(1-x)^{n+1}+\alpha^2-4\alpha x}{(n+\beta)^2(n-1)} \\
&\quad + \lambda\frac{2\alpha n-2\alpha(\alpha+n)(x^{n+1}+(1-x)^n)+\alpha^2x(n^2+1)(1-x)^n}{(n+\beta)^2(n^2-1)}.
\end{aligned}
$$

Corollary 2. *The following identities hold:*

$$\lim_{n\to\infty} n\ B^{\lambda}_{n,\alpha,\beta}(t-x;x;) = \alpha - \beta x;$$
$$\lim_{n\to\infty} n\ B^{\lambda}_{n,\alpha,\beta}((t-x)^2;x) = x(1-x).$$

We obtain the uniform convergence of operators $B^{\lambda}_{n,\alpha,\beta}(f;x)$ by applying well-known Bohman–Korovkin–Popoviciu theorem.

Theorem 1. *Let $C[0,1]$ denote the space of all real-valued continuous functions on $[0,1]$ endowed with the supremum norm. Then*

$$\lim_{n\to\infty} B^{\lambda}_{n,\alpha,\beta}(f;x) = f(x) \quad (f \in C[0,1])$$

uniformly in $[0,1]$.

Proof. It is sufficient to show that

$$\lim_{n\to\infty} ||B^{\lambda}_{n,\alpha,\beta}(t^j;x) - t^j||_{C[0,1]} = 0, \qquad j = 0,1,2$$

as stated in Bohman–Korovkin–Popoviciu theorem. We have the following relations by Lemma 1:

$$\lim_{n\to\infty} ||B^{\lambda}_{n,\alpha,\beta}(t^0;x) - t^0||_{C[0,1]} = 0 \quad \text{and} \quad \lim_{n\to\infty} ||B^{\lambda}_{n,\alpha,\beta}(t;x) - t||_{C[0,1]} = 0.$$

It is easy to show

$$\begin{aligned} B^{\lambda}_{n,\alpha,\beta}(t^2;x) \leq & \frac{n(n+1)x^2 + (1+2\alpha)nx + \alpha^2}{(n+\beta)^2} \\ & + \lambda\left[\frac{2nx + 1 + 4nx^2 + (2n+1)x^{n+1} + (1-x)^{n+1}}{(n+\beta)^2(n-1)} + \frac{\alpha^2 + 4\alpha x}{(n+\beta)^2(n-1)}\right. \\ & \left. + \frac{2\alpha n + 2\alpha(\alpha+n)(x^{n+1} + (1-x)^n) + \alpha^2(n^2+1)x(1-x)^n}{(n+\beta)^2(n^2-1)}\right] \end{aligned}$$

and hence

$$\lim_{n\to\infty} ||B^{\alpha,\beta}_n(t^2;x;\lambda) - t^2||_{C[0,1]} = 0.$$

This implies $B^{\lambda}_{n,\alpha,\beta}(f;x)$ converge uniformly to f on $[0,1]$. □

Recall that the first and second order Ditzian–Totik uniform modulus of smoothness are given by

$$\omega_{\xi}(f,\delta) := \sup_{0<|h|\leq\delta} \sup_{x,x+h\xi(x)\in[0,1]} \{|f(x+h\xi(x)) - f(x)|\}$$

and

$$\omega_2^{\phi}(f,\delta) := \sup_{0<|h|\leq\delta} \sup_{x,x\pm h\phi(x)\in[0,1]} \{|f(x+h\phi(x)) - 2f(x) + f(x-h\phi(x))|\},$$

respectively, where ϕ is an admissible step-weight function on $[a,b]$, that is, $\phi(x) = [(x-a)(b-x)]^{1/2}$ if $x \in [a,b]$ (see [24]). Let

$$K_{2,\phi(x)}(f,\delta) = \inf_{g\in W^2(\phi)} \{||f-g||_{C[0,1]} + \delta||\phi^2 g''||_{C[0,1]} : g \in C^2[0,1]\} \qquad (\delta > 0)$$

be the corresponding K-functional, where

$$W^2(\phi) = \{g \in C[0,1] : g' \in AC[0,1],\ \phi^2 g'' \in C[0,1]\}$$

and

$$C^2[0,1] = \{g \in C[0,1] : g', g'' \in C[0,1]\}.$$

In this case, $g' \in AC[0,1]$ means that g' is absolutely continuous on $[0,1]$. It is known by [25] that there exists an absolute constant $C > 0$, such that

$$C^{-1}\omega_2^{\phi}(f,\sqrt{\delta}) \leq K_{2,\phi(x)}(f,\delta) \leq C\omega_2^{\phi}(f,\sqrt{\delta}). \tag{6}$$

We are now ready to obtain global approximation theorem.

Theorem 2. *Let $\lambda \in [-1,1]$ and $f \in C[0,1]$. Suppose that $\phi(\neq 0)$ such that ϕ^2 is concave. Then*

$$|B_{n,\alpha,\beta}^{\lambda}(f;x) - f(x)| \leq C\omega_2^{\phi}\Big(f, \frac{\delta_n(\alpha,\beta,\lambda;x)}{2\phi(x)}\Big) + \omega_{\xi}\Big(f, \frac{\mu_n(\alpha,\beta,\lambda;x)}{\xi(x)}\Big)$$

for $x \in [0,1]$ and $C > 0$, where $\mu_n(\alpha,\beta,\lambda;x) = B_{n,\alpha,\beta}^{\lambda}(t-x;x)$, $\delta_n(\alpha,\beta,\lambda;x) = (\nu_n(\alpha,\beta,\lambda;x) + \mu_n^2(\alpha,\beta,\lambda;x)(x))^{\frac{1}{2}}$ and $\nu_n(\alpha,\beta,\lambda;x)(x) = B_{n,\alpha,\beta}^{\lambda}((t-x)^2;x)$.

Proof. Consider the operators

$$\begin{aligned}\tilde{B}_{n,\alpha,\beta}^{\lambda}(f;x) &= B_{n,\alpha,\beta}^{\lambda}(f;x) + f(x)\\ &\quad - f\Big(\frac{\alpha-\beta x}{n+\beta} + \lambda\frac{\alpha x(1-x)^n}{n+\beta} + \lambda\frac{1-2x+x^{n+1}+(\alpha-1)(1-x)^{n+1}}{(n+\beta)(n-1)}\Big)\end{aligned} \tag{7}$$

for $\lambda \in [-1,1]$, $x \in [0,1]$. We observe that $\tilde{B}_{n,\alpha,\beta}^{\lambda}(1;x) = 1$ and $\tilde{B}_{n,\alpha,\beta}^{\lambda}(t;x) = x$, that is $\tilde{B}_{n,\alpha,\beta}^{\lambda}(t-x;x) = 0$.

Let $u = \rho x + (1-\rho)t$, $\rho \in [0,1]$. Since ϕ^2 is concave on $[0,1]$, we have $\phi^2(u) \geq \rho\phi^2(x) + (1-\rho)\phi^2(t)$ and hence

$$\frac{|t-u|}{\phi^2(u)} \leq \frac{\rho|x-t|}{\rho\phi^2(x)+(1-\rho)\phi^2(t)} \leq \frac{|t-x|}{\phi^2(x)}. \tag{8}$$

So

$$\begin{aligned}|\tilde{B}_{n,\alpha,\beta}^{\lambda}(f;x) - f(x)| &\leq |\tilde{B}_{n,\alpha,\beta}^{\lambda}(f-g;x)| + |\tilde{B}_{n,\alpha,\beta}^{\lambda}(g;x) - g(x)| + |f(x)-g(x)|\\ &\leq 4\|f-g\|_{C[0,1]} + |\tilde{B}_{n,\alpha,\beta}^{\lambda}(g;x) - g(x)|.\end{aligned} \tag{9}$$

We obtain the following relations by applying the Taylor's formula:

$$\begin{aligned}&|\tilde{B}_{n,\alpha,\beta}^{\lambda}(g;x) - g(x)|\\ &\leq B_{n,\alpha,\beta}^{\lambda}\Big(\Big|\int_x^t |t-u|\,|g''(u)|du\Big|;x\Big) + \Big|\int_x^{x+\mu_n} |x+\mu_n(\alpha,\beta,\lambda;x)-u|\,|g''(u)|\,du\Big|\\ &\leq \|\phi^2 g''\|_{C[0,1]} B_{n,\alpha,\beta}^{\lambda}\Big(\Big|\int_x^t \frac{|t-u|}{\phi^2(u)}du\Big|;x\Big) + \|\phi^2 g''\|_{C[0,1]}\Big|\int_x^{x+\mu_n}\frac{|x+\mu_n(\alpha,\beta,\lambda;x)-u|}{\phi^2(u)}\,du\Big|\\ &\leq \phi^{-2}(x)\|\phi^2 g''\|_{C[0,1]} B_{n,\alpha,\beta}^{\lambda}((t-x)^2;x) + \phi^{-2}(x)\|\phi^2 g''\|_{C[0,1]}\beta_n^2(x).\end{aligned} \tag{10}$$

By using the definition of K-functional together with (6) and the inequalities (9) and (10), we have

$$|\tilde{B}^{\lambda}_{n,\alpha,\beta}(f;x) - f(x)| \leq \phi^{-2}(x)\|\phi^2 g''\|_{C[0,1]}(\nu_n(\alpha,\beta,\lambda;x) + \mu_n^2(\alpha,\beta,\lambda;x)) + 4\|f-g\|_{C[0,1]}$$

$$\leq C\omega_2^{\phi}\left(f, \frac{(\nu_n(\alpha,\beta,\lambda;x) + \mu_n^2(\alpha,\beta,\lambda;x))^{\frac{1}{2}}}{2\phi(x)}\right).$$

Also, by first order Ditzian–Totik uniform modulus of smoothness, we have

$$\begin{aligned}|f(x+\mu_n) - f(x)| &= \left|f\left(x + \xi(x)\frac{\mu_n(\alpha,\beta,\lambda;x)}{\xi(x)}\right) - f(x)\right| \\ &\leq \omega_{\xi}\left(f, \frac{\mu_n(\alpha,\beta,\lambda;x)}{\xi(x)}\right).\end{aligned}$$

Therefore, the following inequalities hold:

$$\begin{aligned}|B^{\lambda}_{n,\alpha,\beta}(f;x) - f(x)| &\leq |\tilde{B}^{\lambda}_{n,\alpha,\beta}(f;x) - f(x)| + |f(x+\mu_n(\alpha,\beta,\lambda;x)) - f(x)| \\ &\leq C\omega_2^{\phi}\left(f, \frac{\delta_n(\alpha,\beta,\lambda;x)}{2\phi(x)}\right) + \omega_{\xi}\left(f, \frac{\mu_n(\alpha,\beta,\lambda;x)}{\xi(x)}\right),\end{aligned}$$

which completes the proof. □

In order to obtain next result, we first recall some concepts and results concerning modulus of continuity and Peetre's K-functional. For $\delta > 0$, the modulus of continuity $w(f,\delta)$ of $f \in C[a,b]$ is given by

$$w(f,\delta) := \sup\{|f(x) - f(y)| : \; x,y \in [a,b], \; |x-y| \leq \delta\}.$$

It is also well known that, for any $\delta > 0$ and each $x \in [a,b]$,

$$|f(x) - f(y)| \leq \omega(f,\delta)\left(\frac{|x-y|}{\delta} + 1\right). \tag{11}$$

For $f \in C[0,1]$, the second-order modulus of smoothness is given by

$$w_2(f,\sqrt{\delta}) := \sup_{0<h\leq\sqrt{\delta}} \; \sup_{x,x+2h\in[0,1]} \{|f(x+2h) - 2f(x+h) + f(x)|\},$$

and the corresponding Peetre's K-functional [26] is

$$K_2(f,\delta) = \inf\{||f-g||_{C[0,1]} + \delta||g''||_{C[0,1]} : g \in W^2[0,1]\},$$

where

$$W^2[0,1] = \{g \in C[0,1] : g', g'' \in C[0,1]\}.$$

It is well-known that the inequality

$$K_2(f,\delta) \leq Cw_2(f,\sqrt{\delta}) \qquad (\delta > 0) \tag{12}$$

holds in which the absolute constant $C > 0$ is independent of δ and f (see [25]).

We are now ready to establish a direct local approximation theorem for operators $B^{\lambda}_{n,\alpha,\beta}(f;x)$ via second order modulus of smoothness and usual modulus of continuity.

Theorem 3. *Assume that $f \in C[0,1]$ and $x \in [0,1]$. Then there exists an absolute constant C such that*

$$|B^{\lambda}_{n,\alpha,\beta}(f;x) - f(x)| \leq C\, w_2\Big(f, \frac{1}{2}\delta_n(\alpha,\beta,\lambda;x)\Big) + w(f,\mu_n(\alpha,\beta,\lambda;x))$$

for the operators $B^{\lambda}_{n,\alpha,\beta}(f;x)$, where $\mu_n(\alpha,\beta,\lambda;x)$ and $\delta_n(\alpha,\beta,\lambda;x)$ are given in Theorem 2.

Proof. Consider the operators $\tilde{B}^{\lambda}_{n,\alpha,\beta}(f;x)$ as defined in Theorem 2. Assume that $t,x \in [0,1]$ and $g \in W^2[0,1]$. The following equality yields by Taylor's expansion formula:

$$g(t) = g(x) + (t-x)g'(x) + \int_x^t (t-u)g''(u)du. \tag{13}$$

If we apply $\tilde{B}^{\lambda}_{n,\alpha,\beta}(\cdot;x)$ to both sides of (13) and keeping in mind these operators preserve constants and linear functions, we obtain

$$\begin{aligned}\tilde{B}^{\lambda}_{n,\alpha,\beta}(g;x) - g(x) &= g'(x)\tilde{B}^{\lambda}_{n,\alpha,\beta}(t-x;x) + \tilde{B}^{\lambda}_{n,\alpha,\beta}\Big(\int_x^t (t-u)g''(u)du;x\Big)\\ &= B^{\lambda}_{n,\alpha,\beta}\Big(\int_x^t (t-u)g''(u)du;x\Big) - \int_x^{x+\mu_n} (x+\mu_n(\alpha,\beta,\lambda;x)-u)g''(u)du.\end{aligned}$$

Therefore,

$$\begin{aligned}|\tilde{B}^{\lambda}_{n,\alpha,\beta}(g;x) - g(x)| &\leq B^{\lambda}_{n,\alpha,\beta}\Big(\Big|\int_x^t |t-u|\,|g''(u)|du\Big|;x\Big)\\ &\quad - \int_x^{x+\mu_n} |x+\mu_n(\alpha,\beta,\lambda;x)-u|\,|g''(u)|\,du\\ &\leq \|g''\|_{C[0,1]}\big(B^{\lambda}_{n,\alpha,\beta}((t-x)^2;x) + (B^{\lambda}_{n,\alpha,\beta}(t-x;x))^2\big).\end{aligned}$$

With the help of (7), one obtains

$$\begin{aligned}\|\tilde{B}^{\lambda}_{n,\alpha,\beta}(g;x)\|_{C[0,1]} &\leq \|B^{\lambda}_{n,\alpha,\beta}(g;x)\|_{C[0,1]} + \|g(x)\|_{C[0,1]} + \|g(x+\mu_n(\alpha,\beta,\lambda;x))\|_{C[0,1]}\\ &\leq \|3g\|_{C[0,1]}.\end{aligned} \tag{14}$$

Now, for $f \in C[0,1]$ and $g \in W^2[0,1]$, using (7) and (14), we get

$$\begin{aligned}|B^{\lambda}_{n,\alpha,\beta}(f;x) - f(x)| &\leq |\tilde{B}^{\lambda}_{n,\alpha,\beta}(f-g;x)| + |\tilde{B}^{\lambda}_{n,\alpha,\beta}(g;x) - g(x)|\\ &\quad + |g(x) - f(x)| + |f(x+\mu_n(\alpha,\beta,\lambda;x)) - f(x)|\\ &\leq \delta_n^2(\alpha,\beta,\lambda;x)\|g''\|_{C[0,1]} + w(f,\mu_n(\alpha,\beta,\lambda;x)) + 4\|f-g\|_{C[0,1]}.\end{aligned}$$

Finally, by assuming the infimum on the right-hand side of the above inequality over all $g \in W^2[0,1]$ togrther with inequality (12), we obtain

$$\begin{aligned}|B^{\lambda}_{n,\alpha,\beta}(f;x) - f(x)| &\leq 4K_2\Big(f, \frac{\delta_n^2(\alpha,\beta,\lambda;x)}{4}\Big) + w(f,\mu_n(\alpha,\beta,\lambda;x))\\ &\leq C\, w_2\Big(f, \frac{1}{2}\delta_n(\alpha,\beta,\lambda;x)\Big) + w(f,\mu_n(\alpha,\beta,\lambda;x)),\end{aligned}$$

which completes the proof. □

In the following theorem, we obtain a local direct estimate of the rate of convergence via Lipschitz-type function involving two parameters for the operators $B_{n,\alpha,\beta}^{\lambda}$. Before proceeding further, let us recall that

$$Lip_M^{(k_1,k_2)}(\eta) := \left\{ f \in C[0,1] : |f(t) - f(x)| \leq M \frac{|t-x|^{\eta}}{(k_1x^2 + k_2x + t)^{\frac{\eta}{2}}};\ x \in (0,1], t \in [0,1] \right\}$$

for $k_1 \geq 0, k_2 > 0$, where $\eta \in (0,1]$ and M is a positive constant (see [27]).

Theorem 4. *If $f \in Lip_M^{(k_1,k_2)}(\eta)$, then*

$$|B_{n,\alpha,\beta}^{\lambda}(f;x) - f(x)| \leq M\sqrt{\frac{\nu_n^{\eta}(\alpha,\beta,\lambda;x)}{(k_1x^2 + k_2x)^{\eta}}}$$

for all $\lambda \in [-1,1]$, $x \in (0,1]$ and $\eta \in (0,1]$, where $\nu_n(\alpha,\beta,\lambda;x)$ is defined in Theorem 2.

Proof. Let $f \in Lip_M^{(k_1,k_2)}(\eta)$ and $\eta \in (0,1]$. First, we are going to show that statement is true for $\eta = 1$. We write

$$\begin{aligned}
|B_{n,\alpha,\beta}^{\lambda}(f;x) - f(x)| &\leq |B_{n,\alpha,\beta}^{\lambda}(|f(t) - f(x)|;x)| + f(x)\, |B_{n,\alpha,\beta}^{\lambda}(1;x) - 1| \\
&\leq \sum_{i=0}^{n} \left| f\left(\frac{i+\alpha}{n+\beta}\right) - f(x) \right| \tilde{b}_{n,i}(x;\lambda) \\
&\leq M \sum_{i=0}^{n} \frac{|\frac{i+\alpha}{n+\beta} - x|}{(k_1x^2 + k_2x + t)^{\frac{1}{2}}} \tilde{b}_{n,i}(x;\lambda)
\end{aligned}$$

for $f \in Lip_M^{(k_1,k_2)}(1)$. By using the relation

$$(k_1x^2 + k_2x + t)^{-1/2} \leq (k_1x^2 + k_2x)^{-1/2} \quad (k_1 \geq 0, k_2 > 0)$$

and applying Cauchy–Schwarz inequality, we obtain

$$\begin{aligned}
|B_{n,\alpha,\beta}^{\lambda}(f;x) - f(x)| &\leq M(k_1x^2 + k_2x)^{-1/2} \sum_{i=0}^{n} \left|\frac{i+\alpha}{n+\beta} - x\right| \tilde{b}_{n,i}(x;\lambda) \\
&= M(k_1x^2 + k_2x)^{-1/2} |B_{n,\alpha,\beta}^{\lambda}(t-x;x)| \\
&\leq M|\nu_n(\alpha,\beta,\lambda;x)|^{1/2}(k_1x^2 + k_2x)^{-1/2}.
\end{aligned}$$

Hence, the statement is true for $\eta = 1$. By the monotonicity of $B_{n,\alpha,\beta}^{\lambda}(f;x)$ and applying Hölder's inequality two times with $a = 2/\eta$ and $b = 2/(2-\eta)$, we can see that the statement is true for $\eta \in (0,1]$ as follows:

$$\begin{aligned}
\left|B_{n,\alpha,\beta}^{\lambda}(f;x) - f(x)\right| &\leq \sum_{i=0}^{n} \left| f\left(\frac{i+\alpha}{n+\beta}\right) - f(x) \right| \tilde{b}_{n,i}(x;\lambda) \\
&\leq \left(\sum_{i=0}^{n} \left| f\left(\frac{i+\alpha}{n+\beta}\right) - f(x) \right|^{\frac{2}{\eta}} \tilde{b}_{n,i}(x;\lambda) \right)^{\frac{\eta}{2}} \left(\sum_{i=0}^{n} \tilde{b}_{n,i}(x;\lambda) \right)^{\frac{2-\eta}{2}} \\
&\leq M \left(\sum_{i=0}^{n} \frac{\left(\frac{i+\alpha}{n+\beta} - x\right)^2 \tilde{b}_{n,i}(x;\lambda)}{\frac{i+\alpha}{n+\beta} + k_1x^2 + k_2x} \right)^{\frac{\eta}{2}}
\end{aligned}$$

$$\leq M(k_1x^2+k_2x)^{-\eta/2}\left\{\sum_{i=0}^{n}\left(\frac{i+\alpha}{n+\beta}-x\right)^2\tilde{b}_{n,i}(x;\lambda)\right\}^{\frac{\eta}{2}}$$
$$\leq M(k_1x^2+k_2x+t)^{-\eta/2}\left[B_n^{\alpha,\beta}((t-x)^2;x;\lambda)\right]^{\frac{\eta}{2}}$$
$$= M\sqrt{\frac{\nu_n^{\eta}(\alpha,\beta,\lambda;x)}{(k_1x^2+k_2x)^{\eta}}}.$$

□

Theorem 5. *The following inequality holds:*

$$|B_{n,\alpha,\beta}^{\lambda}(f;x)-f(x)|\leq|\mu_n(\alpha,\beta,\lambda;x)|\,|f'(x)|+2\sqrt{\nu_n(\alpha,\beta,\lambda;x)}w(f',\sqrt{\nu_n(\alpha,\beta,\lambda;x)}\,)$$

for $f\in C^1[0,1]$ and $x\in[0,1]$, where $\mu_n(\alpha,\beta,\lambda;x)$ and $\nu_n(\alpha,\beta,\lambda;x)$ are defined in Theorem 2.

Proof. We have

$$f(t)-f(x)=(t-x)f'(x)+\int_x^t(f'(u)-f'(x))du \tag{15}$$

for any $t\in[0,1]$ and $x\in[0,1]$. By applying the operators $B_{n,\alpha,\beta}^{\lambda}(\cdot;x)$ to both sides of (15), we have

$$B_{n,\alpha,\beta}^{\lambda}(f(t)-f(x);x)=f'(x)B_{n,\alpha,\beta}^{\lambda}(t-x;x)+B_{n,\alpha,\beta}^{\lambda}\left(\int_x^t(f'(u)-f'(x))du;x\right).$$

The following inequality holds for any $\delta>0$, $u\in[0,1]$ and $f\in C[0,1]$:

$$|f(u)-f(x)|\leq w(f,\delta)\left(\frac{|u-x|}{\delta}+1\right).$$

Thus, we obtain

$$\left|\int_x^t(f'(u)-f'(x))du\right|\leq w(f',\delta)\left(\frac{(t-x)^2}{\delta}+|t-x|\right).$$

Hence

$$|B_{n,\alpha,\beta}^{\lambda}(f;x)-f(x)|\leq|f'(x)|\,|B_{n,\alpha,\beta}^{\lambda}(t-x;x)|$$
$$+w(f',\delta)\left\{\frac{1}{\delta}B_{n,\alpha,\beta}^{\lambda}((t-x)^2;x)+B_{n,\alpha,\beta}^{\lambda}(t-x;x)\right\}. \tag{16}$$

By applying Cauchy–Schwarz inequality on the right hand side of last inequality (16), we have

$$|B_{n,\alpha,\beta}^{\lambda}(f;x)-f(x)|\leq|f'(x)|\,|\mu_n(\alpha,\beta,\lambda;x)|$$
$$+w(f',\delta)\left\{\frac{1}{\delta}\sqrt{B_{n,\alpha,\beta}^{\lambda}((t-x)^2;x)}+1\right\}\sqrt{B_{n,\alpha,\beta}^{\lambda}(|t-x|;x)}.$$

Consequently, we obtain the desired result if we choose δ as $\nu_n^{1/2}(\alpha,\beta,\lambda;x)$. □

3. Voronovskaja-Type Theorems

Here, we prove the following Voronovskaja-type theorems by $B_{n,\alpha,\beta}^{\lambda}(f;x)$.

Theorem 6. *Let $f, f', f'' \in C_B[0,1]$, where $C_B[0,1]$ is the set of all real-valued bounded and continuous functions defined on $[0,1]$. Then, for each $x \in [0,1]$, we have*

$$\lim_{n\to\infty} n\{B^{\lambda}_{n,\alpha,\beta}(f;x) - f(x)\} = (\alpha - \beta x)\, f'(x) + \frac{x(1-x)}{2} f''(x)$$

uniformly on $[0,1]$.

Proof. We first write the following equality by Taylor's expansion theorem of function $f(x)$ in $C_B[0,1]$:

$$f(t) = f(x) + (t-x)f'(x) + \frac{1}{2}(t-x)^2 f''(x) + (t-x)^2\, r_x(t), \tag{17}$$

where $r_x(t)$ is Peano form of the remainder, $r_x \in C[0,1]$ and $r_x(t) \to 0$ as $t \to x$. Applying the operators $B^{\lambda}_{n,\alpha,\beta}(\cdot;x)$ to identity (17), we have

$$B^{\lambda}_{n,\alpha,\beta}(f;x) - f(x) = f'(x) B^{\lambda}_{n,\alpha,\beta}(t-x;x) + \frac{f''(x)}{2} B^{\lambda}_{n,\alpha,\beta}((t-x)^2;x) + B^{\lambda}_{n,\alpha,\beta}((t-x)^2 r_x(t);x).$$

Using Cauchy–Schwarz inequality, we have

$$B^{\lambda}_{n,\alpha,\beta}((t-x)^2 r_x(t);x) \le \sqrt{B^{\lambda}_{n,\alpha,\beta}((t-x)^4;x)}\sqrt{B^{\lambda}_{n,\alpha,\beta}(r_x^2(t);x)}. \tag{18}$$

We observe that $\lim_n B^{\lambda}_{n,\alpha,\beta}(r_x^2(t);x) = 0$ and hence

$$\lim_{n\to\infty} n\{B^{\lambda}_{n,\alpha,\beta}((t-x)^2 r_x(t);x)\} = 0.$$

Thus

$$\lim_{n\to\infty} n\{B^{\lambda}_{n,\alpha,\beta}(f;x) - f(x)\} = \lim_{n\to\infty} n\{B^{\lambda}_{n,\alpha,\beta}(t-x;x)f'(x) + \frac{f''(x)}{2} B^{\lambda}_{n,\alpha,\beta}((t-x)^2;x) + B^{\lambda}_{n,\alpha,\beta}((t-x)^2 r_x(t);x)\}.$$

The result follows immediately by applying the Corollaries 1 and 2. □

For $f \in C[0,1]$ and $\delta > 0$, the Ditzian–Totik modulus of smoothness is given by

$$\omega_\phi(f,\delta) := \sup_{0<|h|\le\delta} \left\{ \left| f\left(x + \frac{h\phi(x)}{2}\right) - f\left(x - \frac{h\phi(x)}{2}\right) \right|, x \pm \frac{h\phi(x)}{2} \in [0,1] \right\},$$

where $\phi(x) = (x(1-x))^{1/2}$, and let

$$K_\phi(f,\delta) = \inf_{g\in W_\phi[0,1]} \{||f-g|| + \delta||\phi g'|| : g \in C^1[0,1]\}$$

be the corresponding Peetre's K-functional, where

$$W_\phi[0,1] = \{g : \ g \in AC_{loc}[0,1],\ \|\phi g'\| < \infty\}$$

and $AC_{loc}[0,1]$ denotes the class of absolutely continuous functions defined on $[a,b] \subset [0,1]$. There exists a constant $C > 0$ such that $K_\phi(f,\delta) \le C\,\omega_\phi(f,\delta)$.

Next, we give a quantitative Voronovskaja-type result for $B_n^{\alpha,\beta}(f;x;\lambda)$.

Theorem 7. *Suppose that $f \in C[0,1]$ such that $f', f'' \in C[0,1]$. Then*

$$\begin{aligned} &\left| B^{\lambda}_{n,\alpha,\beta}(f;x)f(x) - f(x) - \mu_n(\alpha,\beta,\lambda;x)f'(x) - \{\nu_n(\alpha,\beta,\lambda;x)+1\}\frac{f''(x)}{2}\right| \\ &\leq \frac{C}{n}\phi^2(x)\omega_\phi\left(f'', \frac{1}{\sqrt{n}}\right). \end{aligned} \tag{19}$$

for every $x \in [0,1]$ and sufficiently large n, where C is a positive constant, $\mu_n(\alpha,\beta,\lambda;x)$ and $\nu_n(\alpha,\beta,\lambda;x)$ are defined in Theorem 2.

Proof. Consider the following equality

$$f(t) - f(x) - (t-x)f'(x) = \int_x^t (t-u)f''(u)du$$

for $f \in C[0,1]$. It follows that

$$f(t) - f(x) - (t-x)f'(x) - \frac{f''(x)}{2}((t-x)^2+1) \leq \int_x^t (t-u)[f''(u) - f''(x)]du. \tag{20}$$

Applying $B^{\lambda}_{n,\alpha,\beta}(\cdot;x)$ to both sides of (20), we obtain

$$\begin{aligned} &\left| B^{\lambda}_{n,\alpha,\beta}(f;x) - f(x) - B^{\lambda}_{n,\alpha,\beta}((t-x);x)f'(x) - \frac{f''(x)}{2}(B^{\lambda}_{n,\alpha,\beta}((t-x)^2;x) + B^{\lambda}_{n,\alpha,\beta}(1;x))\right| \\ &\leq B^{\lambda}_{n,\alpha,\beta}\left(\left|\int_x^t |t-u|\,|f''(u)-f''(x)|\,du\right|;x\right). \end{aligned} \tag{21}$$

The quantity in the right hand side of (21) can be estimated as

$$\left|\int_x^t |t-u|\,|f''(u)-f''(x)|\,du\right| \leq 2\|f''-g\|(t-x)^2 + 2\|\phi g'\|\phi^{-1}(x)|t-x|^3, \tag{22}$$

where $g \in W_\phi[0,1]$. There exists $C > 0$ such that

$$B^{\lambda}_{n,\alpha,\beta}((t-x)^2;x) \leq \frac{C}{2n}\phi^2(x) \quad \text{and} \quad B^{\lambda}_{n,\alpha,\beta}((t-x)^4;x) \leq \frac{C}{2n^2}\phi^4(x) \tag{23}$$

for sufficiently large n. By taking (21)–(23) into our account and using Cauchy–Schwarz inequality, we have

$$\begin{aligned} &\left| B^{\lambda}_{n,\alpha,\beta}(f;x) - f(x) - B^{\lambda}_{n,\alpha,\beta}((t-x);x)f'(x) - \frac{f''(x)}{2}(B^{\lambda}_{n,\alpha,\beta}((t-x)^2;x) + B^{\lambda}_{n,\alpha,\beta}(1;x))\right| \\ &\leq 2\|f''-g\|B^{\lambda}_{n,\alpha,\beta}((t-x)^2;x) + 2\|\phi g'\|\phi^{-1}(x)B^{\lambda}_{n,\alpha,\beta}(|t-x|^3;x) \\ &\leq \frac{C}{n}x(1-x)\|f''-g\| + 2\|\phi g'\|\phi^{-1}(x)\{B^{\lambda}_{n,\alpha,\beta}((t-x)^2;x)\}^{1/2}\{B^{\lambda}_{n,\alpha,\beta}((t-x)^4;x)\}^{1/2} \\ &\leq \frac{C}{n}\phi^2(x)\left\{\|f''-g\| + n^{-1/2}\|\phi g'\|\right\}. \end{aligned}$$

Finally, by taking infimum over all $g \in W_\phi[0,1]$, this last inequality leads us to the assertion (19) of Theorem 7. □

As an immediate consequence of Theorem 7, we have the following result.

Corollary 3. *If $f \in C[0,1]$ such that $f', f'' \in C[0,1]$, then*

$$\lim_{n\to\infty} n\left|B^{\lambda}_{n,\alpha,\beta}(f;x)f(x) - f(x) - \mu_n(\alpha,\beta,\lambda;x)f'(x) - \{\nu_n(\alpha,\beta,\lambda;x)+1\}\frac{f''(x)}{2}\right| = 0,$$

where $\mu_n(\alpha,\beta,\lambda;x)$ and $\nu_n(\alpha,\beta,\lambda;x)$ are defined in Theorem 2.

4. The Bivariate Case of the Operators $B^{\lambda}_{n,\alpha,\beta}(f;x)$

We construct bivariate version of Stancu-type λ-Bernstein operators defined which was defined in the first section of this manuscript as (5) and study their approximation properties.

For $0 \le \alpha_i \le \beta_i$ $(i = 1,2)$, we defined the bivariate version of Stancu-type λ-Bernstein operators by

$$B^{\lambda,\alpha,\beta}_{n,m}(f;x,y) = \sum_{i_1=0}^{n}\sum_{i_2=0}^{m} f\left(\frac{i_1+\alpha_1}{n+\beta_1}, \frac{i_2+\alpha_2}{m+\beta_2}\right)\tilde{b}_{n,i_1}(\lambda_1;x)\tilde{b}_{m,i_2}(\lambda_2;y) \tag{24}$$

for $(x,y) \in I$ and $f \in C(I)$, where $I = [0,1]\times[0,1]$ and $\tilde{b}_{n,i_1}(\lambda_1;x)$ and $\tilde{b}_{m,i_2}(\lambda_2;x)$ are Bézier bases defined in (4).

We remark that if we take $\lambda_1 = \lambda_2 = 0$ in bivariate λ-Bernstein–Stancu operators, then (24) reduces to the classical bivariate Bernstein–Stancu operators defined in [28]. Also, for $\alpha_1 = \beta_1 = \lambda_1 = 0$ and $\alpha_2 = \beta_2 = \lambda_2 = 0$, the bivariate λ-Bernstein–Stancu operators (24) reduce to classical bivariate Bernstein operators defined in [29].

Lemma 2. *The following equalities hold for bivariate λ-Bernstein–Stancu operators:*

$$\begin{aligned}
B^{\lambda,\alpha,\beta}_{n,m}(1;x,y) &= 1;\\
B^{\alpha,\beta}_{n,m}(s;x,y) &= \frac{\alpha_1+nx}{n+\beta_1} + \lambda_1\left[\frac{1-2x+x^{n+1}+(\alpha_1-1)(1-x)^{n+1}}{(n+\beta_1)(n-1)} + \frac{\alpha_1 x(1-x)^n}{n+\beta_1}\right];\\
B^{\lambda,\alpha,\beta}_{n,m}(t;x,y) &= \frac{\alpha_2+my}{m+\beta_2} + \lambda_2\left[\frac{1-2y+y^{m+1}+(\alpha_2-1)(1-y)^{m+1}}{(m+\beta_2)(m-1)} + \frac{\alpha_2 y(1-y)^m}{m+\beta_2}\right];\\
B^{\lambda,\alpha,\beta}_{n,m}(s^2;x,y) &= \frac{1}{(n+\beta_1)^2}\left\{n(n-1)x^2 + (1+2\alpha_1)nx + \alpha_1^2\right\}\\
&\quad + \lambda_1\left[\frac{2nx-1-4nx^2+(2n+1)x^{n+1}+(1-x)^{n+1}}{(n+\beta_1)^2(n-1)} + \frac{\alpha_1^2-4\alpha_1 x}{(n+\beta_1)^2(n-1)}\right.\\
&\quad \left. + \frac{2\alpha_1 n - 2\alpha_1(\alpha_1+n)(x^{n+1}+(1-x)^n) + \alpha_1^2 x(n^2+1)(1-x)^n}{(n+\beta_1)^2(n^2-1)}\right];\\
B^{\lambda,\alpha,\beta}_{n,m}(t^2;x,y) &= \frac{1}{(m+\beta_2)^2}\left\{m(m-1)y^2 + (1+2\alpha_2)my + \alpha_2^2\right\}\\
&\quad + \lambda_2\left[\frac{2my-1-4my^2+(2m+1)y^{m+1}+(1-y)^{m+1}}{(m+\beta_2)^2(m-1)} + \frac{\alpha_2^2-4\alpha_2 y}{(n+\beta_2)^2(m-1)}\right.\\
&\quad \left. + \frac{2\alpha_2 m - 2\alpha_2(\alpha_2+m)(y^{m+1}+(1-y)^m) + \alpha_2^2 y(m^2+1)(1-y)^m}{(m+\beta_2)^2(m^2-1)}\right].
\end{aligned}$$

Theorem 8. *Let $e_{ij}(x,y) = x^i y^j$, where $0 \le i+j \le 2$. Then, the sequence $B^{\lambda,\alpha,\beta}_{n,m}(f;x,y)$ of operators converges uniformly to f on I for each $f \in C(I)$.*

Proof. It is enough to prove the following condition

$$\lim_{n,m\to\infty} B^{\lambda,\alpha,\beta}_{n,m}(e_{ij};x,y) = e_{ij}$$

converges uniformly on I. With the help of Lemma 2, one can see that

$$\lim_{m,n\to\infty} B_{n,m}^{\lambda,\alpha,\beta}(e_{00};x,y) = e_{00},$$

$$\lim_{n,m\to\infty} B_{n,m}^{\lambda,\alpha,\beta}(e_{10};x,y) = e_{10}, \quad \lim_{n,m\to\infty} B_{n,m}^{\lambda,\alpha,\beta}(e_{01};x,y) = e_{01}$$

and

$$\lim_{n,m\to\infty} B_{n,m}^{\lambda,\alpha,\beta}(e_{02}+e_{20};x,y) = e_{02}+e_{20}.$$

Keeping in mind the above conditions and Korovkin type theorem established by Volkov [30], we obtain

$$\lim_{m,n\to\infty} B_{n,m}^{\lambda,\alpha,\beta}(f;x,y) = f$$

converges uniformly. □

Now, we compute the rate of convergence of operators (24) by means of the modulus of continuity. Recall that the modulus of continuity for bivariate case is defined as

$$\omega(f,\delta) = \sup\left\{|f(s,t)-f(x,y)| : \sqrt{(s-x)^2+(t-y)^2} \le \delta\right\}$$

for $f \in C(I_{ab})$ and for every $(s,t),(x,y) \in I_{ab} = [0,a]\times[0,b]$. The partial moduli of continuity with respect to x and y are defined by

$$\begin{aligned}\omega_1(f,\delta) &= \sup\{|f(x_1,y)-f(x_2,y)| : y\in[0,a] \text{ and } |x_1-x_2|\le\delta\},\\ \omega_2(f,\delta) &= \sup\{|f(x,y_1)-f(x,y_2)| : x\in[0,b] \text{ and } |y_1-y_2|\le\delta\}.\end{aligned}$$

Peetre's K-functional is given by

$$K(f,\delta) = \inf_{g\in C^2(I_{ab})}\left\{\|f-g\|_{C(I_{ab})} + \delta\|g\|_{C^2(I_{ab})}\right\}$$

for $\delta > 0$, where $C^2(I_{ab})$ is the space of functions of f such that f, $\frac{\partial^j f}{\partial x^j}$ and $\frac{\partial^j f}{\partial y^j}$ $(j=1,2)$ in $C(I_{ab})$ [26]. We now give an estimate of the rates of convergence of operators $B_{n,m}^{\lambda,\alpha,\beta}(f;x,y)$.

Theorem 9. *Let $f \in C(I)$. Then*

$$\left|B_{n,m}^{\lambda,\alpha,\beta}(f;x,y) - f(x,y)\right| \le 4\omega\left(f;\nu_n^{1/2}(\alpha,\beta,\lambda;x),\nu_m^{1/2}(\alpha,\beta,\lambda;y)\right)$$

for all $x \in I$, where

$$\nu_n(\alpha,\beta,\lambda;x) = B_{n,m}^{\lambda,\alpha,\beta}\left((s-x)^2;x,y\right) \text{ and } \nu_m(\alpha,\beta,\lambda;y) = B_{n,m}^{\lambda,\alpha,\beta}\left((t-y)^2;x,y\right).$$

Proof. Since (24) is linear and positive, we have

$$\begin{aligned}|B_{n,m}^{\lambda,\alpha,\beta}(f;x,y)-f(x,y)| &\le B_{n,m}^{\lambda,\alpha,\beta}(|f(s,t)-f(x,y)|;x,y)\\ &\le B_{n,m}^{\lambda,\alpha,\beta}\left(\omega\left(f;\sqrt{(s-x)^2+(t-y)^2}\right);x,y\right)\\ &\le \omega\left(f;\sqrt{\nu_n(\alpha,\beta,\lambda;x)},\sqrt{\nu_m(\alpha,\beta,\lambda;y)}\right)\end{aligned}$$

$$\times \left[\frac{1}{\sqrt{v_n(\alpha,\beta,\lambda;x)v_m(\alpha,\beta,\lambda;y)}} B_{n,m}^{\lambda,\alpha,\beta}\left(\sqrt{(s-x)^2+(t-y)^2};x,y\right)\right].$$

The Cauchy–Schwartz inequality gives that

$$\begin{aligned}
&|B_{n,m}^{\lambda,\alpha,\beta}(f;x,y)-f(x,y)| \\
&\leq \omega\left(f;\sqrt{v_n(\alpha,\beta,\lambda;x)},\sqrt{v_m(\alpha,\beta,\lambda;y)}\right) \\
&\times\left[1+\frac{1}{\sqrt{v_n(\alpha,\beta,\lambda;x)v_m(\alpha,\beta,\lambda;y)}}\left\{B_{n,m}^{\lambda,\alpha,\beta}\left((s-x)^2;x,y\right)B_{n,m}^{\lambda,\alpha,\beta}\left((t-y)^2;x,y\right)\right\}^{1/2}\right. \\
&\left.+\frac{\sqrt{B_{n,m}^{\lambda,\alpha,\beta}((s-x)^2;x,y)}}{\sqrt{v_n(\alpha,\beta,\lambda;x)}}+\frac{\sqrt{B_{n,m}^{\lambda,\alpha,\beta}((t-y)^2;x,y)}}{\sqrt{v_m(\alpha,\beta,\lambda;y)}}\right].
\end{aligned}$$

If we choose

$$v_n(\alpha,\beta,\lambda;x)=B_{n,m}^{\lambda,\alpha,\beta}\left((s-x)^2;x,y\right) \quad\text{and}\quad v_m(\alpha,\beta,\lambda;y)=B_{n,m}^{\lambda,\alpha,\beta}\left((t-y)^2;x,y\right)$$

for all $(x,y)\in I$ we complete the proof, where

$$\begin{aligned}
B_{n,m}^{\lambda,\alpha,\beta}\left((s-x)^2;x,y\right) &= B_{n,m}^{\lambda,\alpha,\beta}\left(s^2;x,y\right)-2xB_{n,m}^{\lambda,\alpha,\beta}(s;x,y)+x^2B_{n,m}^{\lambda,\alpha,\beta}(1;x,y) \\
&= \frac{nx(1-x)+(\beta_1x-\alpha_1)^2}{(n+\beta_1)^2} \\
&\quad+\lambda_1\left[\frac{4x^2-2x-2x^{n+2}-2(\alpha_1-1)x(1-x)^{n+1}}{(n+\beta_1)(n-1)}-\frac{2\alpha_1x^2(1-x)^n}{n+\beta_1}\right] \\
&\quad+\lambda_1\frac{2nx-1-4nx^2+(2n+1)x^{n+1}+(1-x)^{n+1}+\alpha_1^2-4\alpha_1x}{(n+\beta_1)^2(n-1)} \\
&\quad+\lambda_1\frac{2\alpha_1n-2\alpha_1(\alpha_1+n)(x^{n+1}+(1-x)^n)+\alpha_1^2x(n^2+1)(1-x)^n}{(n+\beta_1)^2(n^2-1)};
\end{aligned}$$

$$\begin{aligned}
B_{n,m}^{\lambda,\alpha,\beta}\left((t-y)^2;x,y\right) &= \frac{my(1-y)+(\beta_2y-\alpha_2)^2}{(m+\beta_2)^2} \\
&\quad+\lambda_2\left[\frac{4y^2-2y-2y^{m+2}-2(\alpha_2-1)y(1-y)^{m+1}}{(m+\beta_2)(m-1)}-\frac{2\alpha_2y^2(1-y)^n}{m+\beta_2}\right] \\
&\quad+\lambda_2\frac{2my-1-4my^2+(2m+1)y^{m+1}+(1-y)^{m+1}+\alpha_2^2-4\alpha_2y}{(m+\beta_2)^2(m-1)} \\
&\quad+\lambda_2\frac{2\alpha_2m-2\alpha_2(\alpha_2+m)(y^{m+1}+(1-y)^m)+\alpha_2^2y(m^2+1)(1-y)^m}{(m+\beta_2)^2(m^2-1)}.
\end{aligned}$$

□

Theorem 10. *Let $f\in C(I)$. Then, the following inequality holds:*

$$\left|B_{n,m}^{\lambda,\alpha,\beta}(f;x,y)-f(x,y)\right|\leq 2\left[\omega_1\left(f;v_n^{1/2}(\alpha,\beta,\lambda;x)\right)+\omega_2\left(f;v_n^{1/2}(\alpha,\beta,\lambda;y)\right)\right],$$

where $v_n(\alpha,\beta,\lambda;x)$ and $v_m(\alpha,\beta,\lambda;y)$ are defined in Theorem 9.

Proof. By using the definition of partial modulus of continuity and Cauchy–Schwartz inequality, we have

$$\begin{aligned}
&|B_{n,m}^{\lambda,\alpha,\beta}(f;x,y)-f(x,y)| \\
&\le B_{n,m}^{\lambda,\alpha,\beta}(|f(s,t)-f(x,y)|;x,y) \\
&\le B_{n,m}^{\lambda,\alpha,\beta}(|f(s,t)-f(x,t)|;x,y)+B_{n,m}^{\lambda,\alpha,\beta}(|f(x,t)-f(x,y)|;x,y) \\
&\le B_{n,m}^{\lambda,\alpha,\beta}(|\omega_1(f;|s-x|)|;x,y)+B_{n,m}^{\lambda,\alpha,\beta}(|\omega_2(f;|t-y|)|;x,y) \\
&\le \omega_1(f,\nu_n(\alpha,\beta,\lambda;x))\left[1+\frac{1}{\nu_n(\alpha,\beta,\lambda;x)}B_{n,m}^{\lambda,\alpha,\beta}(|s-x|;x,y)\right] \\
&\quad +\omega_2(f,\nu_m(\alpha,\beta,\lambda;y))\left[1+\frac{1}{\nu_m(\alpha,\beta,\lambda;y)}B_{n,m}^{\lambda,\alpha,\beta}(|t-y|;x,y)\right] \\
&\le \omega_1(f,\nu_n^{1/2}(\alpha,\beta,\lambda;x))\left[1+\frac{1}{\nu_n^{1/2}(\alpha,\beta,\lambda;x)}\left(B_{n,m}^{\lambda,\alpha,\beta}\left((s-x)^2;x,y\right)\right)^{1/2}\right] \\
&\quad +\omega_2(f,\nu_n^{1/2}(\alpha,\beta,\lambda;x))\left[1+\frac{1}{\nu_m^{1/2}(\alpha,\beta,\lambda;y)}\left(B_{n,m}^{\lambda,\alpha,\beta}\left((t-y)^2;x,y\right)\right)^{1/2}\right].
\end{aligned}$$

Finally, by choosing $\nu_n(\alpha,\beta,\lambda;x)$ and $\nu_m(\alpha,\beta,\lambda;y)$ as defined in Theorem 9, we obtain desired result. □

We recall that the Lipschitz class $Lip_M(\widehat{\beta}_1,\widehat{\beta}_2)$ for the bivariate is given by

$$|f(s,t)-f(x,y)|\le M|s-x|^{\widehat{\beta}_1}|t-y|^{\widehat{\beta}_2}$$

for $\widehat{\beta}_1,\widehat{\beta}_2\in(0,1]$ and $(s,t),(x,y)\in I_{ab}$.

Theorem 11. *Let $f\in Lip_M(\widehat{\beta}_1,\widehat{\beta}_2)$. Then, for all $(x,y)\in I_{ab}$, we have*

$$|B_{n,m}^{\lambda,\alpha,\beta}(f;x,y)-f(x,y)|\le M\nu_n^{\widehat{\beta}_1/2}(\alpha,\beta,\lambda;x)\nu_m^{\widehat{\beta}_2/2}(\alpha,\beta,\lambda;y),$$

where $\nu_n(\alpha,\beta,\lambda;x)$ and $\nu_m(\alpha,\beta,\lambda;y)$ are defined in Theorem 9.

Proof. We have

$$\begin{aligned}
|B_{n,m}^{\lambda,\alpha,\beta}(f;x,y)-f(x,y)| &\le B_{n,m}^{\lambda,\alpha,\beta}(|f(s,t)-f(x,y)|;x,y) \\
&\le MB_{n,m}^{\lambda,\alpha,\beta}(|s-x|^{\widehat{\beta}_1}|t-y|^{\widehat{\beta}_2};x,y)
\end{aligned}$$

$$= MB_{n,m}^{\lambda,\alpha,\beta}(|s-x|^{\widehat{\beta}_1}|;x,y)B_{n,m}^{\lambda,\alpha,\beta}(|t-y|^{\widehat{\beta}_2};x,y)$$

since $f\in Lip_M(\widehat{\beta}_1,\widehat{\beta}_2)$. Then, by applying the Hölder's inequality for

$$\widehat{p}_1=\frac{2}{\widehat{\beta}_1},\widehat{q}_1=\frac{2}{2-\widehat{\beta}_1}$$

and

$$\widehat{p}_2=\frac{1}{\widehat{\beta}_2},\widehat{q}_2=\frac{2}{2-\widehat{\beta}_2},$$

we obtain

$$\begin{aligned}|B_{n,m}^{\lambda,\alpha,\beta}(f;x,y)-f(x,y)| &\leq M\{B_{n,m}^{\lambda,\alpha,\beta}(|s-x|^2;x,y)\}^{\widehat{\beta}_1/2}\{B_{n,m}^{\lambda,\alpha,\beta}(1;x,y)\}^{\widehat{\beta}_1/2}\\ &\quad\times\{B_{n,m}^{\lambda,\alpha,\beta}(|t-y|^2;x,y)\}^{\widehat{\beta}_2/2}\{B_{n,m}^{\lambda,\alpha,\beta}(1;x,y)\}^{\widehat{\beta}_2/2}\\ &= M\nu_n(\alpha,\beta,\lambda;x)^{\widehat{\beta}_1/2}\nu_m(\alpha,\beta,\lambda;y)^{\widehat{\beta}_2/2}.\end{aligned}$$

This completes the proof. □

Theorem 12. *For $f\in C^1(I)$, the following inequality holds:*

$$|B_{n,m}^{\lambda,\alpha,\beta}(f;x,y)-f(x,y)| \leq \parallel f_x\parallel_{C(I)} \nu_n^{1/2}(\alpha,\beta,\lambda;x)+\parallel f_y\parallel_{C(I)} \nu_m^{1/2}(\alpha,\beta,\lambda;y),$$

where $\nu_n(\alpha,\beta,\lambda;x)$ and $\nu_m(\alpha,\beta,\lambda;y)$ are defined in Theorem 9.

Proof. We have

$$f(t)-f(s)=\int_x^t f_u(u,s)du+\int_y^s f_v(x,v)du$$

for $(s,t)\in I$. Thus, by applying the operators defined in (24) to the above equality, we obtain

$$\begin{aligned}&|B_{n,m}^{\lambda,\alpha,\beta}(f;x,y)-f(x,y)|\\ &\leq B_{n,m}^{\lambda,\alpha,\beta}\left(\left|\int_x^t f_u(u,s)du\right|;x,y\right)+B_{n,m}^{\lambda,\alpha,\beta}\left(\left|\int_y^s f_v(x,v)du\right|;x,y\right).\end{aligned}$$

By taking the following relations into our consideration

$$\left|\int_x^t f_u(u,s)du\right|\leq\parallel f_x\parallel_{C(I_{ab})}|s-x|$$

and

$$\left|\int_y^s f_v(x,v)du\right|\leq f_y\parallel_{C(I_{ab})}|t-y|,$$

one obtains

$$\begin{aligned}&|B_{n,m}^{\lambda,\alpha,\beta}(f;x,y)-f(x,y)|\\ &\leq\parallel f_x\parallel_{C(I)} B_{n,m}^{\lambda,\alpha,\beta}(|s-x|;x,y)+\parallel f_y\parallel_{C(I)} B_{n,m}^{\lambda,\alpha,\beta}(|t-y|;x,y).\end{aligned}$$

Using Cauchy–Schwarz inequality, we have

$$\begin{aligned}&|B_{n,m}^{\lambda,\alpha,\beta}(f;x,y)-f(x,y)|\\ &\leq\parallel f_x\parallel_{C(I)}\{B_{n,m}^{\lambda,\alpha,\beta}\left((s-x)^2;x,y\right)\}^{1/2}\{B_{n,m}^{\lambda,\alpha,\beta}(1;x,y)\}^{1/2}\\ &\quad+\parallel f_y\parallel_{C(I)}\{B_{n,m}^{\lambda,\alpha,\beta}\left((t-y)^2;x,y\right)\}^{1/2}\{B_{n,m}^{\lambda,\alpha,\beta}(1;x,y)\}^{1/2}.\end{aligned}$$

□

Finally, we presents a Voronovskaja-type theorem for $B_{n,n}^{\lambda,\alpha,\beta}(f;x,y)$.

Theorem 13. *Let $f \in C^2(I)$. Then*

$$\lim_{n\to\infty} n\left[B_{n,n}^{\lambda,\alpha,\beta}(f;x,y) - f(x,y)\right] = (\alpha_1 - \beta_1 x)f_x + (\alpha_2 - \beta_2 y)f_y + \frac{x(1-x)}{2}f_{xx} + \frac{y(1-y)}{2}f_{yy}.$$

Proof. Let $(x,y) \in I$ and write the Taylor's formula of $f(s,t)$ as

$$\begin{aligned} f(s,t) &= f(x,y) + f_x(s-x) + f_y(t-y) \\ &\quad + \frac{1}{2}\left\{f_{xx}(s-x)^2 + 2f_{xy}(s-x)(t-y) + f_{yy}(t-y)^2\right\} \\ &\quad + \varepsilon(s,t)\left((s-x)^2 + (t-y)^2\right), \end{aligned} \tag{25}$$

where $(s,t) \in I$ and $\varepsilon(s,t) \longrightarrow 0$ as $(s,t) \longrightarrow (x,y)$. If we apply sequence of operators $B_{n,n}^{\lambda,\alpha,\beta}(\cdot;x,y)$ on (25) keeping in mind linearity of operator, we have

$$\begin{aligned} &B_{n,n}^{\lambda,\alpha,\beta}(f;s,t) - f(x,y) \\ &= f_x(x,y)B_{n,n}^{\lambda,\alpha,\beta}((s-x);x,y) + f_y(x,y)B_{n,n}^{\lambda,\alpha,\beta}((t-y);x,y) \\ &\quad + \frac{1}{2}\left\{f_{xx}B_{n,n}^{\lambda,\alpha,\beta}((s-x)^2;x,y) + 2f_{xy}B_{n,n}^{\lambda,\alpha,\beta}((s-x)(t-y);x,y)\right. \\ &\quad \left.+ f_{yy}B_{n,n}^{\lambda,\alpha,\beta}((t-y)^2;x,y)\right\} + B_{n,n}^{\lambda,\alpha,\beta}\left(\varepsilon(s,t)\left((s-x)^2 + (t-y)^2\right);x,y\right). \end{aligned}$$

Applying limit to both sides of the last equality as $n \to \infty$, we have

$$\begin{aligned} &\lim_{n\to\infty} n(B_{n,n}^{\lambda,\alpha,\beta}(f;s,t) - f(x,y)) \\ &= \lim_{n\to\infty} n\left\{f_x(x,y)B_{n,n}^{\lambda,\alpha,\beta}((s-x);x,y) + f_y(x,y)B_{n,n}^{\lambda,\alpha,\beta}((t-y);x,y)\right\} \\ &\quad + \lim_{n\to\infty} \frac{n}{2}\left\{f_{xx}B_{n,n}^{\lambda,\alpha,\beta}((s-x)^2;x,y)\right. \\ &\quad \left.+ 2f_{xy}B_{n,n}^{\lambda,\alpha,\beta}((s-x)(t-y);x,y) + f_{yy}B_{n,n}^{\lambda,\alpha,\beta}((t-y)^2;x,y)\right\} \\ &\quad + \lim_{n\to\infty} nB_{n,n}^{\lambda,\alpha,\beta}\left(\varepsilon(s,t)\left((s-x)^2 + (t-y)^2\right);x,y\right). \end{aligned}$$

Using Hölder inequality for the last term of above equality, we have

$$\begin{aligned} &B_{n,n}^{\lambda,\alpha,\beta}\left(\varepsilon(s,t)\left((s-x)^2 + (t-y)^2\right);x,y\right) \\ &\le \sqrt{2}\sqrt{B_{n,n}^{\lambda,\alpha,\beta}\left(\varepsilon^2(s,t);x,y\right)} \\ &\quad \times \sqrt{B_{n,n}^{\lambda,\alpha,\beta}\left(\varepsilon(s,t)\left((s-x)^4 + (t-y)^4\right);x,y\right)}. \end{aligned}$$

Since

$$\lim_{n\to\infty} B_{n,n}^{\lambda,\alpha,\beta}\left(\varepsilon^2(s,t);x,y\right) = \varepsilon^2(x,y) = 0$$

we have

$$\lim_{n\to\infty} n\, B_{n,n}^{\lambda,\alpha,\beta}\left(\varepsilon(s,t)\left((s-x)^4 + (t-y)^4\right);x,y\right) = 0. \tag{26}$$

Consequently, we obtain

$$\lim_{n\to\infty} n\, B_{n,n}^{\lambda,\alpha,\beta}((s-x);x,y) = \alpha_1 - \beta_1 x, \tag{27}$$
$$\lim_{n\to\infty} n\, B_{n,n}^{\lambda,\alpha,\beta}((t-y);x,y) = \alpha_2 - \beta_2 y, \tag{28}$$
$$\lim_{n\to\infty} n\, B_{n,n}^{\lambda,\alpha,\beta}((s-x)^2;x,y) = x(1-x), \tag{29}$$
$$\lim_{n\to\infty} n\, B_{n,n}^{\lambda,\alpha,\beta}((t-y)^2;x,y) = y(1-y). \tag{30}$$

Combining (26)–(30), we deduce the desired result. □

Author Contributions: All authors contributed equally in this work.

References

1. Bernstein, S.N. Démonstration du théorème de Weierstrass fondée sur le calcul des probabilités. *Comm. Soc. Math. Kharkow* **1913**, *13*, 1–2.
2. Stancu, D.D. Asupra unei generalizari a polinoamelor lui Bernstein. *Studia Univ. Babes-Bolyai Ser. Math.-Phys.* **1969**, *14*, 31–45.
3. Acar, T.; Mohiuddine, S.A.; Mursaleen, M. Approximation by (p,q)-Baskakov-Durrmeyer-Stancu operators. *Complex Anal. Oper. Theory* **2018**, *12*, 1453–1468. [CrossRef]
4. Baxhaku, B.; Agrawal, P.N. Degree of approximation for bivariate extension of Chlodowsky-type q-Bernstein-Stancu-Kantorovich operators. *Appl. Math. Comput.* **2017**, *306*, 56–72. [CrossRef]
5. Chauhan, R.; Ispir, N.; Agrawal, P.N. A new kind of Bernstein-Schurer-Stancu-Kantorovich-type operators based on q-integers. *J. Inequal. Appl.* **2017**, *2017*, 50. [CrossRef] [PubMed]
6. Mursaleen, M.; Ansari, K.J.; Khan, A. Some approximation results by (p,q)-analogue of Bernstein-Stancu operators. *Appl. Math. Comput.* **2015**, *264*, 392–402. [CrossRef]
7. Cai, Q.-B.; Lian, B.-Y.; Zhou, G. Approximation properties of λ-Bernstein operators, *J. Inequal. Appl.* **2018**, *2018*, 61. [CrossRef] [PubMed]
8. Ye, Z.; Long, X.; Zeng, X.-M. Adjustment algorithms for Bézier curve and surface. In Proceedings of the International Conference on Computer Science and Education, Hefei, China, 24–27 August 2010; pp. 1712–1716.
9. Cai, Q.-B. The Bézier variant of Kantorovich type λ-Bernstein operators. *J. Inequal. Appl.* **2018**, *2018*, 90. [CrossRef] [PubMed]
10. Acu, A.M.; Manav, N.; Sofonea, D.F. Approximation properties of λ-Kantorovich operators. *J. Inequal. Appl.* **2018**, *2018*, 202. [CrossRef]
11. Özger, F. Some general statistical approximation results for λ-Bernstein operators. *arXiv* **2018**, arXiv:1901.01099.
12. Acar, T.; Aral, A. On pointwise convergence of q-Bernstein operators and their q-derivatives. *Numer. Funct. Anal. Optim.* **2015**, *36*, 287–304. [CrossRef]
13. Acar, T.; Aral, A.; Mohiuddine, S.A. On Kantorovich modification of (p,q)-Bernstein operators. *Iran. J. Sci. Technol. Trans. Sci.* **2018**, *42*, 1459–1464. [CrossRef]
14. Acar, T.; Aral, A.; Mohiuddine, S.A. Approximation by bivariate (p,q)-Bernstein-Kantorovich operators. *Iran. J. Sci. Technol. Trans. Sci.* **2018**, *42*, 655–662. [CrossRef]
15. Acar, T.; Aral, A.; Mohiuddine, S.A. On Kantorovich modification of (p,q)-Baskakov operators. *J. Inequal. Appl.* **2016**, *2016*, 98. [CrossRef]
16. Acu, A.M.; Muraru, C. Approximation properties of bivariate extension of q-Bernstein-Schurer-Kantorovich operators. *Results Math.* **2015**, *67*, 265–279. [CrossRef]
17. Mishra, V.N.; Patel, P. On generalized integral Bernstein operators based on q-integers. *Appl. Math. Comput.* **2014**, *242*, 931–944. [CrossRef]
18. Mohiuddine, S.A.; Acar, T.; Alotaibi, A. Construction of a new family of Bernstein-Kantorovich operators. *Math. Methods Appl. Sci.* **2017**, *40*, 7749–7759. [CrossRef]

19. Mohiuddine, S.A.; Acar, T.; Alotaibi, A. Durrmeyer type (p,q)-Baskakov operators preserving linear functions. *J. Math. Inequal.* **2018**, *12*, 961–973. [CrossRef]
20. Mohiuddine, S.A.; Acar, T.; Alghamdi, M.A. Genuine modified Bernstein-Durrmeyer operators. *J. Inequal. Appl.* **2018**, *2018*, 104. [CrossRef] [PubMed]
21. Mursaleen, M.; Ansari, K.J.; Khan, A. On (p,q)-analogue of Bernstein operators. *Appl. Math. Comput.* **2018**, *266* , 874–882; Erratum in *Appl. Math. Comput.* **2016**, *278*, 70–71. [CrossRef]
22. Braha, N.L.; Srivastava, H.M.; Mohiuddine, S.A. A Korovkin's type approximation theorem for periodic functions via the statistical summability of the generalized de la Vallée Poussin mean. *Appl. Math. Comput.* **2014**, *228*, 62–169. [CrossRef]
23. Srivastava, H.M.; Zeng, X.-M. Approximation by means of the Szász-Bézier integral operators. *Int. J. Pure Appl. Math.* **2004**, *14*, 283–294.
24. Ditzian, Z.; Totik, V. *Moduli of Smoothness*; Springer: New York, NY, USA, 1987.
25. DeVore, R.A.; Lorentz, G.G. *Constructive Approximation*; Springer: Berlin, Germany, 1993.
26. Peetre, J. *A Theory of Interpolation of Normed Spaces*; Notas Mat.: Rio de Janeiro, Brazil, 1963.
27. Ozarslan, M.A.; Aktuğlu, H. Local approximation for certain King type operators. *Filomat* **2013**, *27*, 173–181. [CrossRef]
28. Büyükyazıcı, İ.; İbikli, E. The properties of generalized Bernstein polynomials of two variables. *Appl. Math. Comput.* **2004**, *156*, 367–380. [CrossRef]
29. Martinez, F.L. Some properties of two-demansional Bernstein polynomials. *J. Approx. Theory* **1989**, *59*, 300–306. [CrossRef]
30. Volkov, V.J. On the convergence of linear positive operators in the space of continuous functions of two variables. *Doklakad Nauk SSSR* **1957**, *115*, 17–19.

9

An Investigation of the Third Hankel Determinant Problem for Certain Subfamilies of Univalent Functions Involving the Exponential Function

Lei Shi [1], Hari Mohan Srivastava [2,3], Muhammad Arif [4,*], Shehzad Hussain [4] and Hassan Khan [4]

1 School of Mathematics and Statistics, Anyang Normal University, Anyan 455002, Henan, China; shimath@163.com
2 Department of Mathematics and Statistics, University of Victoria, Victoria, BC V8W 3R4, Canada; harimsri@math.uvic.ca
3 Department of Medical Research, China Medical University Hospital, China Medical University, Taichung 40402, Taiwan
4 Department of Mathematics, Abdul Wali Khan University Mardan, Mardan 23200, Pakistan; shehzad873822@gmail.com (S.H.); hassanmath@awkum.edu.pk (H.K.)
* Correspondence: marifmaths@awkum.edu.pk

Abstract: In the current article, we consider certain subfamilies $\mathcal{S}_e^*$ and $\mathcal{C}_e$ of univalent functions associated with exponential functions which are symmetric along real axis in the region of open unit disk. For these classes our aim is to find the bounds of Hankel determinant of order three. Further, the estimate of third Hankel determinant for the family $\mathcal{S}_e^*$ in this work improve the bounds which was investigated recently. Moreover, the same bounds have been investigated for 2-fold symmetric and 3-fold symmetric functions.

Keywords: subordinations; exponential function; Hankel determinant

---

1. Introduction and Definitions

Let the collection of functions f that are holomorphic in $\Delta = \{z \in \mathbb{C} : |z| < 1\}$ and normalized by conditions $f(0) = f'(0) - 1 = 0$ be denoted by the symbol $\mathcal{A}$. Equivalently; if $f \in \mathcal{A}$, then the Taylor-Maclaurin series representation has the form:

$$f(z) = z + \sum_{k=2}^{\infty} a_k z^k \quad (z \in \Delta). \tag{1}$$

Further, let we name by the notation $\mathcal{S}$ the most basic sub-collection of the set $\mathcal{A}$ that are univalent in Δ. The familiar coefficient conjecture for the function $f \in \mathcal{S}$ of the form (1) was first presented by Bieberbach [1] in 1916 and proved by de-Branges [2] in 1985. In 1916-1985, many mathematicians struggled to prove or disprove this conjecture and as result they defined several subfamilies of the set $\mathcal{S}$ of univalent functions connected with different image domains. Now we mention some of them, that is; let the notations $\mathcal{S}^*$, $\mathcal{C}$ and $\mathcal{K}$, shows the families of starlike, convex and close-to-convex functions respectively and are defined as:

$$\mathcal{S}^* = \left\{f \in \mathcal{S} : \frac{zf'(z)}{f(z)} \prec \frac{1+z}{1-z}, \ (z \in \Delta)\right\},$$
$$\mathcal{C} = \left\{f \in \mathcal{S} : \frac{(zf'(z))'}{f'(z)} \prec \frac{1+z}{1-z}, \ (z \in \Delta)\right\},$$
$$\mathcal{K} = \left\{f \in \mathcal{S} : \frac{f'(z)}{g'(z)} \prec \frac{1+z}{1-z}, \text{ for } g(z) \in \mathcal{C}, \ (z \in \Delta)\right\},$$

where the symbol " $\prec$ " denotes the familiar subordinations between analytic functions and is define as; the function h_1 is subordinate to a function h_2, symbolically written as $h_1 \prec h_2$ or $h_1(z) \prec h_2(z)$, if we can find a function w, which is holomorphic in Δ with $w(0) = 0$ & $|w(z)| < 1$ such that $h_1(z) = h_2(w(z))$ $(z \in \Delta)$. Thus, $h_1(z) \prec h_2(z)$ implies $h_1(\Delta) \subset h_2(\Delta)$. In case of univalency of h_1 in Δ, then the following relation holds:

$$h_1(z) \prec h_2(z) \quad (z \in \Delta) \quad \Longleftrightarrow \quad h_1(0) = h_2(0) \quad \text{and} \quad h_1(\Delta) \subset h_2(\Delta).$$

In [3], Padmanabhan and Parvatham in 1985 defined a unified families of starlike and convex functions using familiar convolution with the function $z/(1-z)^a$, for all $a \in \mathbb{R}$. Later on, Shanmugam [4] generalized the idea of paper [3] and introduced the set

$$\mathcal{S}_h^*(\phi) = \left\{f \in \mathcal{A} : \frac{z(f*h)'}{(f*h)} \prec \phi(z), \ (z \in \Delta)\right\},$$

where "$*$" stands for the familiar convolution, ϕ is a convex and h is a fixed function in $\mathcal{A}$. We obtain the families $\mathcal{S}^*(\phi)$ and $\mathcal{C}(\phi)$ when taking $z/(1-z)$ and $z/(1-z)^2$ instead of h in $\mathcal{S}_h^*(\phi)$ respectively. In 1992, Ma and Minda [5] reduced the restriction to a weaker supposition that ϕ is a function, with $Re\phi > 0$ in Δ, whose image domain is symmetric about the real axis and starlike with respect to $\phi(0) = 1$ with $\phi'(0) > 0$ and discussed some properties. The set $\mathcal{S}^*(\phi)$ generalizes various subfamilies of the set $\mathcal{A}$, for example:

1. If $\phi(z) = \frac{1+Az}{1+Bz}$ with $-1 \le B < A \le 1$, then $\mathcal{S}^*[A,B] := \mathcal{S}^*\left(\frac{1+Az}{1+Bz}\right)$ is the set of Janowski starlike functions, see [6]. Further, if $A = 1-2\alpha$ and $B = -1$ with $0 \le \alpha < 1$, then we get the set $\mathcal{S}^*(\alpha)$ of starlike functions of order α.
2. The class $\mathcal{S}_L^* := \mathcal{S}^*(\sqrt{1+z})$ was introduced by Sokól and Stankiewicz [7], consisting of functions $f \in \mathcal{A}$ such that $zf'(z)/f(z)$ lies in the region bounded by the right-half of the lemniscate of Bernoulli given by $|w^2 - 1| < 1$.
3. For $\phi(z) = 1 + \sin z$, the class $\mathcal{S}^*(\phi)$ lead to the class $\mathcal{S}_{\sin}^*$, introduced in [8].
4. The family $\mathcal{S}_e^* := \mathcal{S}^*(e^z)$ was introduced by Mediratta et al. [9] given as:

$$\mathcal{S}_e^* = \left\{f \in \mathcal{S} : \frac{zf'(z)}{f(z)} \prec e^z, \ (z \in \Delta)\right\}, \tag{2}$$

or, equivalently

$$\mathcal{S}_e^* = \left\{f \in \mathcal{S} : \left|\log \frac{zf'(z)}{f(z)}\right| < 1, \ (z \in \Delta)\right\}. \tag{3}$$

They investigated some interesting properties and also links these classes to the familiar subfamilies of the set $\mathcal{S}$. In [9], the authors choose the function $f(z) = z + \frac{1}{4}z^2$ (Figure 1) and then sketch the following figure of the function class $\mathcal{S}_e^*$ by using the form (3) as:

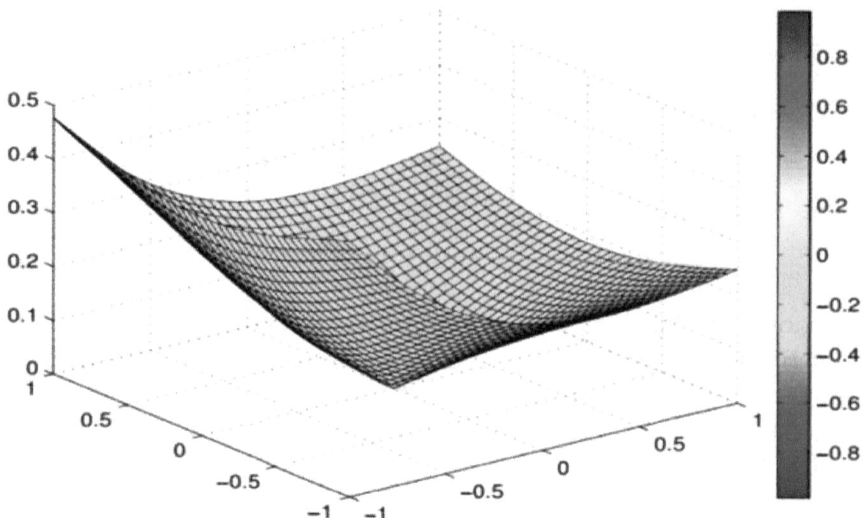

Figure 1. The figure of the function class $\mathcal{S}_1^*$ for $f(z) = z + \frac{1}{4}z^2$.

Similarly, by using Alexandar type relation in [9], we have;

$$\mathcal{C}_e = \left\{ f \in \mathcal{S} : \frac{(zf'(z))'}{f'(z)} \prec e^z, \ (z \in \Delta) \right\}. \tag{4}$$

From the above discussion, we conclude that the families $\mathcal{S}_e^*$ and $\mathcal{C}_e$ considered in this paper are symmetric about the real axis.

For given parameters $q, n \in \mathbb{N} = \{1, 2, \ldots\}$, the Hankel determinant $H_{q,n}(f)$ was defined by Pommerenke [10,11] for a function $f \in \mathcal{S}$ of the form (1) as follows:

$$H_{q,n}(f) = \begin{vmatrix} a_n & a_{n+1} & \ldots & a_{n+q-1} \\ a_{n+1} & a_{n+2} & \ldots & a_{n+q} \\ \vdots & \vdots & \ldots & \vdots \\ a_{n+q-1} & a_{n+q} & \ldots & a_{n+2q-2} \end{vmatrix}. \tag{5}$$

The concept of Hankel determinant is very useful in the theory of singularities [12] and in the study of power series with integral coefficients. For deep insight, the reader is invited to read [13–15]. Specifically, the absolute sharp bound of the functional $H_{2,2}(f) = a_2a_4 - a_3^2$ for each of the sets $\mathcal{S}^*$ and $\mathcal{C}$ were proved by Janteng et al. [16,17] while the exact estimate of this determinant for the family of close-to-convex functions is still unknown (see, [18]). On the other side for the set of Bazilevič functions, the sharp estimate of $|H_{2,2}(f)|$ was given by Krishna et al. [19]. Recently, Srivastava and his coauthors [20] found the estimate of second Hankel determinant for bi-univalent functions involving symmetric q-derivative operator while in [21], the authors discussed Hankel and Toeplitz determinants for subfamilies of q-starlike functions connected with a general form of conic domain. For more literature see [22–29]. The determinant with entries from (1)

$$H_{3,1}(f) = \begin{vmatrix} 1 & a_2 & a_3 \\ a_2 & a_3 & a_4 \\ a_3 & a_4 & a_5 \end{vmatrix}$$

is known as Hankel determinant of order three and the estimation of this determinant $|H_{3,1}(f)|$ is very hard as compared to derive the bound of $|H_{2,2}(f)|$. The very first paper on $H_{3,1}(f)$ visible in 2010 by

Babalola [30] in which he got the upper bound of $H_{3,1}(f)$ for the families of $\mathcal{S}^*$ and $\mathcal{C}$. Later on, many authors published their work regarding $|H_{3,1}(f)|$ for different sub-collections of univalent functions, see [8,31–36]. In 2017, Zaprawa [37] upgraded the results of Babalola [30] by giving

$$|H_{3,1}(f)| \leq \begin{cases} 1, & \text{for} \quad f \in \mathcal{S}^*, \\ \frac{49}{540}, & \text{for} \quad f \in \mathcal{C}, \end{cases}$$

and claimed that these bounds are still not best possible. Further for the sharpness, he examined the subfamilies of $\mathcal{S}^*$ and $\mathcal{C}$ consisting of functions with m-fold symmetry and obtained the sharp bounds. Moreover this determinant was further improved by Kwon et al. [38] and proved $|H_{3,1}(f)| \leq 8/9$ for $f \in \mathcal{S}^*$, yet not best possible. The authors in [39–41] contributed in similar direction by generalizing different classes of univalent functions with respect to symmetric points. In 2018, Kowalczyk et al. [42] and Lecko et al. [43] got the sharp inequalities

$$|H_{3,1}(f)| \leq 4/135, \quad \text{and} \quad |H_{3,1}(f)| \leq 1/9,$$

for the recognizable sets $\mathcal{K}$ and $\mathcal{S}^*(1/2)$ respectively, where the symbol $\mathcal{S}^*(1/2)$ indicates to the family of starlike functions of order $1/2$. Also we would like to cite the work done by Mahmood et al. [44] in which they studied third Hankel determinant for a subset of starlike functions in q-analogue. Additionally Zhang et al. [45] studied this determinant for the set $\mathcal{S}_e^*$ and obtained the bound $|H_{3,1}(f)| \leq 0.565$.

In the present article, our aim is to investigate the estimate of $|H_{3,1}(f)|$ for both the above defined classes $\mathcal{S}_e^*$ and $\mathcal{C}_e$. Moreover, we also study this problem for m-fold symmetric starlike and convex functions associated with exponential function.

2. A Set of Lemmas

Let $\mathcal{P}$ denote the family of all functions p that are analytic in $\mathbb{D}$ with $\Re(p(z)) > 0$ and has the following series representation

$$p(z) = 1 + \sum_{n=1}^{\infty} c_n z^n \; (z \in \Delta). \tag{6}$$

Lemma 1. *If $p \in \mathcal{P}$ and has the form , then*

$$\begin{aligned} |c_n| &\leq 2 \text{ for } n \geq 1, &(7)\\ |c_{n+k} - \mu c_n c_k| &< 2, \text{ for } 0 \leq \mu \leq 1, &(8)\\ |c_m c_n - c_k c_l| &\leq 4 \text{ for } m+n = k+l, &(9)\\ \left|c_{n+2k} - \mu c_n c_k^2\right| &\leq 2(1+2\mu); \text{ for } \mu \in \mathbb{R}, &(10)\\ \left|c_2 - \frac{c_1^2}{2}\right| &\leq 2 - \frac{|c_1|^2}{2}, &(11) \end{aligned}$$

and for complex number λ, we have

$$\left|c_2 - \lambda c_1^2\right| \leq 2\max\{1, |2\lambda - 1|\}. \tag{12}$$

For the inequalities (7), (11), (8), (10), (9) see [46] and (12) is given in [47].

3. Improved Bound of $|H_{3,1}(f)|$ for the Set $\mathcal{S}_e^*$

Theorem 1. *If f belongs to $\mathcal{S}_e^*$, then*

$$|H_{3,1}(f)| \leq 0.50047781.$$

Proof. Let $f \in \mathcal{S}_e^*$. Then we can write (2), in terms of Schwarz function as

$$\frac{zf'(z)}{f(z)} = e^{w(z)}.$$

If $h \in \mathcal{P}$, then it can be written in form of Schwarz function as

$$h(z) = \frac{1+w(z)}{1-w(z)} = 1 + c_1 z + c_2 z^2 + \cdots.$$

From above, we can get

$$w(z) = \frac{h(z)-1}{h(z)+1} = \frac{c_1 z + c_2 z^2 + c_3 z^3 + \cdots}{2 + c_1 z + c_2 z^2 + c_3 z^3 + \cdots}.$$

$$\frac{zf'(z)}{f(z)} = 1 + a_2 z + \left(2a_3 - a_2^2\right) z^2 + \left(3a_4 - 3a_2 a_3 + a_2^3\right) z^3 + \left(4a_5 - 2a_3^2 - 4a_2 a_4 + 4a_2^2 a_3 - a_2^4\right) z^4 = 1 + p_1 z + p_2 z^2 + \cdots. \quad (13)$$

and from the series expansion of w along with some calculations, we have

$$e^{w(z)} = 1 + w(z) + \frac{(w(z))^2}{2!} + \frac{(w(z))^3}{3!} + \frac{(w(z))^4}{4!} + \frac{(w(z))^5}{5!} + \cdots.$$

After some computations and rearranging, it yields

$$e^{w(z)} = 1 + \frac{1}{2}c_1 z + \left(\frac{c_2}{2} - \frac{c_1^2}{8}\right) z^2 + \left(\frac{c_1^3}{48} + \frac{c_3}{2} - \frac{c_1 c_2}{4}\right) z^3 + \left(\frac{1}{384}c_1^4 + \frac{1}{2}c_4 - \frac{1}{8}c_2^2 + \frac{1}{16}c_1^2 c_2 - \frac{1}{4}c_1 c_3\right) z^4 + \cdots. \quad (14)$$

Comparing (13) and (14), we have

$$a_2 = \frac{c_1}{2}, \quad (15)$$

$$a_3 = \frac{1}{4}\left(c_2 + \frac{c_1^2}{4}\right), \quad (16)$$

$$a_4 = \frac{1}{6}\left(c_3 + \frac{c_1 c_2}{4} - \frac{c_1^3}{48}\right), \quad (17)$$

$$a_5 = \frac{1}{4}\left(\frac{c_1^4}{288} + \frac{c_4}{2} + \frac{c_1 c_3}{12} - \frac{c_1^2 c_2}{24}\right). \quad (18)$$

From (5), the Third Hankel determinant can be written as

$$H_{3,1}(f) = -a_2^2 a_5 + 2a_2 a_3 a_4 - a_3^3 + a_3 a_5 - a_4^2.$$

Using (15), (16), (17) and (18), we get

$$H_{3,1}(f) = \frac{35}{27648}c_1^4 c_2 + \frac{53}{6912}c_1^3 c_3 + \frac{c_2 c_4}{32} + \frac{19}{576}c_1 c_2 c_3 - \frac{211}{331776}c_1^6 - \frac{c_2^3}{64} - \frac{3}{128}c_1^2 c_4 - \frac{13}{2304}c_1^2 c_2^2 - \frac{c_3^2}{36}.$$

After rearranging, it yields

$$\begin{aligned} H_{3,1}(f) &= \frac{211}{165888}c_1^4\left(c_2-\frac{c_1^2}{2}\right)+\frac{3}{64}c_4\left(c_2-\frac{c_1^2}{2}\right)-\frac{c_1c_3}{96}\left(c_2-\frac{c_1^2}{2}\right)+\frac{1}{165888}c_1^3(c_3-c_1c_2) \\ &+\frac{407}{165888}c_1^2\left(c_1c_3-c_2^2\right)-\frac{c_3}{36}(c_3-c_1c_2)-\frac{c_2}{64}(c_4-c_1c_3)-\frac{529}{165888}c_1^2c_2^2-\frac{c_2^3}{64}. \end{aligned}$$

Using triangle inequality along with (7), (11), (8) and (9), provide us

$$\begin{aligned} |H_{3,1}(f)| &\leq \frac{211}{165888}|c_1|^4\left(2-\frac{|c_1|^2}{2}\right)+\frac{3}{32}\left(2-\frac{|c_1|^2}{2}\right)+\frac{|c_1|}{48}\left(2-\frac{|c_1|^2}{2}\right)+\frac{1}{82944}|c_1|^3 \\ &+\frac{407}{41472}|c_1|^2+\frac{1}{9}+\frac{1}{16}+\frac{529}{41472}|c_1|^2+\frac{1}{8}. \end{aligned}$$

If we substitute $|c_1|=x\in[0,2]$, we obtain a function of variable x. Therefore, we can write

$$\begin{aligned} |H_{3,1}(f)| &\leq \frac{211}{165888}x^4\left(2-\frac{x^2}{2}\right)+\frac{3}{32}\left(2-\frac{x^2}{2}\right)+\frac{x}{48}\left(2-\frac{x^2}{2}\right)+\frac{1}{82944}x^3 \\ &+\frac{407}{41472}x^2+\frac{1}{9}+\frac{1}{16}+\frac{529}{41472}x^2+\frac{1}{8}. \end{aligned}$$

The above function attains its maximum value at $x=0.64036035$, which is

$$|H_{3,1}(f)|\leq 0.50047781.$$

Thus, the proof is completed. □

4. Bound of $|H_{3,1}(f)|$ for the Set $\mathcal{C}_e$

Theorem 2. *Let f has the form* (1) *and belongs to $\mathcal{C}_e$. Then*

$$|a_2| \leq \frac{1}{2}, \tag{19}$$

$$|a_3| \leq \frac{1}{4}, \tag{20}$$

$$|a_4| \leq \frac{17}{144}, \tag{21}$$

$$|a_5| \leq \frac{7}{96}. \tag{22}$$

The first three inequalities are sharp.

Proof. If $f\in\mathcal{C}_e$, then we can write (4), in form of Schwarz function as

$$1+\frac{zf''(z)}{f'(z)}=e^{w(z)}.$$

From (1), we can write

$$\begin{aligned} 1+\frac{zf''(z)}{f'(z)} &= 1+2a_2z+\left(6a_3-4a_2^2\right)z^2+\left(12a_4-18a_2a_3+8a_2^3\right)z^3 \\ &+\left(20a_5-18a_3^2-32a_2a_4+48a_2^2a_3-16a_2^4\right)z^4+\cdots. \end{aligned} \tag{23}$$

By comparing (23) and (14), we get

$$a_2 = \frac{c_1}{4}, \tag{24}$$

$$a_3 = \frac{1}{12}\left(c_2 + \frac{c_1^2}{4}\right), \tag{25}$$

$$a_4 = \frac{1}{24}\left(\frac{c_1 c_2}{4} + c_3 - \frac{c_1^3}{48}\right), \tag{26}$$

$$a_5 = \frac{1}{20}\left(\frac{c_1^4}{288} + \frac{c_4}{2} + \frac{c_1 c_3}{12} - \frac{c_1^2 c_2}{24}\right). \tag{27}$$

Implementing (7), in (24) and (25), we have

$$|a_2| \leq \tfrac{1}{2} \quad \text{and} \quad |a_3| \leq \tfrac{1}{4}.$$

Reshuffling (26), we have

$$|a_4| = \frac{1}{24}\left|\frac{5}{24}c_1 c_2 + \frac{c_1}{24}\left(c_2 - \frac{c_1^2}{2}\right) + c_3\right|.$$

Application of triangle inequality and (7) and (11) leads us to

$$|a_4| \leq \frac{1}{24}\left\{\frac{5}{12}|c_1| + \frac{|c_1|}{24}\left(2 - \frac{|c_1|^2}{2}\right) + 2\right\}.$$

If we insert $|c_1| = x \in [0, 2]$, then we get

$$|a_4| \leq \frac{1}{24}\left\{\frac{5}{12}x + \frac{x}{24}\left(2 - \frac{x^2}{2}\right) + 2\right\}.$$

The overhead function has a maximum value at $x = 2$, thus

$$|a_4| \leq \frac{17}{144}.$$

Reordering (27), we have

$$|a_5| = \frac{1}{20}\left|\frac{1}{2}\left(c_4 - \frac{c_1^2 c_2}{48}\right) - \frac{c_1^2}{96}\left(c_2 - \frac{c_1^2}{3}\right) + \frac{c_1}{12}\left(c_3 - \frac{c_1 c_2}{4}\right)\right|.$$

By using triangle inequality along with (7), and (8), we get

$$|a_5| \leq \frac{7}{96}.$$

Equalities are obtain if we take

$$f(z) = \int_0^z e^{J(t)} dt = z + \frac{1}{2}z^2 + \frac{1}{4}z^3 + \frac{17}{144}z^4 + \frac{19}{360}z^5 + \cdots \tag{28}$$

where

$$J(t) = \int_0^t \frac{e^x - 1}{x} dx.$$

□

Theorem 3. *If f is of the form (1) belongs to $\mathcal{C}_e$, then*

$$\left|a_3 - \gamma a_2^2\right| \leq \frac{1}{6} \max\left\{1, \frac{3}{2}\left|\gamma - 1\right|\right\}, \tag{29}$$

where γ is a complex number.

Proof. From (24) and (25), we get

$$\left|a_3 - \gamma a_2^2\right| = \left|\frac{c_2}{12} + \frac{c_1^2}{48} - \frac{\gamma}{16}c_1^2\right|.$$

By reshuffling it, provides

$$\left|a_3 - \gamma a_2^2\right| = \frac{1}{12}\left|\left(c_2 - \frac{1}{2}\left(\frac{3\gamma - 1}{2}\right)c_1^2\right)\right|.$$

Application of (12), leads us to

$$\left|a_3 - \gamma a_2^2\right| \leq \max\left\{\frac{1}{6}, \frac{1}{12}\left|3\gamma - 3\right|\right\}.$$

□

Substituting $\gamma = 1$, we obtain the following inequality.

Corollary 1. *If $f \in \mathcal{C}_e$ and has the series represntaion (1), then*

$$\left|a_3 - a_2^2\right| \leq \frac{1}{6}. \tag{30}$$

Theorem 4. *If f has the form (1) belongs to $\mathcal{C}_e$, then*

$$\left|a_2a_3 - a_4\right| \leq \frac{31}{288}. \tag{31}$$

Proof. Using (24), (25) and (26), we have

$$\left|a_2a_3 - a_4\right| = \left|\frac{c_1c_2}{96} + \frac{7}{1152}c_1^3 - \frac{c_3}{24}\right|.$$

By rearranging it, gives

$$\left|a_2a_3 - a_4\right| = \left|-\frac{1}{48}\left(c_3 - \frac{c_1c_2}{2}\right) - \frac{1}{48}\left(c_3 - \frac{7}{24}c_1^3\right)\right|.$$

By applying triangle inequality plus (8) and (10), we get

$$\left|a_2a_3 - a_4\right| \leq \left\{\frac{1}{24} + \frac{19}{288}\right\} = \frac{31}{288}.$$

□

Theorem 5. *Let $f \in \mathcal{C}_e$ be of the form (1). Then*

$$\left|a_2a_4 - a_3^2\right| \leq \frac{3}{64}. \tag{32}$$

Proof. From (24), (25) and (26), we have

$$\left|a_2a_4 - a_3^2\right| = \left|\frac{c_1c_3}{96} - \frac{c_1^4}{1536} - \frac{c_1^2c_2}{1152} - \frac{c_2^2}{144}\right|.$$

By reordering it, yields

$$\left|a_2a_4 - a_3^2\right| = \left|\frac{c_1}{576}\left(c_3 - \frac{c_1c_2}{2}\right) + \frac{c_1}{576}\left(c_3 - \frac{3}{8}c_1^3\right) + \frac{1}{144}\left(c_1c_3 - c_2^2\right)\right|.$$

Application of triangle inequality plus (7), (11), (10) and (9), we obtain

$$\left|a_2a_4 - a_3^2\right| \leq \frac{4}{576} + \frac{7}{576} + \frac{4}{144} = \frac{3}{64}.$$

□

Theorem 6. *If $f \in \mathcal{C}_e$ and has the form (1), then*

$$|H_{3,1}(f)| \leq 0.0234598.$$

Proof. Using (5), the Hankel determinant of order three can be formed as;

$$H_{3,1}(f) = -a_2^2a_5 + 2a_2a_3a_4 - a_3^3 + a_3a_5 - a_4^2.$$

Using (24), (25), (26) and (27), gives us

$$H_{3,1}(f) = \frac{7}{5760}c_1c_2c_3 - \frac{c_3^2}{576} - \frac{c_2^3}{1728} - \frac{173}{6635520}c_1^6 + \frac{23}{276480}c_1^4c_2 + \frac{c_2c_4}{480} - \frac{13}{46980}c_1^2c_2^2 - \frac{c_1^2c_4}{960} + \frac{23}{69120}c_1^3c_3.$$

Now, rearranging it provides

$$\begin{aligned} H_{3,1}(f) = {} & \frac{173}{3317760}c_1^4\left(c_2 - \frac{c_1^2}{2}\right) - \frac{103}{1658880}c_1^2c_2\left(c_2 - \frac{c_1^2}{2}\right) + \frac{c_4}{480}\left(c_2 - \frac{c_1^2}{2}\right) \\ & + \frac{11}{17280}c_1c_2\left(c_3 - \frac{365}{1056}c_1c_2\right) + \frac{c_2}{1728}\left(c_1c_3 - c_2^2\right) - \frac{c_3}{576}\left(c_3 - \frac{23}{120}c_1^3\right). \end{aligned}$$

Application of triangle inequality plus (7), (11), (8), (10) and (9), leads us to

$$|H_{3,1}(f)| \leq \frac{173}{3317760}|c_1|^4\left(2 - \frac{|c_1|^2}{2}\right) + \frac{103}{829440}|c_1|^2\left(2 - \frac{|c_1|^2}{2}\right) + \frac{1}{4320}|c_1| + \frac{83}{8640} + \frac{1}{216}.$$

Now, replacing $|c_1| = x \in [0,2]$, then, we can write

$$|H_{3,1}(f)| \leq \frac{173}{3317760}x^4\left(2 - \frac{x^2}{2}\right) + \frac{103}{829440}x^2\left(2 - \frac{x^2}{2}\right) + \frac{1}{240}\left(2 - \frac{x^2}{2}\right) + \frac{11}{4320}x + \frac{41}{2880}.$$

The above function gets its maximum at $x = 0.7024858$, Therefore, we have

$$|H_{3,1}(f)| \leq 0.02345979.$$

Thus the proof is completed. □

5. Bounds of $|H_{3,1}(f)|$ for 2-Fold and 3-Fold Functions

Let $m \in \mathbb{N} = \{1, 2, \ldots\}$. If a rotation $\triangle$ about the origin through an angle $2\pi/m$ carries $\triangle$ on itself, then such a domain $\triangle$ is called m-fold symmetric. An analytic function f is m-fold symmetric in Δ, if

$$f\left(e^{2\pi i/m}z\right) = e^{2\pi i/m} f(z), \ (z \in \Delta).$$

By $\mathcal{S}^{(m)}$, we define the set of m-fold univalent functions having the following Taylor series form

$$f(z) = z + \sum_{k=1}^{\infty} a_{mk+1} z^{mk+1}, \ \ (z \in \Delta). \tag{33}$$

The sub-families $\mathcal{S}_e^{*(m)}$ and $\mathcal{C}_e^{(m)}$ of $\mathcal{S}^{(m)}$ are the sets of m-fold symmetric starlike and convex functions respectively associated with exponential functions. More intuitively, an analytic function f of the form (33), belongs to the families $\mathcal{S}_e^{*(m)}$ and $\mathcal{C}_e^{(m)}$, if and only if

$$\frac{zf'(z)}{f(z)} = \exp\left(\frac{p(z)-1}{p(z)+1}\right), \ p \in \mathcal{P}^{(m)}, \tag{34}$$

$$1 + \frac{zf''(z)}{f'(z)} = \exp\left(\frac{p(z)-1}{p(z)+1}\right), \ p \in \mathcal{P}^{(m)}. \tag{35}$$

where the set $\mathcal{P}^{(m)}$ is defined by

$$\mathcal{P}^{(m)} = \left\{ p \in \mathcal{P} : p(z) = 1 + \sum_{k=1}^{\infty} c_{mk} z^{mk}, \ \ (z \in \Delta) \right\}. \tag{36}$$

Here we prove some theorems related to 2-fold and 3-fold symmetric functions.

Theorem 7. *If $f \in \mathcal{S}_e^{*(2)}$ and has the form (33), then*

$$|H_{3,1}(f)| \le \frac{1}{8}.$$

Proof. Let $f \in \mathcal{S}_e^{*(2)}$. Then, there exists a function $p \in \mathcal{P}^{(2)}$, such that

$$\frac{zf'(z)}{f(z)} = \exp\left(\frac{p(z)-1}{p(z)+1}\right).$$

Using the series form (33) and (36), when $m = 2$ in the above relation, we can get

$$a_3 = \frac{c_2}{4}, \tag{37}$$

$$a_5 = \frac{c_4}{8}. \tag{38}$$

Now,

$$H_3(f) = a_3 a_5 - a_3^3.$$

Utilizing (37) and (38), we get

$$H_{3,1}(f) = -\frac{c_2^3}{64} + \frac{c_2 c_4}{32}.$$

By rearranging, it yields

$$H_{3,1}(f) = \frac{c_2}{32}\left(c_4 - \frac{c_2^2}{2}\right).$$

Using triangle inequality long with (8) and (7), gives us

$$|H_{3,1}(f)| \leq \frac{1}{8}.$$

Hence, the proof is done. □

Theorem 8. *If $f \in \mathcal{S}_e^{*(3)}$ and has the series form (33), then*

$$|H_{3,1}(f)| \leq \frac{1}{9}.$$

This result is sharp for the function

$$f(z) = \exp\left(\int_0^z \frac{e^{x^3}}{x} dx\right) = z + \frac{1}{3}z^4 + \frac{5}{36}z^7 + \cdots \tag{39}$$

Proof. As, $f \in \mathcal{S}_e^{*(3)}$, therefore there exists a function $p \in \mathcal{P}^{(3)}$, such that

$$\frac{zf'(z)}{f(z)} = \exp\left(\frac{p(z)-1}{p(z)+1}\right).$$

Utilizing the series form (33) and (36), when $m = 3$ in the above relation, we can obtain

$$a_4 = \frac{c_3}{6}.$$

Then,

$$H_{3,1}(f) = -a_4^2 = -\frac{c_3^2}{36}.$$

Utilizing (7) and triangle inequality, we have

$$|H_{3,1}(f)| \leq \frac{1}{9}.$$

Thus the proof is ended. □

Theorem 9. *Let $f \in \mathcal{C}_e^{(2)}$ and has the form given in (33). Then*

$$|H_{3,1}(f)| \leq \frac{1}{120}.$$

Proof. As, $f \in \mathcal{C}_e^{(2)}$, then there exists a function $p \in \mathcal{P}^{(2)}$, such that

$$1 + \frac{zf''(z)}{f'(z)} = \exp\left(\frac{p(z)-1}{p(z)+1}\right).$$

Utilizing the series form (33) and (36), when $m = 2$ in the above relation, we can obtain

$$a_3 = \frac{c_2}{12}, \tag{40}$$

$$a_5 = \frac{c_4}{40}. \tag{41}$$

$$H_{3,1}(f) = a_3 a_5 - a_3^3.$$

Using (40) and (41), we have

$$H_{3,1}(f) = -\frac{c_2^3}{1728} + \frac{c_2 c_4}{480}.$$

Now, reordering the above equation, we obtain

$$H_3(f) = \frac{c_2}{480}\left(c_4 - \frac{5}{18}c_2^2\right).$$

Application of (7), (8) and triangle inequality, leads us to

$$|H_{3,1}(f)| \leq \frac{1}{120}.$$

Thus, the required result is completed. □

Theorem 10. *If $f \in \mathcal{C}_e^{(3)}$ and has the form given in* (33)*, then*

$$|H_{3,1}(f)| \leq \frac{1}{144}. \quad (42)$$

This result is sharp for the function

$$f(z) = \int_0^z e^{I(t)}dt = z + \frac{1}{12}z^4 + \frac{5}{252}z^7 + \cdots \quad (43)$$

where

$$I(t) = \int_0^t \frac{e^{x^3} - 1}{x}dx.$$

Proof. Let, $f \in \mathcal{C}_e^{(3)}$. Then there exists a function $p \in \mathcal{P}^{(3)}$, such that

$$1 + \frac{zf''(z)}{f'(z)} = \exp\left(\frac{p(z)-1}{p(z)+1}\right).$$

Utilizing the series form (33) and (36), when $m = 3$ in the above relation, we can obtain

$$a_4 = \frac{c_3}{24}.$$

Then,

$$H_{3,1}(f) = -\frac{c_3^2}{576}.$$

Implementing (7) and triangle inequality, we have

$$|H_{3,1}(f)| \leq \frac{1}{144}.$$

Hence, the proof is done. □

6. Conclusions

In this article, we studied Hankel determinant $H_{3,1}(f)$ for the families $\mathcal{S}_e^*$ and $\mathcal{C}_e$ whose image domain are symmetric about the real axis. Furthermore, we improve the bound of third Hankel determinant for the family $\mathcal{S}_e^*$. These bounds are also discussed for 2-fold symmetric and 3-fold symmetric functions.

Author Contributions: Conceptualization, L.S. and H.K.; Methodology, M.A.; Software, H.M.S.; Validation, H.K. and M.A.; Formal Analysis, L.S.; Investigation, M.A. and H.M.S; Resources, H.K. and S.H.; Data Curation, S.H.; Writing—Original Draft Preparation, S.H.; Writing—Review and Editing, H.K., M.A. and L.S.; Visualization, M.A.;Supervision, M.A., L.S.; Project Administration, L.S.; Funding Acquisition, L.S.

References

1. Bieberbach, L. *Über dié koeffizienten derjenigen Potenzreihen, welche eine Schlichte Abbildung des Einheitskreises vermitteln*; Reimer in Komm: Berlin, Germany, 1916.
2. De-Branges, L. A proof of the Bieberbach conjecture. *Acta. Math.* **1985**, *154*, 137–152.
3. Padmanabhan, K.S.; Parvatham, R. Some applications of differential subordination. *Bull. Aust. Math. Soc.***1985**, *32*, 321–330.
4. Shanmugam, T.N. Convolution and differential subordination. *Int. J. Math. Math. Sci.* **1989**, *12*, 333–340. [CrossRef]
5. Ma, W.; Minda, D. A unified treatment of some special classes of univalent functions. *Int. J. Math. Math. Sci.* **2011**. [CrossRef]
6. Janowski, W. Extremal problems for a family of functions with positive real part and for some related families. *Ann. Polonici Math.* **1971**, *23*, 159–177. [CrossRef]
7. Sokoł, J.; Stankiewicz, J. Radius of convexity of some subclasses of strongly starlike functions. *Zeszyty Naukowe/Oficyna Wydawnicza al. Powstańców Warszawy* **1996**, *19*, 101–105.
8. Cho, N.E.; Kumar, V.; Kumar, S.S.; Ravichandran, V. Radius problems for starlike functions associated with the sine function. *Bull. Iran. Math. Soc.* **2019**, *45*, 213–232. [CrossRef]
9. Mendiratta, R.; Nagpal, S.; Ravichandran, V. On a subclass of strongly starlike functions associated with exponential function. *Bull. Malays. Math. Sci. Soc.* **2015**, *38*, 365–386. [CrossRef]
10. Pommerenke, C. On the coefficients and Hankel determinants of univalent functions. *J. Lond. Math. Soc.* **1966**, *1*, 111–122. [CrossRef]
11. Pommerenke, C. On the Hankel determinants of univalent functions. *Mathematika* **1967**, *14*, 108–112. [CrossRef]
12. Dienes, P. *The Taylor Series: An Introduction to the Theory of Functions of a Complex Variable*; NewYork-Dover: Mineola, NY, USA, 1957.
13. Cantor, D. G. Power series with integral coefficients. *Bull. Am. Math. Soc.*. **1963**, *69*, 362–366. [CrossRef]
14. Edrei, A. Sur les determinants recurrents et less singularities d'une fonction donee por son developpement de Taylor. *Comput. Math.* **1940**, *7*, 20–88.
15. Polya, G.; Schoenberg, I.J. Remarks on de la Vallee Poussin means and convex conformal maps of the circle. *Pac. J. Math.* **1958**, *8*, 259–334. [CrossRef]
16. Janteng, A.; Halim, S.A.; Darus, M. Coefficient inequality for a function whose derivative has a positive real part. *J. Inequal. Pure Appl. Math.* **2006**, *7*, 1–5.
17. Janteng, A.; Halim, S.A.; Darus, M. Hankel determinant for starlike and convex functions. *Int. J. Math. Anal.* **2007**, *1*, 619–625.
18. Răducanu, D.; Zaprawa, P. Second Hankel determinant for close-to-convex functions. *C. R. Math.* **2017**, *355*, 1063–1071. [CrossRef]
19. Krishna, D.V.; RamReddy, T. Second Hankel determinant for the class of Bazilevic functions. *Stud. Univ. Babes-Bolyai Math.* **2015**, *60*, 413–420.
20. Srivastava, H.M.; Altınkaya, S.; Yalcın, S. Hankel determinant for a subclass of bi-univalent functions defined by using a symmetric q-derivative operator. *Filomath* **2018**, *32*, 503–516. [CrossRef]
21. Srivastava, H.M.; Ahmad, Q.Z.; Khan, N.; Khan, B. Hankel and Toeplitz determinants for a subclass of q-starlike functions associated with a general conic domain. *Mathematics* **2019**, *7*, 181. [CrossRef]
22. Çaglar, M.; Deniz, E.; Srivastava, H.M. Second Hankel determinant for certain subclasses of bi-univalent functions. *Turk. J. Math.* **2017**, *41*, 694–706. [CrossRef]
23. Bansal, D. Upper bound of second Hankel determinant for a new class of analytic functions. *Appl. Math. Lett.* **2013**, *26*, 103–107. [CrossRef]
24. Hayman, W.K. On second Hankel determinant of mean univalent functions. *Proc. Lond. Math. Soc.* **1968**, *3*, 77–94. [CrossRef]
25. Lee, S.K.; Ravichandran, V.; Supramaniam, S. Bounds for the second Hankel determinant of certain univalent functions. *J. Inequal. Appl.* **2013**. [CrossRef]

26. Altınkaya, Ş.; Yalçın, S. Upper bound of second Hankel determinant for bi-Bazilevic functions. *Mediterr. J. Math.* **2016**, *13*, 4081–4090. [CrossRef]
27. Liu, M.S.; Xu, J.F.; Yang,M. Upper bound of second Hankel determinant for certain subclasses of analytic functions. *Abstr. Appl. Anal.* **2014**. [CrossRef]
28. Noonan, J.W.; Thomas, D.K. On the second Hankel determinant of areally mean p-valent functions. *Trans. Am. Math. Soc.* **1976**, *223*, 337–346.
29. Orhan, H.; Magesh, N.; Yamini, J. Bounds for the second Hankel determinant of certain bi-univalent functions. *Turk. J. Math.* **2016**, *40*, 679–687. [CrossRef]
30. Babalola, K.O. On $H_3(1)$ Hankel determinant for some classes of univalent functions. *Inequal. Theory Appl.* **2010**, *6*, 1–7.
31. Arif, M.; Noor, K. I.; Raza, M. Hankel determinant problem of a subclass of analytic functions. *J. Inequal. Appl.* **2012**, *2012*, 2. [CrossRef]
32. Altınkaya, Ş.; Yalçın, S. Third Hankel determinant for Bazilevič functions. *Adv. Math.* **2016**, *5*, 91–96.
33. Bansal, D.; Maharana, S.; Prajapat, J.K. Third order Hankel Determinant for certain univalent functions. *J. Korean Math. Soc.* **2015**, *52*, 1139–1148. [CrossRef]
34. Krishna, D.V.; Venkateswarlu, B.; RamReddy, T. Third Hankel determinant for bounded turning functions of order alpha. *J. Niger. Math. Soc.* **2015**, *34*, 121–127. [CrossRef]
35. Raza, M.; Malik, S.N. Upper bound of third Hankel determinant for a class of analytic functions related with lemniscate of Bernoulli. *J. Inequal. Appl.* **2013**, *2013*, 412. [CrossRef]
36. Shanmugam, G.; Stephen, B.A.; Babalola, K.O. Third Hankel determinant for α-starlike functions. *Gulf J. Math.* **2014**, *2*, 107–113.
37. Zaprawa, P. Third Hankel determinants for subclasses of univalent functions. *Mediterr. J. Math.* **2017**, *14*, 10. [CrossRef]
38. Kwon, O.S.; Lecko, A.; Sim, Y.J. The bound of the Hankel determinant of the third kind for starlike functions. *Bull. Malays. Math. Sci. Soc.* **2019**, *42*, 767–780. [CrossRef]
39. Mahmood, S.; Khan, I.; Srivastava, H.M.; Malik, S.N. Inclusion relations for certain families of integral operators associated with conic regions. *J. Inequal. Appl.* **2019**, 59. [CrossRef]
40. Mahmood, S.; Srivastava, H.M.; Malik, S.N. Some subclasses of uniformly univalent functions with respect to symmetric points. *Symmetry* **2019**, *11*, 287. [CrossRef]
41. Mahmood, S.; Jabeen, M.; Malik, S.N.; Srivastava, H.M.; Manzoor, R.; Riaz, S.M. Some coefficient inequalities of q-starlike functions associated with conic domain defined by q-derivative. *J. Funct. Spaces* **2018**, *1*, 1–13. [CrossRef]
42. Kowalczyk, B.; Lecko, A.; Sim, Y.J. The sharp bound of the Hankel determinant of the third kind for convex functions. *Bull. Aust. Math. Soc.* **2018**, *97*, 435–445. [CrossRef]
43. Lecko, A.; Sim, Y.J.; Śmiarowska, B. The sharp bound of the Hankel determinant of the third kind for starlike functions of order 1/2. *Complex Anal. Oper. Theory* **2018**. [CrossRef]
44. Mahmood, S.; Srivastava, H.M.; Khan, N.; Ahmad, Q.Z.; Khan, B.; Ali, I. Upper bound of the third Hankel determinant for a subclass of q-starlike functions. *Symmetry* **2019**, *11*, 347. [CrossRef]
45. Zhang, H.-Y.; Tang, H.; Niu, X.-M. Third-order Hankel determinant for certain class of analytic functions related with exponential function. *Symmetry* **2018**, *10*, 501. [CrossRef]
46. Pommerenke, C.; Jensen, G. *Univalent Functions*; Vandenhoeck and Ruprecht: Gottingen, Germany, 1975.
47. Keough, F.; Merkes, E. A coefficient inequality for certain subclasses of analytic functions. *Proc. Am. Math. Soc.* **1969**, *20*, 8–12. [CrossRef]

10

The Principle of Differential Subordination and its Application to Analytic and p-Valent Functions Defined by a Generalized Fractional Differintegral Operator

Nak Eun Cho [1,*], Mohamed Kamal Aouf [2] and Rekha Srivastava [3]

1 Department of Applied Mathematics, Pukyong National University, Busan 48513, Korea
2 Department of Mathematics, Faculty of Science, Mansoura University, 35516 Mansoura, Egypt
3 Department of Mathematics and Statistics, University of Victoria, Victoria, BC V8W 3R4, Canada
* Correspondence: necho@pknu.ac.kr

Abstract: A useful family of fractional derivative and integral operators plays a crucial role on the study of mathematics and applied science. In this paper, we introduce an operator defined on the family of analytic functions in the open unit disk by using the generalized fractional derivative and integral operator with convolution. For this operator, we study the subordination-preserving properties and their dual problems. Differential sandwich-type results for this operator are also investigated.

Keywords: analytic function; Hadamard product; differential subordination; differential superordination; generalized fractional differintegral operator

MSC: 30C45; 30C50

---

1. Introduction

Let $\mathcal{H}(\mathbb{D})$ be the family of analytic functions in $\mathbb{D} = \{z \in \mathbb{C} : |z| < 1\}$ and $\mathcal{H}[c, n]$ be the subfamily of $\mathcal{H}(\mathbb{D})$ consisting of functions of the form:

$$f(z) = c + b_n z^n + b_{n+1} z^{n+1} + \cdots \quad (c \in \mathbb{C};\ n \in \mathbb{N} = \{1, 2, \cdots\}).$$

Let $\mathcal{A}(p)$ denote the family of analytic functions in $\mathbb{D} = \{z \in \mathbb{C} : |z| < 1\}$ of the form:

$$f(z) = z^p + \sum_{n=1}^{\infty} b_{p+n} z^{p+n} \ (p \in \mathbb{N};\ f^{(p+1)}(0) \neq 0). \tag{1}$$

For $f, F \in \mathcal{H}(\mathbb{D})$, the function $f(z)$ is said to be subordinate to $F(z)$ or $F(z)$ is superordinate to $f(z)$, written $f \prec F$ or $f(z) \prec F(z)$, if there exists a Schwarz function $\omega(z)$ for $z \in \mathbb{D}$ such that $f(z) = F(\omega(z))$. If $F(z)$ is univalent, then $f(z) \prec F(z)$ if and only if $f(0) = F(0)$ and $f(\mathbb{D}) \subset F(\mathbb{D})$ (see [1,2]).

Let $\phi : \mathbb{C}^2 \times \mathbb{D} \to \mathbb{C}$ and $h(z)$ be univalent in $\mathbb{D}$. If $p(z)$ is analytic in $\mathbb{D}$ and satisfies

$$\phi\left(p(z), zp'(z); z\right) \prec h(z), \tag{2}$$

then $p(z)$ is solution Relation (2). The univalent function $q(z)$ is called a dominant of the solutions of Relation (2) if $p(z) \prec q(z)$ for all $p(z)$ satisfying Relation (2). A univalent dominant $\tilde{q}$ that satisfies

$\tilde{q} \prec q$ for all dominants of Relation (2) is called the best dominant. If $p(z)$ and $\phi(p(z), zp'(z); z)$ are univalent in $\mathbb{D}$ and if $p(z)$ satisfies

$$h(z) \prec \phi(p(z), zp'(z); z), \tag{3}$$

then $p(z)$ is a solution of Relation (3). An analytic function $q(z)$ is called a subordinant of the solutions of Relation (3) if $q(z) \prec p(z)$ for all $p(z)$ satisfying Relation (3). A univalent subordinant $\tilde{q}$ that satisfies $q \prec \tilde{q}$ for all subordinants of Relation (3) is called the best subordinant (see [1,2]).

We now introduce the operator $S_{0,z}^{\lambda,\mu,\eta,p}$ due to Goyal and Prajapat [3] (see also [4]) as follows:

$$S_{0,z}^{\lambda,\mu,\eta,p} f(z) = \begin{cases} \dfrac{\Gamma(p+1-\mu)\Gamma(p+1-\lambda+\eta)}{\Gamma(p+1)\Gamma(p+1-\mu+\eta)} z^{\mu} J_{0,z}^{\lambda,\mu,\eta} f(z) \ (0 \leq \lambda < \eta+p+1;\ z \in \mathbb{D}), \\ \\ \dfrac{\Gamma(p+1-\mu)\Gamma(p+1-\lambda+\eta)}{\Gamma(p+1)\Gamma(p+1-\mu+\eta)} z^{\mu} I_{0,z}^{-\lambda,\mu,\eta} f(z) \ (-\infty < \lambda < 0;\ z \in \mathbb{D}), \end{cases} \tag{4}$$

where $J_{0,z}^{\lambda,\mu,\eta}$ and $I_{0,z}^{-\lambda,\mu,\eta}$ are the generalized fractional derivative and integral operators, respectively, due to Srivastava et al. [5] (see also [6,7]). For $f \in \mathcal{A}(p)$ of form Equation (1), we have

$$\begin{aligned} S_{0,z}^{\lambda,\mu,\eta,p} f(z) &= z^p {}_3F_2(1, 1+p, 1+p+\eta-\mu; 1+p-\mu, 1+p+\eta-\lambda; z) * f(z) \\ &= z^p + \sum_{n=1}^{\infty} \frac{(p+1)_n (p+1-\mu+\eta)_n}{(p+1-\mu)_n (p+1-\lambda+\eta)_n} b_{p+n} z^{p+n} \\ &\quad (p \in \mathbb{N};\ \mu, \eta \in \mathbb{R};\ \mu < p+1;\ -\infty < \lambda < \eta+p+1), \end{aligned} \tag{5}$$

where ${}_qF_s$ $(q \leq s+1;\ q, s \in \mathbb{N}_0 = \mathbb{N} \cup \{0\})$ is the well-known generalized hypergeometric function (for details, see [8,9]), the symbol $*$ stands for convolution of two analytic functions [1] and $(\nu)_n$ is the Pochhammer symbol [8,10].

Setting

$$\begin{aligned} G_{p,\eta,\mu}^{\lambda}(z) &= z^p + \sum_{n=1}^{\infty} \frac{(p+1)_n (p+1-\mu+\eta)_n}{(p+1-\mu)_n (p+1-\lambda+\eta)_n} z^{p+n} \\ &\quad (p \in \mathbb{N};\ \mu, \eta \in \mathbb{R};\ \mu < \min\{p+1, p+1+\eta\};\ -\infty < \lambda < \eta+p+1) \end{aligned} \tag{6}$$

and

$$G_{p,\eta,\mu}^{\lambda}(z) * \left[G_{p,\eta,\mu}^{\lambda,\delta}(z)\right] = \frac{z^p}{(1-z)^{\delta+p}} \ (\delta > -p;\ z \in \mathbb{D}),$$

Tang et al. [11] (see also [12]) defined the operator $H_{p,\eta,\mu}^{\lambda,\delta} : \mathcal{A}(p) \to \mathcal{A}(p)$ by

$$H_{p,\eta,\mu}^{\lambda,\delta} f(z) = \left[G_{p,\eta,\mu}^{\lambda,\delta}(z)\right] * f(z).$$

Then, for $f \in \mathcal{A}(p)$, we have

$$H_{p,\eta,\mu}^{\lambda,\delta} f(z) = z^p + \sum_{n=1}^{\infty} \frac{(\delta+p)_n (p+1-\mu)_n (p+1-\lambda+\eta)_n}{(1)_n (p+1)_n (p+1-\mu+\eta)_n} b_{p+n} z^{p+n}. \tag{7}$$

It is easy to verify that

$$z\left(H_{p,\eta,\mu}^{\lambda,\delta} f(z)\right)' = (\delta+p) H_{p,\eta,\mu}^{\lambda,\delta+1} f(z) - \delta H_{p,\eta,\mu}^{\lambda,\delta} f(z), \tag{8}$$

and

$$z\left(H_{p,\eta,\mu}^{\lambda+1,\delta} f(z)\right)' = (p+\eta-\lambda) H_{p,\eta,\mu}^{\lambda,\delta} f(z) - (\eta-\lambda) H_{p,\eta,\mu}^{\lambda+1,\delta} f(z). \tag{9}$$

Making use of the hypergeometric function in the kernel, Saigo [13] proposed generalizations of fractional calculus of both Riemann–Liouville and Weyl types. The general theory of fractional calculus thus developed was applied to the study for several multiplication properties of fractional integrals [14]. In particular, Owa et al. [15] and Srivastava et al. [5] investigated some distortion theorems involving fractional integrals, and sufficient conditions for fractional integrals of analytic functions in the open unit disk to be starlike or convex. Moreover, the theory of fractional calculus is widely applied to not only pure mathematics but also applied science. For some interesting developments in applied science such as bioengineering and applied physics, the readers may be referred to the works of (for examples) Hassan et al. [16], Magin [17], Martínez-García et al. [18] and Othman and Marin [19].

By using the principle of subordination, Miller et al. [20] investigated subordinations-preserving properties for certain integral operators. In addition, Miller and Mocanu [2] studied some important properties on superordinations as the dual problem of subordinations. Furthermore, the study of the subordinaton-preserving properties and their dual problems for various operators is a significant role in pure and applied mathematics. The aim of the present paper, motivated by the works mentioned above, is to systematically investigate the subordination- and superordination-preserving results of the generalized fractional differintegral operator defined Equation (7) with certain differential sandwich-type theorems as consequences of the results presented here. Our results give interesting new properties, and together with other papers that appeared in the last years could emphasize the perspective of the importance of differential subordinations and generalized fractional differintegral operators. We also note that, in recent years, several authors obtained many interesting results involving various linear and nonlinear operators associated with differential subordinations and their dual problrms (for details, see [21–28]).

For the proofs of our main results, we shall need some definitions and lemmas stated below.

Definition 1 ([1]). *We denote by $\mathcal{Q}$ the set of all functions $q(z)$ that are analytic and injective on $\overline{\mathbb{D}}\backslash E(q)$, where*

$$E(q) = \left\{ \zeta \in \partial\mathbb{D} : \lim_{z\to\zeta} q(z) = \infty \right\},$$

and are $q'(\zeta) \neq 0$ for $\zeta \in \partial\mathbb{D}\backslash E(q)$.

Definition 2 ([2]). *A function $\mathcal{I}(z,t)$ $(z \in \mathbb{D},\ t \geq 0)$ is a subordination chain if $\mathcal{I}(.,t)$ is analytic and univalent in $\mathbb{D}$ for all $t \geq 0$, $\mathcal{I}(z,.)$ is continuously differentiable on $[0,\infty)$ for all $z \in \mathbb{D}$ and $\mathcal{I}(z,s) \prec \mathcal{I}(z,t)$ for all $0 \leq s \leq t$.*

Lemma 1 ([29]). *Let $H : \mathbb{C}^2 \to \mathbb{C}$ satisfy*

$$\Re\{H(i\sigma;\tau)\} \leq 0$$

for all real σ, τ with $\tau \leq -n\left(1+\sigma^2\right)/2$ and $n \in \mathbb{N}$. If $p(z) = 1 + p_n z^n + p_{n+1} z^{n+1} + \cdots$ is analytic in $\mathbb{D}$ and

$$\Re\{H(p(z); zp'(z))\} > 0 \ (z \in \mathbb{D}),$$

then $\Re\{p(z)\} > 0$ for $z \in \mathbb{D}$.

Lemma 2 ([30]). *Let $\kappa, \gamma \in \mathbb{C}$ with $\kappa \neq 0$ and let $h \in H(\mathbb{D})$ with $h(0) = c$. If $\Re\{\kappa h(z) + \gamma\} > 0\,(z \in \mathbb{D})$, then the solution of the differential equation:*

$$q(z) + \frac{zq'(z)}{\kappa q(z) + \gamma} = h(z) \ \ (z \in \mathbb{D};\ q(0) = c)$$

is analytic in $\mathbb{D}$ and satisfies $\Re\{\kappa q(z) + \gamma\} > 0$ for $z \in \mathbb{D}$.

Lemma 3 ([1]). *Suppose that $p \in \mathcal{Q}$ with $q(0) = a$ and $q(z) = a + q_n z^n + q_{n+1} z^{n+1} + \cdots$ is analytic in $\mathbb{D}$ with $q(z) \neq a$ and $n \geq 1$. If $q(z)$ is not subordinate to $p(z)$, then there exists two points $z_0 = r_0 e^{i\theta} \in \mathbb{D}$ and $\xi_0 \in \partial\mathbb{D} \backslash E(q)$ such that*

$$q(z_0) = p(\xi_0) \text{ and } z_0 q'(z_0) = m\xi_0 p'(\xi_0) \ (m \geq n).$$

Lemma 4 ([2]). *Let $q \in \mathcal{H}[c,1]$ and $\varphi : \mathbb{C}^2 \to \mathbb{C}$. In addition, let $\varphi(q(z), zq'(z)) = h(z)$. If $\mathcal{I}(z,t) = \varphi(q(z), tzq'(z))$ is a subordination chain and $q \in \mathcal{H}[c,1] \cap \mathcal{Q}$, then*

$$h(z) \prec \varphi(p(z), zp'(z)),$$

implies that $q(z) \prec p(z)$. Moreover, if $\varphi(q(z), zq'(z)) = h(z)$ has a univalent solution $q \in \mathcal{Q}$, then q is the best subordinant.

Lemma 5 ([31]). *The function $\mathcal{I}(z,t) : \mathbb{D} \times [0,\infty) \longrightarrow \mathbb{C}$ of the form*

$$\mathcal{I}(z,t) = a_1(t)z + \cdots (a_1(t) \neq 0;\ t \geq 0)$$

and $\lim_{t\to\infty} |a_1(t)| = \infty$ is a subordination chain if and only if

$$\Re\left\{\frac{z\frac{\partial \mathcal{I}(z,t)}{\partial z}}{\frac{\partial \mathcal{I}(z,t)}{\partial t}}\right\} > 0 \ (z \in \mathbb{D};\ t \geq 0)$$

and

$$|\mathcal{I}(z,t)| \leq K_0 |a_1(t)| \ (t \geq 0)$$

for constants $K_0 > 0$ and r_0 ($|z| < r_0 < 1$).

2. Main Results

Throughout this paper, we assume that $p \in \mathbb{N}$, $\alpha, \beta > 0$, $\delta > -p$, $\mu, \eta \in \mathbb{R}$, $\mu < \min\{p+1, p+1+\eta\}$, $-\infty < \lambda < \eta + p + 1$, $H_{p,\eta,\mu}^{\lambda,\delta} f(z)/z^p \neq 0$ for $f \in \mathcal{A}(p)$ and all the powers are understood as principal values.

Theorem 1. *Suppose that $f, g \in \mathcal{A}(p)$ and*

$$\Re\left\{1 + \frac{z\phi''(z)}{\phi'(z)}\right\} > -\rho \tag{10}$$

$$\left(\phi(z) = (1-\alpha)\left[\frac{H_{p,\eta,\mu}^{\lambda,\delta} g(z)}{z^p}\right]^\beta + \alpha\left[\frac{H_{p,\eta,\mu}^{\lambda,\delta+1} g(z)}{H_{p,\eta,\mu}^{\lambda,\delta} g(z)}\right]\left[\frac{H_{p,\eta,\mu}^{\lambda,\delta} g(z)}{z^p}\right]^\beta;\ z \in \mathbb{D}\right),$$

where ρ is given by

$$\rho = \frac{\alpha^2 + \beta^2(\delta+p)^2 - \left|\alpha^2 - \beta^2(\delta+p)^2\right|}{4\alpha\beta(\delta+p)}. \tag{11}$$

Then,

$$(1-\alpha)\left[\frac{H_{p,\eta,\mu}^{\lambda,\delta} f(z)}{z^p}\right]^\beta + \alpha\left[\frac{H_{p,\eta,\mu}^{\lambda,\delta+1} f(z)}{H_{p,\eta,\mu}^{\lambda,\delta} f(z)}\right]\left[\frac{H_{p,\eta,\mu}^{\lambda,\delta} f(z)}{z^p}\right]^\beta \prec \phi(z) \tag{12}$$

implies that

$$\left[\frac{H_{p,\eta,\mu}^{\lambda,\delta} f(z)}{z^p}\right]^\beta \prec \left[\frac{H_{p,\eta,\mu}^{\lambda,\delta} g(z)}{z^p}\right]^\beta \tag{13}$$

and $\left[\frac{H_{p,\eta,\mu}^{\lambda,\delta}g(z)}{z^p}\right]^{\beta}$ *is the best dominant.*

Proof. We define two functions $\Phi(z)$ and $\Psi(z)$ by

$$\Phi(z) = \left[\frac{H_{p,\eta,\mu}^{\lambda,\delta}f(z)}{z^p}\right]^{\beta} \text{ and } \Psi(z) = \left[\frac{H_{p,\eta,\mu}^{\lambda\delta}g(z)}{z^p}\right]^{\beta} \ (z \in \mathbb{D}). \tag{14}$$

Firstly, we will show that, if

$$q(z) = 1 + \frac{z\Psi''(z)}{\Psi'(z)} \ (z \in \mathbb{D}), \tag{15}$$

then

$$\Re\{q(z)\} > 0 \ (z \in \mathbb{D}).$$

From the definitions of $\Psi(z)$ and $\phi(z)$ with Equation (8), we have

$$\phi(z) = \Psi(z) + \frac{\alpha}{\beta(\delta+p)} z\Psi'(z). \tag{16}$$

Differentiation both sides of Equation (16) with respect to z yields

$$\phi'(z) = \Psi'(z) + \frac{\alpha\left[z\Psi''(z) + \Psi'(z)\right]}{\beta(\delta+p)}. \tag{17}$$

From Equations (15) and (17), we easily obtain

$$1 + \frac{z\phi''(z)}{\phi'(z)} = q(z) + \frac{zq'(z)}{q(z) + \frac{\beta(\delta+p)}{\alpha}} = h(z) \ (z \in \mathbb{D}). \tag{18}$$

It follows from Relations (10) and (18) that

$$\Re\left\{h(z) + \frac{\beta(\delta+p)}{\alpha}\right\} > 0 \ (z \in \mathbb{D}). \tag{19}$$

Furthermore, by means of Lemma 2, we deduce that Equation (18) has a solution $q \in \mathcal{H}(\mathbb{D})$ with $h(0) = q(0) = 1$. Let

$$H(u,v) = u + \frac{v}{u + \frac{\beta(\delta+p)}{\alpha}} + \rho, \tag{20}$$

where ρ is given by Equation (11). From Equations (18) and (19), we have

$$\Re\{H(q(z); zq'(z))\} > 0 \ (z \in \mathbb{D}).$$

Now, we will show that

$$\Re\{H(i\sigma;\tau)\} \leq 0 \left(\sigma \in \mathbb{R};\ \tau \leq -\frac{1+\sigma^2}{2}\right). \tag{21}$$

From Equation (20), we obtain

$$\begin{aligned}\Re\{\mathcal{H}(i\sigma;\tau)\} &= \Re\left\{i\sigma+\frac{\tau}{\frac{\beta(\delta+p)}{\alpha}+i\sigma}+\rho\right\}\\ &= \rho+\frac{\frac{\beta(\delta+p)\tau}{\alpha}}{\left|\frac{\beta(\delta+p)}{\alpha}+i\sigma\right|^2}\leq-\frac{E_\rho(\sigma)}{2\left|\frac{\beta(\delta+p)}{\alpha}+i\sigma\right|^2},\end{aligned}$$

where

$$E_\rho(\sigma)=\left(\frac{\beta(\delta+p)}{\alpha}-2\rho\right)\sigma^2-2\left(\frac{\beta(\delta+p)}{\alpha}\right)^2\rho+\frac{\beta(\delta+p)}{\alpha}. \tag{22}$$

For ρ given by Equation (11), since the coefficient of σ^2 in $E_\rho(\sigma)$ of Equation (22) is positive or equal to zero and $E_\rho(\sigma)\geq 0$, we obtain that $\Re\{H(i\sigma;\tau)\}\leq 0$ for all $\sigma\in\mathbb{R}$ and $\tau\leq-\frac{1+\sigma^2}{2}$. Thus, by applying Lemma 1, we obtain that

$$\Re\{q(z)\}>0\ (z\in\mathbb{D}).$$

Moreover, $\Psi'(0)\neq 0$ since $g^{(p+1)}(0)\neq 0$. Hence, $\Psi(z)$ defined by Equation (14) is convex (univalent) in $\mathbb{D}$. Next, we verify that the Condition (12) implies that

$$\Phi(z)\prec\Psi(z)$$

for $\Phi(z)$ and $\Psi(z)$ given by Equation (14). Without loss of generality, we assume that $\Psi(z)$ is analytic, univalent on $\overline{\mathbb{D}}$ and

$$\Psi'(\xi)\neq 0\ (|\xi|=1).$$

Let us consider the function $\mathcal{I}(z,t)$ defined by

$$\mathcal{I}(z,t)=\Psi(z)+\frac{\alpha(1+t)}{\beta(\delta+p)}z\Psi'(z)\ (0\leq t<\infty;\ z\in\mathbb{D}). \tag{23}$$

Then, we see easily that

$$\left.\frac{\partial\mathcal{I}(z,t)}{\partial z}\right|_{z=0}=\Psi'(0)\left(1+\frac{\alpha}{\beta(\delta+p)}(1+t)\right)\neq 0\ (0\leq t<\infty;\ z\in\mathbb{D}).$$

This shows that

$$\mathcal{I}(z,t)=a_1(t)z+\cdots$$

satisfies the restrictions $\lim_{t\to\infty}|a_1(t)|=\infty$ and $a_1(t)\neq 0\ (0\leq t<\infty)$. In addition, we obtain

$$\Re\left\{\frac{z\frac{\partial\mathcal{I}(z,t)}{\partial z}}{\frac{\partial\mathcal{I}(z,t)}{\partial t}}\right\}=\Re\left\{\frac{\beta(\delta+p)}{\alpha}+(1+t)\left(1+\frac{z\Psi''(z)}{\Psi'(z)}\right)\right\}>0$$
$$(0\leq t<\infty;\ z\in\mathbb{D}),$$

since $\Psi(z)$ is convex and $\Re\left(\frac{\beta(\delta+p)}{\alpha}\right)>0$. Moreover, we have

$$\left|\frac{\mathcal{I}(z,t)}{a_1(t)}\right|=\left|\frac{\Psi(z)+\frac{\alpha(1+t)}{\beta(\delta+p)}z\Psi'(z)}{\Psi(0)\left(1+\frac{\alpha(1+t)}{\beta(\delta+p)}\right)}\right| \tag{24}$$

and also the function $\Psi(z)$ may be written by

$$\Psi(z) = \Psi(0) + \Psi'(0)\psi(z) \quad (z \in \mathbb{D}), \tag{25}$$

where $\psi(z)$ is a normalized univalent function in $\mathbb{D}$. We note that, for the function $\psi(z)$, we have the following sharp growth and distortion results [32]:

$$\frac{r}{(1+r)^2} \leq |\psi(z)| \leq \frac{r}{(1-r)^2} \quad (|z| = r < 1) \tag{26}$$

and

$$\frac{1-r}{(1+r)^3} \leq \psi'(z) \leq \frac{1+r}{(1-r)^3} \quad (|z| = r < 1). \tag{27}$$

Hence, by applying Equations (25), (26) and (27) to Equation (24), we can find easily an upper bound for the right-hand side of Equation (24). Thus, the function $\mathcal{I}(z,t)$ satisfies the second condition of Lemma 5, which proves that $\mathcal{I}(z,t)$ is a subordination chain. From the definition of subordination chain, we note that

$$\phi(z) = \Psi(z) + \frac{\alpha}{\beta(\delta+p)} z\Psi'(z) = \mathcal{I}(z,0)$$

and

$$\mathcal{I}(z,0) \prec \mathcal{I}(z,t) \ (0 \leq t < \infty),$$

which implies that

$$\mathcal{I}(\xi,t) \notin \mathcal{I}(\mathbb{D},0) = \phi(\mathbb{D}) \ (0 \leq t < \infty; \ \xi \in \partial\mathbb{D}). \tag{28}$$

If $\Phi(z)$ is not subordinate to $\Psi(z)$, by Lemma 3, we see that there exist two points $z_0 \in \mathbb{D}$ and $\xi_0 \in \partial\mathbb{D}$ satisfying

$$\phi(z_0) = \Psi(\xi_0) \text{ and } z_0\Phi'(z_0) = (1+t)\,\xi_0\Psi'(\xi_0) \ (0 \leq t < \infty). \tag{29}$$

Hence, by using Relations (12), (14), (23) and (29), we obtain

$$\begin{aligned}
\mathcal{I}(\xi_0,t) &= \Psi(\xi_0) + \frac{\alpha}{\beta(\delta+p)}(1+t)\,\xi_0\Psi'(\xi_0) \\
&= \Phi(z_0) + \frac{\alpha}{\beta(\delta+p)} z_0\Phi'(z_0) \\
&= (1-\alpha)\left[\frac{H_{p,\eta,\mu}^{\lambda,\delta} f(z_0)}{z_0^p}\right]^\beta + \alpha\left[\frac{H_{p,\eta,\mu}^{\lambda,\delta+1} f(z_0)}{H_{p,\eta,\mu}^{\lambda,\delta} f(z_0)}\right]\left[\frac{H_{p,\eta,\mu}^{\lambda,\delta} f(z_0)}{z_0^p}\right]^\beta \in \phi(\mathbb{D}).
\end{aligned}$$

This Contradicts (28). Thus, we conclude that $\Phi(z) \prec \Psi(z)$. If we consider $\Phi = \Psi$, then we know that Ψ is the best dominant. Therefore, we complete the proof of Theorem 1. □

Remark 1. *The function $\Psi'(z) \neq 0$ for $z \in \mathbb{D}$ in Theorem 1 under the assumption*

$$\Re\{q(z)\} = 1 + \Re\left\{\frac{z\Psi''(z)}{\Psi'(z)}\right\} > 0 \ \ (z \in \mathbb{D}). \tag{30}$$

In fact, if $\Psi'(z)$ has a zero of order m at $z = z_1 \in \mathbb{D}\backslash\{0\}$, then we may write

$$\Psi(z) = (z - z_1)^m \Psi_1(z) \quad (m \in \mathbb{N}),$$

where $\Psi_1(z)$ is analytic in $\mathbb{D}\backslash\{0\}$ and $\Psi_1(z_1) \neq 0$. Then, we have

$$q(z) = 1 + \frac{z\Psi''(z)}{\Psi'(z)} = 1 + \frac{mz}{z - z_1} + \frac{z\Psi_1'(z)}{\Psi_1(z)}. \tag{31}$$

Thus, choosing $z \to z_1$ suitably, the real part of the right-hand side of Equation (31) can take any negative infinite values, which contradicts hypothesis Equation (30). In addition, it is obvious that $\Psi'(0) \neq 0$ since $g^{(p+1)}(0) \neq 0$.

Using similar methods given in the proof of Theorem 1, we have the following result.

Theorem 2. *Suppose that $f, g \in \mathcal{A}(p)$ and*

$$\Re\left\{1+\frac{z\psi''(z)}{\psi'(z)}\right\} > -\sigma \tag{32}$$

$$\left(\psi(z) = (1-\alpha)\left[\frac{H_{p,\eta,\mu}^{\lambda+1,\delta}g(z)}{z^p}\right]^\beta + \alpha\left[\frac{H_{p,\eta,\mu}^{\lambda,\delta}g(z)}{H_{p,\eta,\mu}^{\lambda+1,\delta}g(z)}\right]\left[\frac{H_{p,\eta,\mu}^{\lambda+1,\delta}g(z)}{z^p}\right]^\beta; z \in \mathbb{D}\right),$$

where σ is given by

$$\sigma = \frac{\alpha^2+\beta^2(p+\eta-\lambda)^2 - \left|\alpha^2-\beta^2(p+\eta-\lambda)^2\right|}{4\alpha\beta(p+\eta-\lambda)}. \tag{33}$$

Then,

$$(1-\alpha)\left[\frac{H_{p,\eta,\mu}^{\lambda+1,\delta}f(z)}{z^p}\right]^\beta + \alpha\left[\frac{H_{p,\eta,\mu}^{\lambda,\delta}f(z)}{H_{p,\eta,\mu}^{\lambda+1,\delta}f(z)}\right]\left[\frac{H_{p,\eta,\mu}^{\lambda+1,\delta}f(z)}{z^p}\right]^\beta \prec \psi(z) \tag{34}$$

implies that

$$\left[\frac{H_{p,\eta,\mu}^{\lambda+1,\delta}f(z)}{z^p}\right]^\beta \prec \left[\frac{H_{p,\eta,\mu}^{\lambda+1,\delta}g(z)}{z^p}\right]^\beta \tag{35}$$

and $\left[\frac{H_{p,\eta,\mu}^{\lambda+1,\delta}g(z)}{z^p}\right]^\beta$ is the best dominant.

Next, we derive the dual result of Theorem 1.

Theorem 3. *Suppose that $f, g \in \mathcal{A}(p)$ and*

$$\Re\left\{1+\frac{z\phi''(z)}{\phi'(z)}\right\} > -\rho$$

$$\left(\phi(z) = (1-\alpha)\left[\frac{H_{p,\eta,\mu}^{\lambda,\delta}g(z)}{z^p}\right]^\beta + \alpha\left[\frac{H_{p,\eta,\mu}^{\lambda,\delta+1}g(z)}{H_{p,\eta,\mu}^{\lambda,\delta}g(z)}\right]\left[\frac{H_{p,\eta,\mu}^{\lambda,\delta}g(z)}{z^p}\right]^\beta; z \in \mathbb{D}\right),$$

where ρ is given by Equation (11). If

$$(1-\alpha)\left[\frac{H_{p,\eta,\mu}^{\lambda,\delta}f(z)}{z^p}\right]^\beta + \alpha\left[\frac{H_{p,\eta,\mu}^{\lambda,\delta+1}f(z)}{H_{p,\eta,\mu}^{\lambda,\delta}f(z)}\right]\left[\frac{H_{p,\eta,\mu}^{\lambda,\delta}f(z)}{z^p}\right]^\beta$$

is univalent in $\mathbb{D}$ and $\left[\frac{H_{p,\eta,\mu}^{\lambda,\delta}f(z)}{z^p}\right]^\beta \in \mathcal{H}[1,1] \cap \mathcal{Q}$, then

$$\phi(z) \prec (1-\alpha)\left[\frac{H_{p,\eta,\mu}^{\lambda,\delta}f(z)}{z^p}\right]^\beta + \alpha\left[\frac{H_{p,\eta,\mu}^{\lambda,\delta+1}f(z)}{H_{p,\eta,\mu}^{\lambda,\delta}f(z)}\right]\left[\frac{H_{p,\eta,\mu}^{\lambda,\delta}f(z)}{z^p}\right]^\beta \tag{36}$$

implies that

$$\left[\frac{H_{p,\eta,\mu}^{\lambda,\delta}g(z)}{z^p}\right]^\beta \prec \left[\frac{H_{p,\eta,\mu}^{\lambda,\delta}f(z)}{z^p}\right]^\beta \tag{37}$$

and $\left[\frac{H_{p,\eta,\mu}^{\lambda,\delta}g(z)}{z^p}\right]^\beta$ *is the best subordinant.*

Proof. By using the functions $\Phi(z)$, $\Psi(z)$ and $q(z)$ given by Equations (14) and (15), we have

$$\phi(z) = \Psi(z) + \frac{\alpha}{\beta(\delta+p)} z\Psi'(z) = \varphi(\Psi(z), z\Psi'(z)) \tag{38}$$

and

$$\Re\{q(z)\} > 0 \ (z \in \mathbb{D}).$$

Next, we will show that $\Psi(z) \prec \Phi(z)$. To derive this, we consider the function $\mathcal{I}(z,t)$ defined by

$$\mathcal{I}(z,t) = \Psi(z) + \frac{\alpha}{\beta(\delta+p)} tz\Psi'(z) \quad (0 \leq t < \infty;\ z \in \mathbb{D}).$$

Then, we see that

$$\left.\frac{\partial \mathcal{I}(z,t)}{\partial z}\right|_{z=0} = \Psi'(0)\left(1 + \frac{\alpha}{\beta(\delta+p)} t\right) \neq 0 \ (0 \leq t < \infty;\ z \in \mathbb{D}),$$

which shows that

$$\mathcal{I}(z,t) = a_1(t) z + \cdots$$

satisfies $\lim_{t\to\infty} |a_1(t)| = \infty$ and $a_1(t) \neq 0 \ (0 \leq t < \infty)$. Furthermore, we obtain

$$\Re\left\{\frac{z\frac{\partial \mathcal{I}(z,t)}{\partial z}}{\frac{\partial \mathcal{I}(z,t)}{\partial t}}\right\} = \Re\left\{\frac{\beta(\delta+p)}{\alpha} + t\left(1 + \frac{z\Psi''(z)}{\Psi'(z)}\right)\right\} > 0$$
$$(0 \leq t < \infty;\ z \in \mathbb{D}).$$

By using a similar method as in the proof of Theorem 1, we can prove the second inequality of Lemma 5. Hence, $\mathcal{I}(z,t)$ is a subordination chain. Therefore, by means of Lemma 4, we see that Relation (36) must imply given by Relation (37). Moreover, since Equation (38) has a univalent solution Ψ, it is the best subordinant. Therefore, we complete the proof. □

Using similar techniques given in the proof of Theorem 3, we have the following result.

Theorem 4. *Suppose that* $f, g \in \mathcal{A}(p)$ *and*

$$\Re\left\{1 + \frac{z\psi''(z)}{\psi'(z)}\right\} > -\sigma$$

$$\left(\psi(z) = (1-\alpha)\left[\frac{H_{p,\eta,\mu}^{\lambda+1,\delta}g(z)}{z^p}\right]^\beta + \alpha\left[\frac{H_{p,\eta,\mu}^{\lambda,\delta}g(z)}{H_{p,\eta,\mu}^{\lambda+1,\delta}g(z)}\right]\left[\frac{H_{p,\eta,\mu}^{\lambda+1,\delta}g(z)}{z^p}\right]^\beta; z \in \mathbb{D}\right),$$

where σ *is given by Equation* (33). *If*

$$(1-\alpha)\left[\frac{H_{p,\eta,\mu}^{\lambda+1,\delta}f(z)}{z^p}\right]^\beta + \alpha\left[\frac{H_{p,\eta,\mu}^{\lambda,\delta}f(z)}{H_{p,\eta,\mu}^{\lambda+1,\delta}f(z)}\right]\left[\frac{H_{p,\eta,\mu}^{\lambda+1,\delta}f(z)}{z^p}\right]^\beta$$

is univalent in $\mathbb{D}$ *and* $\left[\frac{H_{p,\eta,\mu}^{\lambda+1,\delta}f(z)}{z^p}\right]^\beta \in \mathcal{H}[1,1] \cap \mathcal{Q}$, *then*

$$\psi(z) \prec (1-\alpha)\left[\frac{H_{p,\eta,\mu}^{\lambda+1,\delta}f(z)}{z^p}\right]^\beta + \alpha\left[\frac{H_{p,\eta,\mu}^{\lambda,\delta}f(z)}{H_{p,\eta,\mu}^{\lambda+1,\delta}f(z)}\right]\left[\frac{H_{p,\eta,\mu}^{\lambda+1,\delta}f(z)}{z^p}\right]^\beta \tag{39}$$

implies that

$$\left[\frac{H_{p,\eta,\mu}^{\lambda+1,\delta}g(z)}{z^p}\right]^\beta \prec \left[\frac{H_{p,\eta,\mu}^{\lambda+1,\delta}f(z)}{z^p}\right]^\beta \tag{40}$$

and $\left[\frac{H_{p,\eta,\mu}^{\lambda+1,\delta}g(z)}{z^p}\right]^\beta$ *is the best subordinant.*

If we combine Theorems 1 and 3, and Theorems 2 and 4, then we have the unified sandwich-type results, respectively.

Theorem 5. *Suppose that* $f, g_j \in \mathcal{A}(p)$ $(j = 1, 2)$ *and*

$$\Re\left\{1 + \frac{z\phi_j''(z)}{\phi_j'(z)}\right\} > -\rho \tag{41}$$

$$\left(\phi_j(z) = (1-\alpha)\left[\frac{H_{p,\eta,\mu}^{\lambda,\delta}g_j(z)}{z^p}\right]^\beta + \alpha\left[\frac{H_{p,\eta,\mu}^{\lambda,\delta+1}g_j(z)}{H_{p,\eta,\mu}^{\lambda,\delta}g_j(z)}\right]\left[\frac{H_{p,\eta,\mu}^{\lambda,\delta}g_j(z)}{z^p}\right]^\beta ;\ z \in \mathbb{D}\right),$$

where ρ *is given by Equation* (11). *If*

$$(1-\alpha)\left[\frac{H_{p,\eta,\mu}^{\lambda,\delta}f(z)}{z^p}\right]^\beta + \alpha\left[\frac{H_{p,\eta,\mu}^{\lambda,\delta+1}f(z)}{H_{p,\eta,\mu}^{\lambda,\delta}f(z)}\right]\left[\frac{H_{p,\eta,\mu}^{\lambda,\delta}f(z)}{z^p}\right]^\beta$$

is univalent in $\mathbb{D}$ *and* $\left[\frac{H_{p,\eta,\mu}^{\lambda,\delta}f(z)}{z^p}\right]^\beta \in \mathcal{H}[1,1] \cap \mathcal{Q}$, *then*

$$\phi_1(z) \prec (1-\alpha)\left[\frac{H_{p,\eta,\mu}^{\lambda,\delta}f(z)}{z^p}\right]^\beta + \alpha\left[\frac{H_{p,\eta,\mu}^{\lambda,\delta+1}f(z)}{H_{p,\eta,\mu}^{\lambda,\delta}f(z)}\right]\left[\frac{H_{p,\eta,\mu}^{\lambda,\delta}f(z)}{z^p}\right]^\beta \prec \phi_2(z) \tag{42}$$

implies that

$$\left[\frac{H_{p,\eta,\mu}^{\lambda,\delta}g_1(z)}{z^p}\right]^\beta \prec \left[\frac{H_{p,\eta,\mu}^{\lambda,\delta}f(z)}{z^p}\right]^\beta \prec \left[\frac{H_{p,\eta,\mu}^{\lambda,\delta}g_2(z)}{z^p}\right]^\beta. \tag{43}$$

Moreover, $\left[\frac{H_{p,\eta,\mu}^{\lambda,\delta}g_1(z)}{z^p}\right]^\beta$ *and* $\left[\frac{H_{p,\eta,\mu}^{\lambda,\delta}g_2(z)}{z^p}\right]^\beta$ *are the best subordinant and the best dominant, respectively.*

Theorem 6. *Suppose that* $f, g_j \in \mathcal{A}(p)$ $(j = 1, 2)$ *and*

$$\Re\left\{1 + \frac{z\psi_j''(z)}{\psi_j'(z)}\right\} > -\sigma \tag{44}$$

$$\left(\psi_j(z) = (1-\alpha)\left[\frac{H_{p,\eta,\mu}^{\lambda+1,\delta}g_j(z)}{z^p}\right]^\beta + \alpha\left[\frac{H_{p,\eta,\mu}^{\lambda,\delta}g_j(z)}{H_{p,\eta,\mu}^{\lambda+1,\delta}g_j(z)}\right]\left[\frac{H_{p,\eta,\mu}^{\lambda+1,\delta}g_j(z)}{z^p}\right]^\beta ;\ z \in \mathbb{D}\right),$$

where σ is given by Equation (33). If

$$(1-\alpha)\left[\frac{H_{p,\eta,\mu}^{\lambda+1,\delta}f(z)}{z^p}\right]^\beta+\alpha\left[\frac{H_{p,\eta,\mu}^{\lambda,\delta}f(z)}{H_{p,\eta,\mu}^{\lambda+1,\delta}f(z)}\right]\left[\frac{H_{p,\eta,\mu}^{\lambda+1,\delta}f(z)}{z^p}\right]^\beta$$

is univalent in $\mathbb{D}$ *and* $\left[\frac{H_{p,\eta,\mu}^{\lambda+1,\delta}f(z)}{z^p}\right]^\beta\in\mathcal{H}[1,1]\cap\mathcal{Q}$, *then*

$$\psi_1(z)\prec(1-\alpha)\left[\frac{H_{p,\eta,\mu}^{\lambda+1,\delta}f(z)}{z^p}\right]^\beta+\alpha\left[\frac{H_{p,\eta,\mu}^{\lambda,\delta}f(z)}{H_{p,\eta,\mu}^{\lambda+1,\delta}f(z)}\right]\left[\frac{H_{p,\eta,\mu}^{\lambda+1,\delta}f(z)}{z^p}\right]^\beta\prec\psi_2(z) \tag{45}$$

implies that

$$\left[\frac{H_{p,\eta,\mu}^{\lambda+1,\delta}g_1(z)}{z^p}\right]^\beta\prec\left[\frac{H_{p,\eta,\mu}^{\lambda+1,\delta}f(z)}{z^p}\right]^\beta\prec\left[\frac{H_{p,\eta,\mu}^{\lambda+1,\delta}g_2(z)}{z^p}\right]^\beta. \tag{46}$$

Moreover, $\left[\frac{H_{p,\eta,\mu}^{\lambda+1,\delta}g_1(z)}{z^p}\right]^\beta$ *and* $\left[\frac{H_{p,\eta,\mu}^{\lambda+1,\delta}g_2(z)}{z^p}\right]^\beta$ *are the best subordinant and the best dominant, respectively.*

We note that the assumption of Theorem 5, which states that

$$(1-\alpha)\left[\frac{H_{p,\eta,\mu}^{\lambda,\delta}f(z)}{z^p}\right]^\beta+\alpha\left[\frac{H_{p,\eta,\mu}^{\lambda,\delta+1}f(z)}{H_{p,\eta,\mu}^{\lambda,\delta}f(z)}\right]\left[\frac{H_{p,\eta,\mu}^{\lambda,\delta}f(z)}{z^p}\right]^\beta \text{ and } \left[\frac{H_{p,\eta,\mu}^{\lambda,\delta}f(z)}{z^p}\right]^\beta$$

needs to be univalent in $\mathbb{D}$, may be exchanged by a different condition.

Corollary 1. *Suppose that* $f,g_j\in\mathcal{A}(p)$ $(j=1,\ 2)$ *and*

$$\Re\left\{1+\frac{z\phi_j''(z)}{\phi_j'(z)}\right\}>-\rho$$

$$\left(\phi_j(z)=(1-\alpha)\left[\frac{H_{p,\eta,\mu}^{\lambda,\delta}g_j(z)}{z^p}\right]^\beta+\alpha\left[\frac{H_{p,\eta,\mu}^{\lambda,\delta+1}g_j(z)}{H_{p,\eta,\mu}^{\lambda,\delta}g_j(z)}\right]\left[\frac{H_{p,\eta,\mu}^{\lambda,\delta}g_j(z)}{z^p}\right]^\beta;\ z\in\mathbb{D}\right)$$

and

$$\Re\left\{1+\frac{z\chi''(z)}{\chi'(z)}\right\}>-\rho, \tag{47}$$

$$\left(\chi(z)=(1-\alpha)\left[\frac{H_{p,\eta,\mu}^{\lambda,\delta}f(z)}{z^p}\right]^\beta+\alpha\left[\frac{H_{p,\eta,\mu}^{\lambda,\delta+1}f(z)}{H_{p,\eta,\mu}^{\lambda,\delta}f(z)}\right]\left[\frac{H_{p,\eta,\mu}^{\lambda,\delta}f(z)}{z^p}\right]^\beta;\ z\in\mathbb{D}\right),$$

where ρ is given by Equation (11). Then,

$$\phi_1(z)\prec(1-\alpha)\left[\frac{H_{p,\eta,\mu}^{\lambda,\delta}f(z)}{z^p}\right]^\beta+\alpha\left[\frac{H_{p,\eta,\mu}^{\lambda,\delta+1}f(z)}{H_{p,\eta,\mu}^{\lambda,\delta}f(z)}\right]\left[\frac{H_{p,\eta,\mu}^{\lambda,\delta}f(z)}{z^p}\right]^\beta\prec\phi_2(z)$$

implies that

$$\left[\frac{H_{p,\eta,\mu}^{\lambda,\delta}g_1(z)}{z^p}\right]^\beta\prec\left[\frac{H_{p,\eta,\mu}^{\lambda,\delta}f(z)}{z^p}\right]^\beta\prec\left[\frac{H_{p,\eta,\mu}^{\lambda,\delta}g_2(z)}{z^p}\right]^\beta.$$

Proof. To derive Corollary 1, we need to show that the Restriction (47) implies the univalence of $\chi(z)$. Noting that $0\le\rho<1/2$, it follows that $\chi(z)$ is close-to-convex function in $\mathbb{D}$ (see [33]) and so $\chi(z)$

is univalent in $\mathbb{D}$. In addition, by applying the similar methods given in the proof of Theorem 1, we see that the function $\Phi(z)$ defined by Equation (14) is convex (univalent) in $\mathbb{D}$. Therefore, by using Theorem 5, we get the desired result. □

Using similar methods given in the proof of Corollary 1 with Theorem 6, we obtain the following corollary.

Corollary 2. *Suppose that* $f, g_j \in \mathcal{A}(p)$ $(j = 1, 2)$ *and*

$$\Re\left\{1 + \frac{z\psi_j''(z)}{\psi_j'(z)}\right\} > -\sigma$$

$$\left(\psi_j(z) = (1-\alpha)\left[\frac{H_{p,\eta,\mu}^{\lambda+1,\delta} g_j(z)}{z^p}\right]^{\beta} + \alpha\left[\frac{H_{p,\eta,\mu}^{\lambda,\delta} g_j(z)}{H_{p,\eta,\mu}^{\lambda,\delta} g_j(z)}\right]\left[\frac{H_{p,\eta,\mu}^{\lambda,\delta} g_j(z)}{z^p}\right]^{\beta}; z \in \mathbb{D}\right)$$

and

$$\Re\left\{1 + \frac{z\Upsilon''(z)}{\Upsilon'(z)}\right\} > -\rho,$$

$$\left(\Upsilon(z) = (1-\alpha)\left[\frac{H_{p,\eta,\mu}^{\lambda+1,\delta} f(z)}{z^p}\right]^{\beta} + \alpha\left[\frac{H_{p,\eta,\mu}^{\lambda,\delta} f(z)}{H_{p,\eta,\mu}^{\lambda+1,\delta} f(z)}\right]\left[\frac{H_{p,\eta,\mu}^{\lambda+1,\delta} f(z)}{z^p}\right]^{\beta}; z \in \mathbb{D}\right),$$

where σ *is given by* (33). *Then,*

$$\psi_1(z) \prec (1-\alpha)\left[\frac{H_{p,\eta,\mu}^{\lambda+1,\delta} f(z)}{z^p}\right]^{\beta} + \alpha\left[\frac{H_{p,\eta,\mu}^{\lambda,\delta} f(z)}{H_{p,\eta,\mu}^{\lambda+1,\delta} f(z)}\right]\left[\frac{H_{p,\eta,\mu}^{\lambda+1,\delta} f(z)}{z^p}\right]^{\beta} \prec \psi_2(z)$$

implies that

$$\left[\frac{H_{p,\eta,\mu}^{\lambda,\delta} g_1(z)}{z^p}\right]^{\beta} \prec \left[\frac{H_{p,\eta,\mu}^{\lambda,\delta} f(z)}{z^p}\right]^{\beta} \prec \left[\frac{H_{p,\eta,\mu}^{\lambda,\delta} g_2(z)}{z^p}\right]^{\beta}.$$

3. Conclusions

Various applications of fractional calculus have an immense impact on the study of pure mathematic and applied science. In the present paper, we obtain new results on subordinations and superordinations for a wide class of operators defined by generalized fractional derivative operators and generalized fractional integral operators. Furthermore, the differential sandwich-type theorems are also discussed for these operators.

Author Contributions: Investigation, N.E.C. and R.S.; Supervision, R.S.; Writing—original draft, M.K.A.; Writing—review and editing, N.E.C.

References

1. Miller, S.S.; Mocanu, P.T. *Differential Subordinations: Theory and Applications, Series on Monographs and Textbooks in Pure and Applied Mathematics*; Marcel Dekker: New York, NY, USA; Basel, Switzerland, 2000; Volume 225.
2. Miller, S.; Mocanu, P.T. Subordinants of differential superordinations. *Complex Var. Theory Appl.* **2003**, *48*, 815–826. [CrossRef]
3. Goyal, G.P.; Prajapat, J.K. A new class of analytic p-valent functions with negative coefficients and fractional calculus operators. *Tamsui Oxf. J. Math. Sci.* **2004**, *20*, 175–186.

4. Prajapat, J.K.; Aouf, M.K. Majorization problem for certain class of p-valently analytic function defined by generalized fractional differintegral operator. *Comput. Math. Appl.* **2012**, *63*, 42–47. [CrossRef]
5. Srivastava, H.M.; Saigo, M.; Owa, S. A class of distortion theorems involving certain operators of fractional calculus. *J. Math. Anal. Appl.* **1988**, *131*, 412–420. [CrossRef]
6. Owa, S. On the distortion theorems I. *Kyungpook Math. J.* **1978**, *18*, 53–59.
7. Prajapat, J.K.; Raina, R.K.; Srivastava, H.M. Some inclusion properties for certain subclasses of strongly starlike and strongly convex functions involving a family of fractional integral operators. *Integr. Transforms Spec. Funct.* **2007**, *18*, 639–651. [CrossRef]
8. Owa, S.; Srivastava, H.M. Univalent and starlike generalized hypergeometric functions. *Can. J. Math.* **1987**, *39*, 1057–1077. [CrossRef]
9. Srivastava, H.M.; Owa, S. Some characterizations and distortions theorems involving fractional calculus, generalized hypergeometric functions, Hadamard products, linear operators and certain subclasses of analytic functions. *Nagoya Math. J.* **1987**, *106*, 1–28. [CrossRef]
10. Kanas, S.; Srivastava, H.M. Linear operators associated with k-uniformly convex functions. *Integr. Transforms Spec. Funct.* **2000**, *9*, 121–132. [CrossRef]
11. Tang, H.; Deng, G.-T.; Li, S.-H.; Aouf, M.K. Inclusion results for certain subclasses of spiral-like multivalent functions involving a generalized fractional differintegral operator. *Integr. Transforms Spec. Funct.* **2013**, *24*, 873–883. [CrossRef]
12. Seoudy, T.M.; Aouf, M.K. Subclasses of p-valent functions of bounded boundary rotation involving the generalized fractional differintegral operator. *Comptes Rendus Math.* **2013**, *351*, 787–792. [CrossRef]
13. Saigo, M. A remark on integral operators involving the Gauss hypergeometric functions. *Math. Rep. Coll. Gen. Ed. Kyushu Univ.* **1978**, *11*, 135–143.
14. Srivastava, H.M.; Saigo, M. Multiplication of fractional calculus operators and boundary value problems involving the Euler-Darboux equation. *J. Math. Anal. Appl.* **1987**, *121*, 325–369. [CrossRef]
15. Owa, S.; Saigo, M.; Srivastava, H.M. Some characterization theorems for starlike and convex functions involving a certain fractional integral operator. *J. Math. Anal. Appl.* **1989**, *140*, 419–426. [CrossRef]
16. Hassan, M.; Marin, M.; Ellahi, R.; Alamri, S.Z. Exploration of convective heat transfer and flow characteristics synthesis by Cu–Ag/water hybrid-nanofluids. *Heat Transf. Res.* **2018**, *49*, 1837–1848. [CrossRef]
17. Richard, L. *Magin, Fractional Calculus in Bioengineering*; Begell House: Redding, CA, USA, 2006.
18. Martínez-García, M.; Gordon, T.; Shu, L. Extended crossover model for human-control of fractional order plants. *IEEE Access* **2017**, *5*, 27622–27635. [CrossRef]
19. Othman, M.I.; Marin, M. Effect of thermal loading due to laser pulse on thermoelastic porous medium under G-N theory. *Results Phys.* **2017**, *7*, 3863–3872. [CrossRef]
20. Miller, S.S.; Mocanu, P.T.; Reade, M.O. Subordination-preserving integral operators. *Trans. Am. Math. Soc.* **1984**, *283*, 605–615. [CrossRef]
21. Aouf, M.K.; Mostafa, A.O.; Zayed, H.M. Subordination and superordination properties of p-valent functions defined by a generalized fractional differintegral operator. *Quaest. Math.* **2016**, *39*, 545–560. [CrossRef]
22. Srivastava, H.M.; Hussain, S.; Raziq, A.; Raza, M. The Fekete-Szegö functional for a subclass of analytic functions associated with quasi-subordination. *Carpath. J. Math.* **2018**, *34*, 103–113.
23. Srivastava, H.M.; Mostafa, A.O.; Aouf, M.K.; Zayed, H.M. Basic and fractional q-calculus and associated Fekete-Szegö problem for p-valently q-starlike functions and p-valently q-convex functions of complex order. *Miskolc Math. Notes* **2019**, *20*, 489–509. [CrossRef]
24. Srivastava, H.M.; Prajapati, A.; Gochhayat, P. Third-order differential subordination and differential superordination results for analytic functions involving the Srivastava-Attiya operator. *Appl. Math. Inf. Sci.* **2018**, *12*, 469–481. [CrossRef]
25. Srivastava, H.M.; Răducanu, D.; Zaprawa, P. A certain subclass of analytic functions defined by means of differential subordination. *Filomat* **2016**, *30*, 3743–3757. [CrossRef]
26. Tang, H.; Srivastava, H.M.; Deng, G.-T. Some families of analytic functions in the upper half-plane and their associated differential subordination and differential superordination properties and problems. *Appl. Math. Inf. Sci.* **2017**, *11*, 1247–1257. [CrossRef]
27. Tang, H.; Srivastava, H.M.; Deng, G.-T.; Li, S.-H. Second-order differential superordination for analytic functions in the upper half-plane. *J. Nonlinear Sci. Appl.* **2017**, *10*, 5271–5280. [CrossRef]

28. Xu, Q.-H.; Xiao, H.-G.; Srivastava, H.M. Some applications of differential subordination and the Dziok-Srivastava convolution operator. *Appl. Math. Comput.* **2014**, *230*, 496–508. [CrossRef]
29. Miller, S.S.; Mocanu, P.T. Differential subordinations and univalent functions. *Mich. Math. J.* **1981**, *28*, 157–172. [CrossRef]
30. Miller, S.S.; Mocanu, P.T. Univalent solutions of Briot-Bouquet differential equations. *J. Differ. Equ.* **1985**, *56*, 297–309. [CrossRef]
31. Pommerenke, C; Jensen, G. *Univalent Functions*; Vandenhoeck and Ruprecht: Gottingen, Germany, 1975.
32. Hallenbeck, D.J.; MacGregor, T.H. *Linear Problems and Convexity Techniques in Geometric Function Theory*; Pitman: London, UK, 1984.
33. Kaplan, W. Close-to-convex schlicht functions. *Mich. Math. J.* **1952**, *2*, 169–185. [CrossRef]

11

Statistically and Relatively Modular Deferred-Weighted Summability and Korovkin-Type Approximation Theorems

Hari Mohan Srivastava [1,2,*], Bidu Bhusan Jena [3], Susanta Kumar Paikray [3] and Umakanta Misra [4]

1 Department of Mathematics and Statistics, University of Victoria, Victoria, BC V8W 3R4, Canada
2 Department of Medical Research, China Medical University Hospital, China Medical University, Taichung 40402, Taiwan
3 Department of Mathematics, Veer Surendra Sai University of Technology, Burla, Odisha 768018, India; bidumath.05@gmail.com (B.B.J.); skpaikray_math@vssut.ac.in (S.K.P.)
4 Department of Mathematics, National Institute of Science and Technology, Palur Hills, Golanthara, Odisha 761008, India; umakanta_misra@yahoo.com
* Correspondence: harimsri@math.uvic.ca

Abstract: The concept of statistically deferred-weighted summability was recently studied by Srivastava et al. (Math. Methods Appl. Sci. **41** (2018), 671–683). The present work is concerned with the deferred-weighted summability mean in various aspects defined over a modular space associated with a generalized double sequence of functions. In fact, herein we introduce the idea of relatively modular deferred-weighted statistical convergence and statistically as well as relatively modular deferred-weighted summability for a double sequence of functions. With these concepts and notions in view, we establish a theorem presenting a connection between them. Moreover, based upon our methods, we prove an approximation theorem of the Korovkin type for a double sequence of functions on a modular space and demonstrate that our theorem effectively extends and improves most (if not all) of the previously existing results. Finally, an illustrative example is provided here by the generalized bivariate Bernstein–Kantorovich operators of double sequences of functions in order to demonstrate that our established theorem is stronger than its traditional and statistical versions.

Keywords: statistical convergence; P-convergent; statistically and relatively modular deferred-weighted summability; relatively modular deferred-weighted statistical convergence; Korovkin-type approximation theorem; modular space; convex space; $\mathcal{N}$-quasi convex modular; $\mathcal{N}$-quasi semi-convex modular

MSC: 40A05; 41A36; 40G15

---

1. Introduction, Preliminaries, and Motivation

The gradual evolution on sequence spaces results in the development of statistical convergence. It is more general than the ordinary convergence in the sense that the ordinary convergence of a sequence requires that almost all elements are to satisfy the convergence condition, that is, every element of the sequence needs to be in some neighborhood (arbitrarily small) of the limit. However, such restriction is relaxed in statistical convergence, where set having a few elements that are not in the neighborhood of the limit is discarded subject to the condition that the natural density of the set is zero, and at the same time the condition of convergence is valid for the other majority of the elements. In the year 1951, Fast [1] and Steinhaus [2] independently studied the term statistical convergence for single real sequences; it is a generalization of the concept of ordinary convergence. Actually, a root of the notion of statistical convergence can be detected by Zygmund (see [3], p. 181), where he used the term "almost convergence", which turned out to be equivalent to the concept of statistical convergence. We also find such concepts in random graph theory (see [4,5]) in the sense that almost convergence means convergence with probability 1, whereas in statistical convergence the probability is not necessarily 1. Mathematically, a sequence of random variables $\{X_n\}$ is statistically convergent (converges in probability) to a random variable X if $\lim_{n\to\infty} P(|Xn - X| \geqq \epsilon) = 0$, for all $\epsilon > 0$ (arbitrarily small); and almost convergent to X if $P(\lim_{n\to\infty} X_n = X) = 1$.

For different results concerning statistical versions of convergence as well as of the summability of single sequences, we refer to References [1,2,6].

Let $\mathbb{N}$ be the set of natural numbers and let $\mathcal{H} \subseteq \mathbb{N}$. Also let

$$\mathcal{H}_n = \{k : k \leqq n, \text{ and } k \in \mathcal{H}\}$$

and suppose that $|\mathcal{H}_n|$ is the cardinality of $\mathcal{H}_n$. Then, the *natural density* of $\mathcal{H}$ is defined by

$$\delta(\mathcal{H}) = \lim_{n\to\infty} \frac{|\mathcal{H}_n|}{n} = \lim_{n\to\infty} \frac{1}{n}\{k : k \leqq n \text{ and } k \in \mathcal{H}\},$$

provided that the limit exists.

A sequence (x_n) is *statistically convergent* to ℓ if for every $\epsilon > 0$,

$$\mathcal{H}_\epsilon = \{k : k \in \mathbb{N} \quad \text{and} \quad |x_k - \ell| \geqq \epsilon\}$$

has zero natural (asymptotic) density (see [1,2]). That is, for every $\epsilon > 0$,

$$\delta(\mathcal{H}_\epsilon) = \lim_{n\to\infty} \frac{|\mathcal{H}_\epsilon|}{n} = \lim_{n\to\infty} \frac{1}{n}|\{k : k \leqq n \quad \text{and} \quad |x_k - \ell| \geqq \epsilon\}| = 0.$$

Here, we write

$$\text{stat} \lim_{n\to\infty} x_n = \ell.$$

As an extension of statistical versions of convergence, the idea of weighted statistical convergence of single sequences was presented by Karakaya and Chishti [7], and it has been further generalized by various authors (see [8–12]). Moreover, the concept of deferred weighted statistical convergence was studied and introduced by Srivastava et al. [13] (see also [14–19]).

In the year 1900, Pringsheim [20] studied the convergence of double sequences. Recall that a double sequence $(x_{m,n})$ is convergent (or P-convergent) to a number ℓ if for given $\epsilon > 0$ there exists $n_0 \in \mathbb{N}$ such that $|x_{m,n} - \ell| < \epsilon$, whenever $m, n \geqq n_0$ and is written as $P \lim x_{m,n} = \ell$. Likewise, $(x_{m,n})$ is bounded if there exists a positive number $\mathcal{K}$ such that $|x_{m,n}| \leqq \mathcal{K}$. In contrast to the case of single sequences, here we note that a convergent double sequence is not necessarily bounded. We further recall that, a double sequence $(x_{m,n})$ is non-increasing in *Pringsheim's sense* if $x_{m+1,n} \leqq x_{m,n}$ and $x_{m,n+1} \leqq x_{m,n}$.

Let $\mathcal{H} \subset \mathbb{N} \times \mathbb{N}$ be the set of integers and let $\mathcal{H}(i,j) = \{(m,n) : m \leqq i \text{ and } n \leqq j\}$. The *double natural density* of $\mathcal{H}$ denoted by $\delta(\mathcal{H})$ is given by

$$\delta(\mathcal{H}) = P \lim_{i,j} \frac{1}{ij}|\mathcal{H}(i,j)|,$$

provided the limit exists. A double sequence $(x_{m,n})$ of real numbers is statistically convergent to ℓ in the *Pringsheim sense* if, for each $\epsilon > 0$

$$\delta(\mathcal{H}_\epsilon(i,j)) = 0,$$

where

$$\delta(\mathcal{H}_\epsilon(i,j)) = \frac{1}{ij}\{(m,n) : m \leqq i, n \leqq j \text{ and } |x_{m,n} - \ell| \geqq \epsilon\}.$$

Here, we write

$$\text{stat}^2 \lim_{m,n} x_{m,n} = \ell.$$

Note that every P-convergent double sequence is stat^2-convergent to the same limit, but the converse is not necessarily true.

Example 1. *Suppose we consider a double sequence* $x = (x_{m,n})$ *as*

$$x_{m,n} = \begin{cases} \sqrt{nm} & (m = k^2,\ n = l^2;\ \forall\, k, l \in \mathbb{N}), \\ \frac{1}{nm} & \textit{otherwise}. \end{cases}$$

It is trivially seen that, in the ordinary sense $(x_{m,n})$ *is not P-convergent; however,* 0 *is its statistical limit.*

Let $\mathcal{I} = [0, \infty) \subseteq \mathbb{R}$, and let the Lebesgue measure v be defined over $\mathcal{I}$. Let $\mathcal{I}^2 = [0, \infty) \times [0, \infty)$ and suppose that $X(\mathcal{I}^2)$ is the space of all measurable real-valued functions defined over $\mathcal{I}^2$ equipped with the equality almost everywhere. Also, let $C(\mathcal{I}^2)$ be the space of all continuous real-valued functions and suppose that $C^\infty(\mathcal{I}^2)$ is the space of all functions that are infinitely differentiable on $\mathcal{I}^2$. We recall here that a functional $\omega : X(\mathcal{I}^2) \to [0, \infty)$ is a *modular* on $X(\mathcal{I}^2)$ such that it satisfies the following conditions:

(i) $\omega(f) = 0$ if and only if $f = 0$, almost everywhere in $\mathcal{I}$ $(\forall\, f \in \mathcal{I}')$,
(ii) $\omega(\alpha f + \beta g) \leqq \omega(f) + \omega(g), \forall\, f, g \in X(\mathcal{I}^2)$ and for any $\alpha, \beta \geqq 0$ with $\alpha + \beta = 1$,
(iii) $\omega(-f) = \omega(f)$, for each $f \in X(\mathcal{I}^2)$, and
(iv) ω is continuous on $[0, \infty)$.

Also, we further recall that a modular ω is

- $\mathcal{N}$*-Quasi convex* if there exists a constant $\mathcal{N} \geqq 1$ satisfying

$$\omega(\alpha f + \beta g) \leqq \mathcal{N} \alpha \omega(\mathcal{N} f) + \mathcal{N} \beta \omega(\mathcal{N} g)$$

for every $f, g \in X(\mathcal{I}^2)$, $\alpha, \beta \geqq 0$ such that $\alpha + \beta = 1$. Also, in particular, for $\mathcal{N} = 1$, ω is simply called *convex*; and

- $\mathcal{N}$*-Quasi semi-convex* if there exists a constant $\mathcal{N} \geqq 1$ such that

$$\omega(\lambda f) \leqq \mathcal{N} \lambda \omega(\mathcal{N} f)$$

holds for all $f \in X(\mathcal{I}^2)$ and $\lambda \in (0, 1]$.

Also, it is trivial that every $\mathcal{N}$-Quasi semi-convex modular is $\mathcal{N}$-Quasi convex. The above concepts were initially studied by Bardaro et al. [21,22].

We now appraise some suitable subspaces of vector space $X(\mathcal{I}^2)$ under the modular ω as follows:

$$L^\omega(\mathcal{I}^2) = \{f \in X(\mathcal{I}^2) : \lim_{\lambda \to 0^+} \omega(\lambda f) = 0\}$$

and

$$E^\omega(\mathcal{I}^2) = \{f \in L^\omega(\mathcal{I}^2) : \omega(\lambda f) < +\infty,\ \forall\ \lambda > 0\}.$$

Here, $L^\omega(\mathcal{I}^2)$ is known as the modular space generated by ω and $E^\omega(\mathcal{I}^2)$ is known as the space of the finite elements of $L^\omega(\mathcal{I}^2)$. Also, it is trivial that whenever ω is $\mathcal{N}$-Quasi semi-convex,

$$\{f \in X(\mathcal{I}^2) : \omega(\lambda f) < +\infty,\ \forall\ \lambda > 0\}$$

coincides with $L^{\omega}(\mathcal{I}^2)$. Moreover, for a *convex modular* ω in $X(\mathcal{I}^2)$, the *F-norm* is given by the formula:

$$\|f\|_{\omega} = \inf\left\{\lambda > 0 : \omega\left(\frac{f}{\lambda}\right) \leqq 1\right\}.$$

The notion of modular was introduced in [23] and also widely discussed in [22].

In the year 1910, Moore [24] introduced the idea of the relatively uniform convergence of a sequence of functions. Later, along similar lines it was modified by Chittenden [25] for a sequence of functions defined over a closed interval $I = [a, b] \subseteq \mathbb{R}$.

We recall here the definition of uniform convergence relative to a scale function as follows.

A sequence of functions (f_n) defined over $[a, b]$ is *relatively uniformly convergent* to a limit function f if there exists a non-zero scale function σ defined over $[a, b]$, such that for each $\epsilon > 0$ there exists an integer n_{ϵ} and for every $n > n_{\epsilon}$,

$$\left|\frac{f_n(x) - f(x)}{\sigma(x)}\right| \leqq \epsilon$$

holds uniformly for all $x \in [a, b] \subseteq \mathbb{R}$.

Now, to see the importance of relatively uniform convergence (ordinary and statistical) over classical uniform convergence, we present the following example.

Example 2. *For all $n \in \mathbb{N}$, we define $f_n : [0, 1] \to \mathbb{R}$ by*

$$f_n(x) = \begin{cases} \frac{nx}{1+n^2x^2} & (0 < x \leqq 1), \\ 0 & (x = 0). \end{cases}$$

It is not difficult to see that the sequence (f_n) of functions is neither classically nor statistically uniformly convergent in $[0, 1]$; however, it is convergent uniformly to $f = 0$ relative to a scale function

$$\sigma(x) = \begin{cases} \frac{1}{x} & (0 < x \leqq 1) \\ 0 & (x = 0) \end{cases}$$

on $[0, 1]$. Here, we write

$$f_n \rightrightarrows f = 0 \quad ([0, 1]; \sigma).$$

In the middle of the twentieth century, H. Bohman [26] and P. P. Korovkin [27] established some approximation results by using positive linear operators. Later, some Korovkin-type approximation results with different settings were extended to several functional spaces, such as Banach space and Musielak–Orlicz space etc. Bardaro, Musielak, and Vinti [22] studied generalized nonlinear integral operators in connection with some approximation results over a modular space. Furthermore, Bardaro and Mantellini [28] proved some approximation theorems defined over a modular space by positive linear operators. They also established a conventional Korovkin-type theorem in a multivariate modular function space (see [21]). In the year 2015, Orhan and Demirci [29] established a result on statistical approximation by double sequences of positive linear operators on modular space. Demirci and Burçak [30] introduced the idea of A-statistical relative modular convergence of positive linear operators. Moreover, Demirci and Orhan [31] established some results on statistically relatively approximation on modular spaces. Recently, Srivastava et al. [13] established some approximation results on Banach space by using deferred weighted statistical convergence. Subsequently, they also introduced deferred weighted equi-statistical convergence to prove some approximation theorems (see [17]). Very recently, Md. Nasiruzzaman et al. [32] proved Dunkl-type generalization of Szász-Kantorovich operators via post-quantum calculus, and consequently, Srivastava et al. [33]

established the construction of Stancu-type Bernstein operators based on Bézier bases with shape parameter λ.

Motivated essentially by the above-mentioned results, in this paper we introduce the idea of relatively modular deferred-weighted statistical convergence and statistically as well as relatively modular deferred-weighted summability for double sequences of functions. We also establish an inclusion relation between them. Moreover, based upon our proposed methods, we prove a Korovkin-type approximation theorem for a double sequence of functions defined over a modular space and demonstrate that our result is a non-trivial generalization of some well-established results.

2. Relatively Modular Deferred-Weighted Mean

Let (a_n) and (b_n) be sequences of non-negative integers satisfying the conditions: (i) $a_n < b_n$ $(n \in \mathbb{N})$ and (ii) $\lim_{n\to\infty} b_n = \infty$. Note that (i) and (ii) are the regularity conditions for the proposed deferred weighted mean (see Agnew [34]). Now, for the double sequence $(f_{m,n})$ of functions, we define the deferred weighted summability mean $(N_D(f_{m,n}))$ as

$$N_D(f_{m,n}) = \frac{1}{T_m S_n} \sum_{u,v=a_n+1}^{b_m,b_n} t_u s_v f_{u,v}(x), \tag{1}$$

where (s_n) and (t_n) are the sequences of non-negative real numbers satisfying

$$S_n = \sum_{v=a_n+1}^{b_n} s_v \text{ and } T_m = \sum_{u=a_n+1}^{b_m} t_v.$$

Definition 1. *A double sequence $(f_{m,n})$ of functions belonging to $L^\omega(\mathcal{I}^2)$ is relatively modular deferred weighted $(N_D(f_{m,n}))$-summable to a function f on $L^\omega(\mathcal{I}^2)$ if and only if there exists a non-negative scale function $\sigma \in X(\mathcal{I}^2)$ such that*

$$P \lim_{m,n\to\infty} \omega\left(\lambda\left(\frac{N_D(f_{m,n}) - f}{\sigma}\right)\right) = 0 \textit{ for some } \lambda_0 > 0.$$

Here, we write

$$\mathcal{N}_D \lim_{m,n} \left\| \frac{f_{m,n} - f}{\sigma} \right\|_\omega = 0 \textit{ for some } \lambda_0 > 0.$$

Definition 2. *A double sequence $(f_{m,n})$ of functions belonging to $L^\omega(\mathcal{I}^2)$ is relatively F-norm (locally convex) deferred weighted summable (or relatively strong deferred weighted summable) to f if and only if*

$$P \lim_{m,n\to\infty} \omega\left(\lambda\left(\frac{N_D(f_{m,n}) - f}{\sigma}\right)\right) = 0 \textit{ for some } \lambda > 0.$$

Here, we write

$$\mathcal{F}\,\mathcal{N}_D \lim_{m,n} \left\| \frac{f_{m,n} - f}{\sigma} \right\|_\omega = 0 \textit{ for some } \lambda_0 > 0.$$

It can be promptly seen that, Definitions 1 and 2 are identical if and only if the modular ω fairly holds the Δ_2-condition, that is, there exists a constant $\mathcal{M} > 0$ such that $\omega(2f) \leqq \mathcal{M}\omega(f)$ for every $f \in X(\mathcal{I}^2)$. Precisely, relatively strong summability of the double sequence $(f_{m,n})$ to f is identical to the condition

$$P \lim_{m,n} \omega\left(2^n \lambda\left(\frac{N_D(f_{m,n}) - f}{\sigma}\right)\right) = 0,$$

$\forall\ n \in \mathbb{N}$ and some $\lambda > 0$. Thus, if $(f_{m,n})$ is relatively modular deferred weighted $(N_D(f_{m,n}))$-summable to f, then by Definition 1 there exists a $\lambda > 0$ such that

$$P \lim_{m,n\to\infty} \omega\left(\lambda\left(\frac{N_D(f_{m,n}) - f}{\sigma}\right)\right) = 0.$$

Clearly, under Δ_2-condition, we have

$$\omega\left(2^n\lambda\left(\frac{N_D(f_{m,n}) - f}{\sigma}\right)\right) \leqq \mathcal{M}^n\omega\left(\lambda\left(\frac{N_D(f_{m,n}) - f}{\sigma}\right)\right).$$

This implies that

$$P \lim_{m,n} \omega\left(2^n\lambda\left(\frac{N_D(f_{m,n}) - f}{\sigma}\right)\right) = 0.$$

Definition 3. *A double sequence $(f_{m,n})$ of functions belonging to $L^\omega(\mathcal{I}^2)$ is relatively modular deferred-weighted $(N_D(f_{m,n}))$ statistically convergent to a function $f \in L^\omega(\mathcal{I}^2)$ if there exists a non-zero scale function $\sigma \in X(\mathcal{I}^2)$ such that, for every $\epsilon > 0$, the following set:*

$$P \lim_{m,n} \frac{1}{T_m S_n}\left\{(u,v) : u \leqq T_m, v \leqq S_m \text{ and } \omega\left(\lambda_0\left(\frac{t_u s_v|f_{u,v} - f|}{\sigma}\right)\right) \geqq \epsilon\right\} \text{ for some } \lambda_0 > 0$$

has zero relatively deferred-weighted density, that is,

$$P \lim_{m,n} \frac{1}{T_m S_n}\left|\left\{(u,v) : u \leqq T_m, v \leqq S_m \text{ and } \omega\left(\lambda_0\left(\frac{t_u s_v|f_{u,v} - f|}{\sigma}\right)\right) \geqq \epsilon\right\}\right| = 0 \text{ for some } \lambda_0 > 0.$$

Here, we write

$$stat_{N_D} \lim_{m,n}\left\|\frac{f_{m,n} - f}{\sigma}\right\|_\omega = 0.$$

Moreover, $(f_{m,n})$ is relatively F-norm (locally convex) deferred-weighted $(N_D(f_{m,n}))$ statistically convergent (or relatively strong deferred-weighted $(N_D(f_{m,n}))$ statistically convergent) to a function $f \in X(\mathcal{I}^2)$ if and only if

$$P \lim_{m,n} \frac{1}{T_m S_n}\left|\left\{(u,v) : u \leqq T_m, v \leqq S_m \text{ and } \omega\left(\lambda_0\left(\frac{t_u s_v|f_{u,v} - f|}{\sigma}\right)\right) \geqq \epsilon\right\}\right| = 0 \text{ for some } \lambda > 0,$$

where $\sigma \in X(\mathcal{I}^2)$ is a non-zero scale function and $\epsilon > 0$.

Here, we write

$$\mathcal{F}\ stat_{N_D} \lim_{m,n}\left\|\frac{f_{m,n} - f}{\sigma}\right\|_\omega = 0.$$

Definition 4. *A double sequence $(f_{m,n})$ of functions belonging to $L^\omega(\mathcal{I}^2)$ is statistically and relatively modular deferred-weighted $(N_D(f_{m,n}))$-summable to a function $f \in L^\omega(\mathcal{I}^2)$ if there exists a non-zero scale function $\sigma \in X(\mathcal{I}^2)$ such that, for every $\epsilon > 0$, the following set:*

$$P \lim_{m,n} \frac{1}{m,n}\left\{(u,v) : u \leqq m, v \leqq m \text{ and } \omega\left(\lambda_0\left(\frac{N_D(f_{m,n}) - f}{\sigma}\right)\right) \geqq \epsilon\right\} \text{ for some } \lambda_0 > 0$$

has zero relatively deferred-weighted density, that is,

$$P \lim_{m,n} \frac{1}{mn}\left|\left\{(u,v) : u \leqq m, v \leqq n \text{ and } \omega\left(\lambda_0\left(\frac{N_D(f_{m,n}) - f}{\sigma}\right)\right) \geqq \epsilon\right\}\right| = 0 \text{ for some } \lambda_0 > 0.$$

Here, we write

$$N_D stat \lim_{m,n} \left\| \frac{f_{m,n} - f}{\sigma} \right\|_\omega = 0.$$

Furthermore, $(f_{m,n})$ is statistically and relatively F-norm (locally convex) deferred-weighted $(N_D(f_{m,n}))$-summable (or statistically and relatively strong deferred-weighted $(N_D(f_{m,n}))$-summable) to a function $f \in X(\mathcal{I}^2)$ if and only if

$$P \lim_{m,n} \frac{1}{m,n} \left| \left\{ (u,v) : u \leqq m, v \leqq n \text{ and } \omega\left(\lambda_0 \left(\frac{N_D(f_{m,n}) - f}{\sigma}\right)\right) \geqq \epsilon \right\} \right| = 0 \text{ for some } \lambda > 0,$$

where $\sigma \in X(\mathcal{I}^2)$ is a non-zero scale function and $\epsilon > 0$.

Here, we write

$$\mathcal{F}\ N_D stat \lim_{m,n} \left\| \frac{f_{m,n} - f}{\sigma} \right\|_\omega = 0.$$

Remark 1. *If we put $a_n = 0$, $b_n = n$, $b_m = m$, and $t_m = s_n = 1$ in Definition 3, then it reduces to relatively modular statistical convergence (see [31]).*

Next, for our present study on a modular space we have the assumptions as follows:

- If $\omega(f) \leqq \omega(g)$ for $|f| \leqq |g|$, then ω is monotone;
- If $\chi \in L^\omega(\mathcal{I}^2)$ with $\mu(A) < \infty$, where A is a measurable subset of $\mathcal{I}^2$, then ω is finite;
- If ω is finite and for each $\epsilon > 0$, $\lambda > 0$, there exists a $\delta > 0$ and $\omega(\lambda\chi_B) < \epsilon$ for any measurable subset $B \subset \mathcal{I}^2$ such that $\mu(B) < \delta$, then ω is absolutely finite;
- If $\chi_{\mathcal{I}^2} \in E^w(\mathcal{I}^2)$, then ω is strongly finite;
- If for each $\epsilon > 0$ there exists a $\delta > 0$ such that $\omega(\alpha f \chi_B) < \epsilon$ $(\alpha > 0)$, where B is a measurable subset of $\mathcal{I}^2$ with $\mu(B) < \delta$ and for each $f \in X(\mathcal{I}^2)$ with $\omega(f) < +\infty$, then ω is absolutely continuous.

It is clearly observed from the above assumptions that if a modular ω is finite and monotone, then $C(\mathcal{I}^2) \subset L^\omega(\mathcal{I}^2)$. Also, if ω is strongly finite and monotone, then $C(\mathcal{I}^2) \subset E^\omega(\mathcal{I}^2)$. Furthermore, if ω is absolutely continuous, monotone, and absolutely finite, then $\overline{C^\infty(\mathcal{I}^2)} = L^\omega(\mathcal{I}^2)$, where the closure $\overline{C^\infty(\mathcal{I}^2)}$ is compact over the modular space.

Now we establish the following theorem by demonstrating an inclusion relation between relatively deferred-weighted statistical convergence and statistically as well as relatively deferred-weighted summability over a modular space.

Theorem 1. *Let ω be a strongly finite, monotone, and $\mathcal{N}$-Quasi convex modular on $L^\omega(\mathcal{I}^2)$. If a double sequence $(f_{m,n})$ of functions belonging to $L^\omega(\mathcal{I}^2)$ is bounded and relatively modular deferred-weighted statistically convergent to a function $f \in L^\omega(\mathcal{I}^2)$, then it is statistically and relatively modular deferred weighted summable to the function f, but not conversely.*

Proof. Assume that $(f_{m,n}) \in L^\omega(\mathcal{I}^2) \cap \ell_\infty$. Let us set

$$\mathcal{H}_\epsilon = \left\{ (u,v) : u \leqq m, v \leqq n \text{ and } \omega\left(\lambda_0 \left(\frac{f_{u,v} - f}{\sigma}\right)\right) \geqq \epsilon \text{ for some } \lambda_0 > 0 \right\}$$

and

$$\mathcal{H}_\epsilon^c = \left\{ (u,v) : u \leqq m, v \leqq n \text{ and } \omega\left(\lambda_0 \left(\frac{f_{u,v} - f}{\sigma}\right)\right) > \epsilon \text{ for some } \lambda_0 > 0 \right\}.$$

From the regularity condition of our proposed mean, we have

$$P \lim_{u,v} \frac{1}{T_m S_n} \sum_{u,v=a_n+1}^{b_m,b_n} t_u s_v = 0. \tag{2}$$

Thus, we obtain

$$\begin{aligned}
\omega\left(\lambda_0\left(N_D\left(\frac{f_{m,n}-f}{\sigma}\right)\right)\right) &= \omega\left(\lambda_0\left(\frac{1}{T_mS_n}\sum_{u,v=a_n+1}^{b_m,b_n} t_u s_v\left(\frac{f_{u,v}-f}{\sigma}\right)\right)\right)\\
&\leqq \omega\Bigg(\frac{\lambda_0}{T_mS_n}\sum_{\substack{u,v=a_n+1,\\(u,v)\in\mathcal{H}_\epsilon}}^{b_m,b_n} t_u s_v\left|\frac{f_{u,v}-f}{\sigma}\right| + \frac{\lambda_0}{T_mS_n}\sum_{\substack{u=0,v=b_n+1,\\(u,v)\in\mathcal{H}_\epsilon}}^{b_m,\infty} t_u s_v\left|\frac{f_{u,v}-f}{\sigma}\right|\\
&+ \frac{\lambda_0}{T_mS_n}\sum_{\substack{u=b_m+1,v=0,\\(u,v)\in\mathcal{H}_\epsilon}}^{\infty,b_n} t_u s_v\left|\frac{f_{u,v}-f}{\sigma}\right| + \frac{\lambda_0}{T_mS_n}\sum_{\substack{u=b_m+1,v=b_n+1,\\(u,v)\in\mathcal{H}_\epsilon}}^{\infty,\infty} t_u s_v\left|\frac{f_{u,v}-f}{\sigma}\right|\\
&+ \omega\Bigg(\frac{\lambda_0}{T_mS_n}\sum_{\substack{u,v=a_n+1,\\(u,v)\in\mathcal{H}_\epsilon^c}}^{b_m,b_n} t_u s_v\left|\frac{f_{u,v}-f}{\sigma}\right| + \frac{\lambda_0}{T_mS_n}\sum_{\substack{u=0,v=b_n+1,\\(u,v)\in\mathcal{H}_\epsilon^c}}^{b_m,\infty} t_u s_v\left|\frac{f_{u,v}-f}{\sigma}\right|\\
&+ \frac{\lambda_0}{T_mS_n}\sum_{\substack{u=b_m+1,v=0;\\(u,v)\in\mathcal{H}_\epsilon^c}}^{\infty,b_n} t_u s_v\left|\frac{f_{u,v}-f}{\sigma}\right| + \frac{\lambda_0}{T_mS_n}\sum_{\substack{u=b_m+1,v=b_n+1\\(u,v)\in\mathcal{H}_\epsilon^c}}^{\infty,\infty} t_u s_v\left|\frac{f_{u,v}-f}{\sigma}\right|\\
&+ \mathcal{K}\left|\frac{1}{T_mS_n}\sum_{u,v=a_n+1}^{\infty,\infty} t_u s_v - 1\right|\Bigg),
\end{aligned}$$

where

$$\mathcal{K} = \sup_{x,y}\left|\frac{f(x,y)}{\sigma}\right|.$$

Further, ω being $\mathcal{N}$-Quasi convex modular, monotone, and strongly finite on $L^\omega(\mathcal{I}^2)$, it follows that

$$\begin{aligned}
\omega\left(\lambda_0\left(N_D\left(\frac{f_{m,n}-f}{\sigma}\right)\right)\right) &\leqq 3\omega\left(\frac{9\lambda_0|\mathcal{H}_\epsilon|G}{T_mS_n}\sum_{\substack{u,v=a_n+1,\\(u,v)\in\mathcal{H}_\epsilon}}^{b_m,b_n} t_u s_v\right)\\
&+ \epsilon\omega\left(\frac{9\lambda_0|\mathcal{H}_\epsilon|}{T_mS_n}\sum_{\substack{u,v=a_n+1,\\(u,v)\in\mathcal{H}_\epsilon}}^{b_m,b_n} t_u s_v\right) + \omega\left(\frac{9\lambda_0 G b_m b_n}{T_mS_n}\sum_{u,v=a_n+1}^{b_m,b_n} t_u s_v\right)\\
&+ \omega\left(\frac{9\lambda_0 G b_m}{T_mS_n}\sum_{u=0,v=a_n+1}^{b_m,\infty} t_u s_v\right) + \omega\left(\frac{9\lambda_0 G b_n}{T_mS_n}\sum_{u=a_n+1,v=0}^{\infty,b_m} t_u s_v\right)\\
&+ \epsilon\omega\left(\frac{9\lambda_0}{T_mS_n}\sum_{u,v=a_n+1}^{\infty,\infty} t_u s_v\right) + \omega\left(\frac{9\lambda_0\mathcal{K}}{T_mS_n}\sum_{u,v=a_n+1}^{\infty,\infty} t_u s_v - 1\right),
\end{aligned}$$

where $G = \max\left|\frac{f_{u,v}-f(x,y)}{\sigma}\right|$, $\forall\, u,v \in \mathbb{N}$ and $(x,y) \in \mathcal{I}^2$. In the last inequality, considering P limit as $m,n \to \infty$ under the regularity conditions of deferred weighted mean and by using (2), we obtain

$$P\lim_{m,n}\omega\left(\lambda_0\left(\frac{N_D(f_{m,n})-f}{\sigma}\right)\right) = 0.$$

This implies that $(f_{m,n})$ is relatively modular deferred weighted $N_D(f_{m,n})$-summable to a function f. Hence,

$$P\lim_{m,n}\frac{1}{m,n}\left|\left\{(u,v) : u \leqq m, v \leqq m \text{ and } \omega\left(\lambda_0\left(\frac{N_D(f_{m,n})-f}{\sigma}\right)\right) \geqq \epsilon\right\}\right| = 0 \text{ for some } \lambda_0 > 0.$$

Next, to see that the converse part of the theorem is not necessarily true, we consider the following example.

Example 3. *Suppose that $\mathcal{I} = [0,1]$ and let $\varphi : [0,\infty) \to [0,\infty)$ be a continuous function with $\varphi(0) = 0$, $\varphi(u) > 0$ for $u > 0$ and $\lim_{u\to\infty} \varphi(u) = \infty$. Let $f \in X(\mathcal{I}^2)$ be a measurable real-valued function, and consider the functional ω^φ on $X(\mathcal{I}^2)$ defined by*

$$\omega^\varphi(f) = \int_0^1 \int_0^1 \varphi(|f_{m,n}(x,y)|)dxdy \quad (f \in X(\mathcal{I}^2).$$

φ being convex, ω^φ is modular convex on $X(\mathcal{I}^2)$, which satisfies the above assumptions. Consider $L^\omega_\varphi(\mathcal{I}^2)$ as the Orlicz space produced by φ of the form:

$$L^\omega_\varphi(\mathcal{I}^2) = \{f \in X(\mathcal{I}^2) : \omega^\varphi(\lambda(f)) < +\infty \text{ for some } \lambda > 0\}.$$

For all $m,n \in \mathbb{N}$, we consider a double sequence of functions $f_{m,n} : [0,1] \times [0,1] \to \mathbb{R}$ defined by

$$f_{m,n}(x,y) = \begin{cases} 1, & (m,n) \in \mathfrak{U} \times \mathfrak{U} \text{ and } (x,y) \in (0,\frac{1}{m}] \times (0,\frac{1}{n}], \\ 0, & \{(m,n) \in \mathfrak{V} \times \mathfrak{V} \text{ and } (x,y) \in (\frac{1}{m},0] \times (\frac{1}{n},1]; \\ & (m,n) \in \mathfrak{U} \times \mathfrak{V} \text{ or } (m,n) \in \mathfrak{V} \times \mathfrak{U} \text{ or } (x,y) \in (0,0)\}, \end{cases}$$

where the set of all odd and even numbers are $\mathfrak{U}$ and $\mathfrak{V}$, respectively.

We have

$$\omega\lambda(N_D(f_{m,n})) = \omega\left(\frac{\lambda_0}{S_m T_n} \sum_{u,v=a_n+1}^{b_m,b_n} t_u s_v\right),$$

and this implies

$$\omega\lambda(N_D(f_{m,n})) = \lambda_0 \begin{cases} \int_0^{1/b_m} \int_0^{1/b_n} dxdy, & (m,n) \in \mathfrak{U} \times \mathfrak{U} \text{ and } (x,y) \in (0,\frac{1}{m}] \times (0,\frac{1}{n}], \\ 0, & \{(m,n) \in \mathfrak{V} \times \mathfrak{V} \text{ and } (x,y) \in (\frac{1}{m},0] \times (\frac{1}{n},1]; \\ & (m,n) \in \mathfrak{U} \times \mathfrak{V} \text{ or } (m,n) \in \mathfrak{V} \times \mathfrak{U} \text{ or } (x,y) \in (0,0)\}. \end{cases}$$

Clearly, $(f_{m,n})$ is relatively modular deferred weighted summable to $f = 0$, with respect to a non-zero scale function $\sigma(x,y)$ such that

$$\sigma(x,y) = \begin{cases} 1, & (x,y) = (0,0) \\ \frac{1}{xy}, & (x,y) \in (0,1] \times (0,1]. \end{cases}$$

That is,

$$P \lim_{m,n} \omega\left(\lambda_0\left(\frac{N_D(f_{m,n}) - f}{\sigma}\right)\right) = 0 \text{ for some } \lambda_0 > 0.$$

Thus, we have

$$P \lim_{m,n} \frac{1}{m,n} \left|\left\{(u,v) : u \leqq m, v \leqq m \text{ and } \omega\left(\lambda_0\left(\frac{N_D(f_{m,n}) - f}{\sigma}\right)\right) \geqq \epsilon\right\}\right| = 0 \text{ for some } \lambda_0 > 0.$$

On the other hand, it is not relatively modular deferred-weighted statistically convergent to the function $f = 0$, that is,

$$P \lim_{m,n} \frac{1}{T_m S_n} \left| \left\{ (u,v) : u \leqq T_m, v \leqq S_m \text{ and } \omega \left(\lambda_0 \left(\frac{t_u s_v |f_{u,v} - f|}{\sigma} \right) \right) \geqq \epsilon \right\} \right| \neq 0 \text{ for some } \lambda_0 > 0.$$
□

3. A Korovkin-Type Theorem in Modular Space

In this section, we extend here the result of Demirci and Orhan [31] by using the idea of the statistically and relatively modular deferred-weighted summability of a double sequence of positive linear operators defined over a modular space.

Let ω be a finite modular and monotone over $X(\mathcal{I}^2)$. Suppose E is a set such that $C^\infty(\mathcal{I}^2) \subset E \subset L^\omega(\mathcal{I}^2)$. We can construct such a subset E when ω is monotone and finite. We also assume $L = \{\mathcal{L}_{m,n}\}$ as the sequence of positive linear operators from E in to $X(\mathcal{I}^2)$, and there exists a subset $X_L \subset E$ containing $C^\infty(\mathcal{I}^2)$. Let $\sigma \in X(\mathcal{I}^2)$ be an unbounded function with $|\sigma(x,y)| \neq 0$, and R is a positive constant such that

$$N_D\text{stat} \limsup_{m,n} \omega \left(\lambda \left(\frac{\Upsilon_{m,n}(f)}{\sigma} \right) \right) \leqq R\omega(\lambda f) \tag{3}$$

holds for each $f \in X_L, \lambda > 0$ and

$$\Upsilon_{m,n}(f;x,y) = \frac{1}{T_m S_n} \sum_{u,v=a_n+1}^{b_m,b_n} t_u s_v \mathcal{T}_{m,n}(f;x,y).$$

We denote here the value of $\mathcal{L}_{m,n}(f)$ at a point $(x,y) \in \mathcal{I}^2$ by $\mathcal{L}_{m,n}(f(x^*,y^*);x,y)$, or briefly by $\mathcal{L}_{m,n}(f;x,y)$. We now prove the following theorem.

Theorem 2. *Let (a_n) and (b_n) be the sequences of non-negative integers and let ω be an $\mathcal{N}$-Quasi semi-convex modular, absolutely continuous, strongly finite, and monotone on $X(\mathcal{I}^2)$. Assume that $L = \{\mathcal{L}_{m,n}\}$ is a double sequence of positive linear operators from E in to $X(\mathcal{I}^2)$ that satisfy the assumption (3) for every $f \in X_L$ and suppose that $\sigma_i(x,y)$ is an unbounded function such that $|\sigma_i(x,y)| \geqq u_i > 0$ $(i = 0,1,2,3)$. Assume further that*

$$N_D\text{stat} \lim_{m,n} \left\| \frac{\mathcal{L}_{m,n}(f_i;x,y) - f(x,y)}{\sigma} \right\|_\omega = 0 \text{ for each } \lambda > 0 \text{ and } i = 0,1,2,3, \tag{4}$$

where

$$f_0(x,y) = 1, \ f_1(x,y) = x, \ f_2(x,y) = y \text{ and } f_3(x,y) = x^2 + y^2.$$

Then, for every $f \in L^\omega(\mathcal{I}^2)$ and $g \in C^\infty(\mathcal{I}^2)$ with $f - g \in X_L$,

$$N_D\text{stat} \lim_{m,n} \left\| \frac{\mathcal{L}_{m,n}(f;x,y) - f(x,y)}{\sigma} \right\|_\omega = 0 \text{ for every } \lambda_0 > 0, \tag{5}$$

where $\sigma(x,y) = \max\{|\sigma_i(x,y)| : i = 0,1,2,3\}$.

Proof. First we claim that,

$$N_D\text{stat} \lim_{m,n} \left\| \frac{\mathcal{L}_{m,n}(g;x,y) - g(x,y)}{\sigma} \right\|_\omega = 0 \text{ for every } \lambda_0 > 0. \tag{6}$$

In order to justify our claim, we assume that $g \in C(\mathcal{I}^2) \cap E$. Since g is continuous on $\mathcal{I}^2$, for given $\epsilon > 0$, there exists a number $\delta > 0$ such that for every $(x^*,y^*), (x,y) \in \mathcal{I}^2$ with $|x^* - x| < \delta$ and $|y^* - y| < \delta$, we have

$$|g(x^*,y^*) - g(x,y)| < \epsilon. \tag{7}$$

Also, for all $(x^*, y^*), (x, y) \in \mathcal{I}^2$ with $|x^* - x| > \delta$ and $|x^* - x| > \delta$, we have

$$|g(x^*, y^*) - g(x, y)| < \frac{2\mathcal{A}}{\delta^2}\left([\varphi_1(x^*, x)]^2 + [\varphi_2(y^*, y)]^2\right), \tag{8}$$

where

$$\varphi_1(x^*, x) = (x^* - x), \quad \varphi_2(y^*, y) = (y^* - y), \quad \text{and } \mathcal{A} = \sup_{x,y \in \mathcal{I}^2} |g(x, y)|.$$

From Equations (7) and (8), we obtain

$$|g(x^*, y^*) - g(x, y)| < \epsilon + \frac{2\mathcal{A}}{\delta^2}\left([\varphi_1(x^*, x)]^2 + [\varphi_2(y^*, y)]^2\right).$$

This implies that

$$-\epsilon - \frac{2\mathcal{A}}{\delta^2}\left([\varphi_1(x^*, x)]^2 + [\varphi_2(y^*, y)]^2\right) < g(x^*, y^*) - g(x, y) < \epsilon + \frac{2\mathcal{A}}{\delta^2}\left([\varphi_1(x^*, x)]^2 + [\varphi_2(y^*, y)]^2\right). \tag{9}$$

Now $\mathcal{L}_{m,n}(g_0; x, y)$ being linear and monotone, by applying the operator $\mathcal{L}_{m,n}(g_0; x, y)$ to this inequality (9), we fairly have

$$\begin{aligned}\mathcal{L}_{m,n}(g_0; x, y)\left(-\epsilon - \frac{2\mathcal{A}}{\delta^2}\left([\varphi_1(x^*, x)]^2 + [\varphi_2(y^*, y)]^2\right)\right) &< \mathcal{L}_{m,n}(g_0; x, y)(g(x^*, y^*) - g(x, y)) \\ &< \mathcal{L}_{m,n}(g_0; x, y)\left(\epsilon + \frac{2\mathcal{A}}{\delta^2}\left([\varphi_1(x^*, x)]^2 + [\varphi_2(y^*, y)]^2\right)\right).\end{aligned} \tag{10}$$

Note that x, y is fixed, and so also $g(x, y)$ is a constant number. This implies that

$$\begin{aligned}-\epsilon\mathcal{L}_{m,n}(g_0; x, y) - \frac{2\mathcal{A}}{\delta^2}\mathcal{L}_{m,n}\left([\varphi_1(x^*, x)]^2 + [\varphi_2(y^*, y)]^2; x, y\right) &< \mathcal{L}_{m,n}(g; x, y) - g(x, y)\mathcal{L}_{m,n}(g_0; x, y) \\ &< \epsilon\mathcal{L}_{m,n}(g_0; x, y) + \frac{2\mathcal{A}}{\delta^2}\mathcal{L}_{m,n}([\varphi_1(x^*, x)]^2 + [\varphi_2(y^*, y)]^2; x, y).\end{aligned} \tag{11}$$

However,

$$\mathcal{L}_{m,n}(g; x, y) - g(x, y) = [\mathcal{L}_{m,n}(g; x, y) - g(x, y)\mathcal{L}_{m,n}(g_0; x, y)] + g(x, y)[\mathcal{L}_{m,n}(g_0; x, y) - g_0(x, y)]. \tag{12}$$

Now, using (11) and (12), we have

$$\begin{aligned}|\mathcal{L}_{m,n}(g; x, y) - g(x, y)| \leqq& \left|\epsilon\mathcal{L}_{m,n}(g_0; x, y) + \frac{2\mathcal{A}}{\delta^2}\mathcal{L}_{m,n}\left([\varphi_1(x^*, x)]^2 + [\varphi_2(y^*, y)]^2; x, y\right)\right| \\ &+ \mathcal{A}[\mathcal{L}_{m,n}(g_0; x, y) - g_0(x, y)].\end{aligned} \tag{13}$$

Next,

$$\begin{aligned}|\mathcal{L}_{m,n}(g; x, y) - g(x, y)| = \epsilon + (\epsilon + \mathcal{A})[\mathcal{L}_{m,n}(g_0; x, y) - g_0(x, y)] &- \frac{4\mathcal{A}}{\delta^2}|g_1(x, y)|[\mathcal{L}_{m,n}(g_1; x, y) - g_1(x, y)] \\ + \frac{2\mathcal{A}}{\delta^2}[\mathcal{L}_{m,n}(g_3; x, y) - g_3(x, y)] &- \frac{4\mathcal{A}}{\delta^2}|g_2(x, y)|[\mathcal{L}_{m,n}(g_2; x, y) - g_2(x, y)] \\ + \frac{2\mathcal{A}}{\delta^2}|g_3(x, y)|[\mathcal{L}_{m,n}(g_0; x, y) - g_0(x, y)].&\end{aligned}$$

Since the choice of ϵ is arbitrarily small, we can easily write

$$
\begin{aligned}
|\mathcal{L}_{m,n}(g;x,y)-g(x,y)| \leqq \epsilon &+ \left(\epsilon+\frac{2\mathcal{A}}{\delta^2}+\mathcal{A}\right)|\mathcal{L}_{m,n}(g_0;x,y)-g_0(x,y)| \\
&+ \frac{4\mathcal{A}}{\delta^2}|g_1(x,y)||\mathcal{L}_{m,n}(g_1;x,y)-g_1(x,y)| + \frac{2\mathcal{A}}{\delta^2}|\mathcal{L}_{m,n}(g_3;x,y)-g_3(x,y)| \\
&- \frac{4\mathcal{A}}{\delta^2}|g_2(x,y)||\mathcal{L}_{m,n}(g_2;x,y)-g_2(x,y)|.
\end{aligned} \tag{14}
$$

Now multiplying $\frac{1}{\sigma(x,y)}$ to both sides of (14), we have, for any $\lambda > 0$

$$
\begin{aligned}
\lambda\left|\frac{\mathcal{L}_{m,n}(g;x,y)-g(x,y)}{\sigma(x,y)}\right| \leqq \frac{\lambda\epsilon}{\sigma(x,y)} &+ \lambda\mathcal{B}\Bigg\{\left|\frac{\mathcal{L}_{m,n}(g_0;x,y)-g_0(x,y)}{\sigma(x,y)}\right| \\
&+\left|\frac{\mathcal{L}_{m,n}(g_1;x,y)-g_1(x,y)}{\sigma(x,y)}\right| + \left|\frac{\mathcal{L}_{m,n}(g_3;x,y)-g_3(x,y)}{\sigma(x,y)}\right| \\
&-\left|\frac{\mathcal{L}_{m,n}(g_2;x,y)-g_2(x,y)}{\sigma(x,y)}\right|\Bigg\},
\end{aligned} \tag{15}
$$

where $\mathcal{B} = \max\left(\epsilon+\frac{2\mathcal{A}}{\delta^2}+\mathcal{A}, \frac{4\mathcal{A}}{\delta^2}, \frac{2\mathcal{A}}{\delta^2}\right)$ and $g_1(x,y)$, $g_2(x,y)$ are constants for $\forall\ (x,y)$.

Next, applying the modular ω to the above inequality, also ω being $\mathcal{N}$-Quasi semi-convex, strongly finite, monotone, and $\sigma(x,y) = \max\{|\sigma_i(x,y)\ \ (i=0,1,2,3)|\}$, we have

$$
\begin{aligned}
\omega\left(\lambda\left(\frac{\mathcal{L}_{m,n}(g;x,y)-g(x,y)}{\sigma(x,y)}\right)\right) \leqq \omega\left(\frac{5\lambda\epsilon}{\sigma(x,y)}\right) &+ \omega\left(5\lambda\mathcal{B}\left(\frac{\mathcal{L}_{m,n}(g_0;x,y)-g_0(x,y)}{\sigma_0(x,y)}\right)\right) \\
&+\omega\left(5\lambda\mathcal{B}\left(\frac{\mathcal{L}_{m,n}(g_1;x,y)-g_1(x,y)}{\sigma_1(x,y)}\right)\right) \\
&+\omega\left(5\lambda\mathcal{B}\left(\frac{\mathcal{L}_{m,n}(g_3;x,y)-g_3(x,y)}{\sigma_2(x,y)}\right)\right) \\
&-\omega\left(5\lambda\mathcal{B}\left(\frac{\mathcal{L}_{m,n}(g_2;x,y)-g_2(x,y)}{\sigma_3(x,y)}\right)\right).
\end{aligned} \tag{16}
$$

Now, replacing $\mathcal{L}_{m,n}(f;x,y)$ by

$$
\frac{1}{S_m T_n}\sum_{u,v=a_n+1}^{b_m,b_n} s_u t_v \mathcal{T}_{u,v}(g;x,y) = \Upsilon_{m,n}(f;x,y)
$$

and then by $\Psi(f;x,y)$ in (16), for a given $\kappa > 0$ there exists $\epsilon > 0$, such that $\omega\left(\frac{5\lambda\epsilon}{\sigma}\right) < \kappa$. Then, by setting

$$
\Psi = \left\{(m,n) : \omega\left(\lambda\left(\frac{\Upsilon_{m,n}(g)-g}{\sigma}\right)\right) \geqq \kappa\right\}
$$

and for $i = 0,1,2$,

$$
\Psi_i = \left\{(m,n) : \omega\left(\lambda\left(\frac{\Upsilon_{m,n}(g_i)-g}{\sigma_i}\right)\right) \geqq \frac{\kappa-\omega\left(\frac{5\lambda\epsilon}{\sigma}\right)}{4\mathcal{B}}\right\},
$$

we obtain

$$
\Psi \leqq \sum_{i=0}^{3}\Psi_i.
$$

Clearly,

$$\frac{\|\Psi\|_\omega}{mn} \leqq \sum_{i=0}^{3} \frac{\|\Psi_i\|_\omega}{mn}. \tag{17}$$

Now, by the assumption under (4) as well as by Definition 4, the right-hand side of (17) tends to zero as $m,n \to \infty$. Clearly, we get

$$\lim_{m,n\to\infty} \frac{\|\Psi\|_\omega}{mn} = 0 \ (\kappa > 0),$$

which justifies our claim (6). Hence, the implication (6) is fairly obvious for each $g \in C^\infty(\mathcal{I}^2)$.

Now let $f \in L^\omega(\mathcal{I}^2)$ such that $f - g \in X_L$ for every $g \in C^\infty(\mathcal{I}^2)$. Also, ω is absolutely continuous, monotone, strongly and absolutely finite on $X(\mathcal{I}^2)$. Thus, it is trivial that the space $C^\infty(\mathcal{I}^2)$ is modularly dense in $L^\omega(\mathcal{I}^2)$. That is, there exists a sequence $(g_{i,j}) \in C^\infty(\mathcal{I}^2)$ provided that $\omega(3\lambda_0^* g) < +\infty$ and

$$P \lim_{i,j} \omega(3\lambda_0^*(g_{i,j} - f)) = 0 \ \text{ for some } \lambda_0^*. \tag{18}$$

This implies that for each $\epsilon > 0$ there exist two positive integers $\bar{i}$ and $\bar{j}$ such that

$$\omega(3\lambda_0^*(g_{i,j} - f)) < \epsilon \ \text{ whenever } \ i \geqq \bar{i} \text{ and } j \geqq \bar{j}.$$

Further, since the operators $\Upsilon_{m,n}$ are positive and linear, we have that

$$\begin{aligned}\lambda_0^*|\Upsilon_{m,n}(f;x,y) - f(x,y)| \leqq{}& \lambda_0^*|\Upsilon_{m,n}(f - g_{\bar{i},\bar{j}};x,y)| + \lambda_0^*|\Upsilon_{m,n}(g_{\bar{i},\bar{j}};x,y) - g_{\bar{i},\bar{j}}(x,y)| \\ &+ \lambda_0^*|g_{\bar{i},\bar{j}}(x,y) - f(x,y)|\end{aligned}$$

holds true for each $m,n \in \mathbb{N}$ and $x,y \in \mathcal{I}$. Applying the monotonicity of modular ω and further multiplying $\frac{1}{\sigma(x,y)}$ to both sides of the above inequality, we have

$$\begin{aligned}\omega\left(\lambda_0^*\left(\frac{\Upsilon_{m,n}(f;x,y) - f(x,y)}{\sigma}\right)\right) \leqq{}& \omega\left(3\lambda_0^*\left(\frac{\Upsilon_{m,n}(f - g_{\bar{i},\bar{j}})}{\sigma}\right)\right) \\ &+ \omega\left(3\lambda_0^*\left(\frac{\Upsilon_{m,n}(g_{\bar{i},\bar{j}}) - g_{\bar{i},\bar{j}}}{\sigma}\right)\right) + \omega\left(3\lambda_0^*\left(\frac{g_{\bar{i},\bar{j}} - f}{\sigma}\right)\right).\end{aligned}$$

Thus, for $|\sigma(x,y)| \geqq M > 0$ ($M = \max\{M_i : i = 0,1,2,3\}$), we can write

$$\begin{aligned}\omega\left(\lambda_0^*\left(\frac{\Upsilon_{m,n}(f) - f}{\sigma}\right)\right) \leqq{}& \omega\left(3\lambda_0^*\left(\frac{\Upsilon_{m,n}(f - g_{\bar{i},\bar{j}})}{\sigma}\right)\right) \\ &+ \omega\left(3\lambda_0^*\left(\frac{\Upsilon_{m,n}(g_{\bar{i},\bar{j}}) - g_{\bar{i},\bar{j}}}{\sigma}\right)\right) + \omega\left(\frac{3\lambda_0^*}{M}\left(g_{\bar{i},\bar{j}} - f\right)\right).\end{aligned} \tag{19}$$

Then, it follows from (18) and (19) that

$$\omega\left(\lambda_0^*\left(\frac{\Upsilon_{m,n}(f) - f}{\sigma}\right)\right) \leqq \epsilon + \omega\left(3\lambda_0^*\left(\frac{\Upsilon_{m,n}(f - g_{\bar{i},\bar{j}})}{\sigma}\right)\right) + \omega\left(3\lambda_0^*\left(\frac{\Upsilon_{m,n}(g_{\bar{i},\bar{j}}) - g_{\bar{i},\bar{j}}}{\sigma}\right)\right). \tag{20}$$

Now, taking statistical limit superior as $m,n \to \infty$ on both sides of (20) and also using (3), we deduce that

$$P\ \limsup_{m,n} \omega\left(\lambda_0^*\left(\frac{\Upsilon_{m,n}(f)-f}{\sigma}\right)\right) \leqq \epsilon + R\omega\left(3\lambda_0^*(f-g_{\bar{i},\bar{j}})\right) + P\ \limsup_{m,n} \omega\left(3\lambda_0^*\left(\frac{\Upsilon_{m,n}(g_{\bar{i},\bar{j}})-g_{\bar{i},\bar{j}}}{\sigma}\right)\right).$$

Thus, it implies that

$$P\ \limsup_{m,n} \omega\left(\lambda_0^*\left(\frac{\Upsilon_{m,n}(f)-f}{\sigma}\right)\right) \leqq \epsilon + \epsilon R + P\ \limsup_{m,n} \omega\left(3\lambda_0^*\left(\frac{\Upsilon_{m,n}(g_{\bar{i},\bar{j}})-g_{\bar{i},\bar{j}}}{\sigma}\right)\right). \tag{21}$$

Next, by (4), for some $\lambda_0^* > 0$, we obtain

$$P\ \limsup_{m,n} \omega\left(3\lambda_0^*\left(\frac{\Upsilon_{m,n}(g_{\bar{i},\bar{j}})-g_{\bar{i},\bar{j}}}{\sigma}\right)\right) = 0. \tag{22}$$

Clearly from (21) and (22), we get

$$P\ \limsup_{m,n} \omega\left(\lambda_0^*\left(\frac{\Upsilon_{m,n}(f)-f}{\sigma}\right)\right) \leqq \epsilon(1+R).$$

Since $\epsilon > 0$ is arbitrarily small, the right-hand side of the above inequality tends to zero. Hence,

$$P\ \limsup_{m,n} \omega\left(\lambda_0^*\left(\frac{\Upsilon_{m,n}(f)-f}{\sigma}\right)\right) = 0,$$

which completes the proof. □

Next, one can get the following theorem as an immediate consequence of Theorem 2 in which the modular ω satisfies the Δ_2-condition.

Theorem 3. *Let $(\mathcal{L}_{m,n})$, (a_n), (b_n), σ and ω be the same as in Theorem 2. If the modular ω satisfies the Δ_2-condition, then the following assertions are identical:*

(a) $N_D stat\ \lim_{m,n} \left\| \frac{\mathcal{L}_{m,n}(f_i;x,y)-f(x,y)}{\sigma} \right\|_\omega = 0$ *for each* $\lambda > 0$ *and* $i = 0,1,2,3$;

(b) $N_D stat\ \lim_{m,n} \left\| \frac{\mathcal{L}_{m,n}(f;x,y)-f(x,y)}{\sigma} \right\|_\omega = 0$ *for each* $\lambda > 0$ *such that any function* $f \in L^\omega(\mathcal{I}^2)$ *provided that* $f - g \in X_L$ *for each* $g \in C^\infty(\mathcal{I}^2)$.

Next, by using the definitions of relatively modular deferred-weighted statistical convergence given in Definition 3 and statistically as well as relatively modular deferred-weighted summability given in Definition 4, we present the following corollaries in view of Theorem 2.

Let $a_n = 0$ and $b_n = n$, $b_m = m$, then Equation (3) reduces to

$$\text{stat}_N\ \limsup_{m,n} \omega\left(\lambda\left(\frac{\mathfrak{L}_{m,n}(f)}{\sigma}\right)\right) \leqq R\omega(\lambda f) \tag{23}$$

for each $f \in X_L$ and $\lambda > 0$, where R is a constant.

Moreover, if we replace stat_N limit by Nstat limit, then Equation (3) reduces to

$$N\text{stat}\ \limsup_{m,n} \omega\left(\lambda\left(\frac{\Omega_{m,n}(f)}{\sigma}\right)\right) \leqq R\omega(\lambda f). \tag{24}$$

Corollary 1. *Let ω be an $\mathcal{N}$-Quasi semi-convex modular, strongly finite, monotone, and absolutely continuous on $X(\mathcal{I}^2)$. Also, let $(\mathfrak{L}_{m,n})$ be a double sequence of positive linear operators from E in to $X(\mathcal{I}^2)$ satisfying the* *assumption (23) for every* X_L *and* $\sigma_i(x,y)$ *be an unbounded function such that* $|\sigma_i(x,y)| \geqq u_i > 0$ $(i = 0,1,2,3)$*. Suppose that*

$$stat_N \lim_{m,n} \left\| \frac{\mathfrak{L}_{m,n}(f_i;x,y) - f(x,y)}{\sigma} \right\|_\omega = 0 \text{ for each } \lambda > 0 \text{ and } i = 0,1,2,3,$$

where

$$f_0(x,y) = 1, \ \ f_1(x,y) = x, \ \ f_2(x,y) = y \text{ and } f_3(x,y) = x^2 + y^2.$$

Then, for every $f \in L^\omega(\mathcal{I}^2)$ *and* $g \in C^\infty(\mathcal{I}^2)$ *with* $f - g \in X_L$,

$$stat_N \lim_{m,n} \left\| \frac{\mathfrak{L}_{m,n}(f;x,y) - f(x,y)}{\sigma} \right\|_\omega = 0 \text{ for each } \lambda_0 > 0,$$

where

$$\sigma(x,y) = \max\{|\sigma_i(x,y)| : i = 0,1,2,3\}. \tag{25}$$

Corollary 2. *Let* ω *be an* $\mathcal{N}$*-Quasi semi-convex modular, absolutely continuous, monotone, and strongly finite on* $X(\mathcal{I}^2)$*. Also, let* $\Omega_{m,n}$ *be a double sequence of positive linear operators from E in to* $X(\mathcal{I}^2)$ *satisfying the assumption (24) for every* X_L *and* $\sigma_i(x,y)$ *be an unbounded function such that* $|\sigma_i(x,y)| \geqq u_i > 0$ $(i = 0,1,2,3)$*. Suppose that*

$$Nstat \lim_{m,n} \left\| \frac{\Omega_{m,n}(f_i;x,y) - f(x,y)}{\sigma} \right\|_\omega = 0 \text{ for each } \lambda > 0 \text{ and } i = 0,1,2,3,$$

where

$$f_0(x,y) = 1, \ \ f_1(x,y) = x, \ \ f_2(x,y) = y \text{ and } f_3(x,y) = x^2 + y^2.$$

Then, for every $f \in L^\omega(\mathcal{I}^2)$ *and* $g \in C^\infty(\mathcal{I}^2)$ *with* $f - g \in X_L$,

$$Nstat \lim_{m,n} \left\| \frac{\Omega_{m,n}(f;x,y) - f(x,y)}{\sigma} \right\|_\omega = 0 \text{ for every } \lambda_0 > 0,$$

where σ *is given by (25).*

Note that for $a_n = 0$, $b_n = n$, $b_m = m$, and $s_m = 1 = t_n$, Equation (3) reduces to

$$\text{stat} \limsup_{m,n} \omega\left(\lambda\left(\mathfrak{L}^*_{m,n}(f)\right)\right) \leqq R\omega(\lambda f) \tag{26}$$

for each $f \in X_L$ and $\lambda > 0$, where R is a positive constant.

Also, if we replace statistically convergent limit by the statistically summability limit, then Equation (3) reduces to

$$\text{stat} \limsup_{m,n} \omega\left(\lambda\left(\Lambda_{m,n}(f)\right)\right) \leqq R\omega(\lambda f). \tag{27}$$

Now, we present the following corollaries in view of Theorem 2 as the generalization of the earlier results of Demirci and Orhan [31].

Corollary 3. *Let* ω *be an* $\mathcal{N}$*-Quasi semi-convex modular, absolutely continuous, monotone, and strongly finite on* $X(\mathcal{I}^2)$*. Also, let* $(\mathfrak{L}^*_{m,n})$ *be a double sequence of positive linear operators from E in to* $X(\mathcal{I}^2)$ *satisfying the assumption (26) for every* X_L *and* $\sigma_i(x,y)$ *be an unbounded function such that* $|\sigma_i(x,y)| \geqq u_i > 0$ $(i = 0,1,2,3)$*. Suppose that*

$$stat \lim_{m,n} \left\| \frac{\mathfrak{L}^*_{m,n}(f_i; x, y) - f(x,y)}{\sigma} \right\|_\omega = 0 \text{ for every } \lambda > 0 \text{ and } i = 0,1,2,3,$$

where

$$f_0(x,y) = 1, \ \ f_1(x,y) = x, \ \ f_2(x,y) = y \text{ and } f_3(x,y) = x^2 + y^2.$$

Then, for every $f \in L^\omega(\mathcal{I}^2)$ *and* $g \in C^\infty(\mathcal{I}^2)$ *with* $f - g \in X_L$,

$$stat \lim_{m,n} \left\| \frac{\mathfrak{L}^*_{m,n}(f; x, y) - f(x,y)}{\sigma} \right\|_\omega = 0 \text{ for every } \lambda_0 > 0,$$

where σ *is given by (25).*

Corollary 4. *Let* ω *be an* $\mathcal{N}$*-Quasi semi-convex modular, monotone, absolutely continuous, and strongly finite on* $X(\mathcal{I}^2)$*. Also, let* $(\Lambda_{m,n})$ *be a double sequence of positive linear operators from E in to* $X(\mathcal{I}^2)$ *satisfying the assumption (27) for every* X_L *and* $\sigma_i(x,y)$ *be an unbounded function such that* $|\sigma_i(x,y)| \geqq u_i > 0$ $(i = 0,1,2,3)$*. Suppose that*

$$stat \lim_{m,n} \left\| \frac{\Lambda_{m,n}(f_i; x, y) - f(x,y)}{\sigma} \right\|_\omega = 0 \text{ for every } \lambda > 0 \text{ and } i = 0,1,2,3,$$

where

$$f_0(x,y) = 1, \ \ f_1(x,y) = x, \ \ f_2(x,y) = y \text{ and } f_3(x,y) = x^2 + y^2.$$

Then, for every $f \in L^\omega(\mathcal{I}^2)$ *and* $g \in C^\infty(\mathcal{I}^2)$ *with* $f - g \in X_L$,

$$stat \lim_{m,n} \left\| \frac{\Lambda_{m,n}(f; x, y) - f(x,y)}{\sigma} \right\|_\omega = 0 \text{ for every } \lambda_0 > 0,$$

where σ *is given by (25).*

4. Application of Korovkin-Type Theorem

In this section, by presenting a further example, we demonstrate that our proposed Korovkin-type approximation results in modular space are stronger than most (if not all) of the previously existing results in view of the corollaries provided in this paper.

Let $\mathcal{I} = [0,1]$ and φ, ω^φ, and $L^\omega_\varphi(\mathcal{I}^2)$ be as given in Example 3. Also, recall the *bivariate Bernstein–Kantorovich* operators (see [35]), $\mathbb{B} = \{B_{m,n}\}$ on the space $L^\omega_\varphi(\mathcal{I}^2)$ given by

$$B_{m,n}(f; x, y) = \sum_{i,j=0}^{m,n} p^{(m,n)}_{i,j}(x,y)(m+1)(n+1) \times \int_{\frac{i}{m+1}}^{\frac{i+1}{m+1}} \int_{\frac{j}{n+1}}^{\frac{j+1}{n+1}} f(s,t)dsdt \tag{28}$$

for $x, y \in \mathcal{I}$ and

$$p^{(m,n)}_{i,j}(x,y) = \binom{m}{i}\binom{n}{j} x^i y^j (1-x)^{m-i}(1-y)^{n-j}.$$

Also, we have

$$\sum_{i,j=0}^{m,n} p^{(m,n)}_{i,j}(x,y) = 1. \tag{29}$$

Clearly, we observe that

$$B_{m,n}(1;x,y) = 1,$$
$$B_{m,n}(s;x,y) = \frac{mx}{m+1} + \frac{1}{2(m+1)},$$
$$B_{m,n}(t;x,y) = \frac{ny}{n+1} + \frac{1}{2(n+1)}$$

and

$$B_{m,n}(t^2+s^2;x,y) = \frac{m(m-1)x^2}{(m+1)^2} + \frac{2mx}{(m+1)^2} + \frac{1}{3(m+1)^2}\frac{n(n-1)y^2}{(n+1)^2} + \frac{2ny}{(n+1)^2} + \frac{1}{3(n+1)^2}.$$

It is further observed that $B_{m,n} : L_\varphi^\omega(\mathcal{I}^2) \to L_\varphi^\omega(\mathcal{I}^2)$. Recall [28] (Lemma 5.1) and [29] (Example 1). Now because of (29), we have from *Jensen inequality*, for each $f \in L_\varphi^\omega(\mathcal{I}^2)$ and $m,n \in \mathbb{N}$, there exists a constant M such that

$$\omega^\varphi\left(\frac{B_{m,n}(f;x,y)}{\sigma}\right) \leqq M\omega^\varphi(f).$$

We now present an illustrative example for the validity of the operators $(\mathcal{L}_{m,n})$ for our Theorem 2.

Example 4. *Let $\mathcal{L}_{m,n} : L^\omega(\mathcal{I}^2) \to L^\omega(\mathcal{I}^2)$ be defined by*

$$\mathcal{L}_{m,n}(f;x,y) = (1+f_{m,n})B_{m,n}(f;x,y), \tag{30}$$

where $(f_{m,n})$ is a sequence defined as in Example 3. Then, we have

$$\mathcal{L}_{m,n}(1;x,y) = 1 + f_{m,n}(x,y),$$
$$\mathcal{L}_{m,n}(1;x,y) = 1 + f_{m,n}(x,y)\cdot\left[\frac{mx}{m+1} + \frac{1}{2(m+1)}\right],$$
$$\mathcal{L}_{m,n}(1;x,y) = 1 + f_{m,n}(x,y)\cdot\left[\frac{ny}{n+1} + \frac{1}{2(n+1)}\right]$$

and

$$\mathcal{L}_{m,n}(1;x,y) = 1 + f_{m,n}(x,y) \cdot\left[\frac{m(m-1)x^2}{(m+1)^2} + \frac{2mx}{(m+1)}^2 + \frac{1}{3(m+1)^2}\frac{n(n-1)y^2}{(n+1)^2} + \frac{2ny}{(n+1)^2} + \frac{1}{3(n+1)^2}\right].$$

We thus obtain

$$N_D stat \lim_{m,n}\left\|\frac{\mathcal{L}_{m,n}(1;x,y)-1}{\sigma}\right\|_\omega = 0,$$
$$N_D stat \lim_{m,n}\left\|\frac{\mathcal{L}_{m,n}(s;x,y)-s}{\sigma}\right\|_\omega = 0,$$
$$N_D stat \lim_{m,n}\left\|\frac{\mathcal{L}_{m,n}(t;x,y)-t}{\sigma}\right\|_\omega = 0,$$
$$N_D stat \lim_{m,n}\left\|\frac{\mathcal{L}_{m,n}(s^2+t^2;x,y)-s^2+t^2}{\sigma}\right\|_\omega = 0.$$

This means that the operators $\mathcal{L}_{m,n}(f;x,y)$ fulfil the conditions (4). Hence, by Theorem 2 we have

$$N_D stat \lim_{m,n} \left\| \frac{\mathcal{L}_{m,n}(f;x,y) - f(x,y)}{\sigma} \right\|_{\omega} - 0 \text{ for every } \lambda_0 > 0.$$

However, since $(f_{m,n})$ is not relatively modular weighted statistically convergent, the result of Demirci and Orhan ([31], p. 1173, Theorem 1) is not fairly true under the operators defined by us in (30). Furthermore, since $(f_{m,n})$ is statistically and relatively modular deferred-weighted summable, we therefore conclude that our Theorem 2 works for the operators which we have considered here.

5. Concluding Remarks and Observations

In the concluding section of our study, we put forth various supplementary remarks and observations concerning several outcomes which we have established here.

Remark 2. *Let $(f_{m,n})_{m,n\in\mathbb{N}}$ be a sequence of functions given in Example 3. Then, since*

$$N_D stat \lim_{m\to\infty} f_{m,n} = 0 \text{ on } [0,1]\times[0,1],$$

we have

$$N_D stat \lim_{m\to\infty} \|\mathcal{L}_{m,n}(f_i;x,y) - f_i(x,y)\|_{\omega} = 0 \quad (i = 0,1,2,3). \tag{31}$$

Thus, we can write (by Theorem 2)

$$N_D stat \lim_{m\to\infty} \|\mathcal{L}_m(f;x,y) - f(x,y)\|_{\omega} = 0, \ (i = 0,1,2,3), \tag{32}$$

where

$$f_0(x,y) = 1, \ f_1(x,y) = x, \ f_2(x,y) = y \text{ and } f_3(x,y) = x^2 + y^2.$$

Moreover, as $(f_{m,m})$ is not classically convergent it therefore does not converge uniformly in modular space. Thus, the traditional Korovkin-type approximation theorem will not work here under the operators defined in (30). Therefore, this application evidently demonstrates that our Theorem 2 is a non-trivial extension of the conventional Korovkin-type approximation theorem (see [27]).

Remark 3. *Let $(f_{m,n})_{m,n\in\mathbb{N}}$ be a sequence as considered in Example 3. Then, since*

$$N_D stat \lim_{m\to\infty} f_{m,n} = 0 \text{ on } [0,1]\times[0,1],$$

(31) fairly holds true. Now under condition (31) and by applying Theorem 2, we have that the condition (32) holds true. Moreover, since $(f_{m,n})$ is not relatively modular statistically Cesàro summable, Theorem 1 of Demirci and Orhan (see [31], p. 1173, Theorem 1) does not hold fairly true under the operators considered in (30). Hence, our Theorem 2 is a non-trivial generalization of Theorem 1 of Demirci and Orhan (see [31], p. 1173, Theorem 1) (see also [29]). Based on the above facts, we conclude here that our proposed method has effectively worked for the operators considered in (30), and therefore it is stronger than the traditional and statistical versions of the Korovkin-type approximation theorems established earlier in References [27,29,31].

Author Contributions: Writing—review and editing, H.M.S.; Investigation, B.B.J.; Supervision, S.K.P.; Visualization, U.M.

References

1. Fast, H. Sur la convergence statistique. *Colloq. Math.* **1951**, *2*, 241–244. [CrossRef]
2. Steinhaus, H. Sur la convergence ordinaire et la convergence asymptotique. *Colloq. Math.* **1951**, *2*, 73–74.
3. Zygmund, A. *Trigonometric Series*, 3rd ed.; Cambridge University Press: Cambridge, UK, 2002.
4. Shang, Y. Estrada and $\mathcal{L}$-Estrada indices of edge-independent random graphs. *Symmetry* **2015**, *7*, 1455–1462. [CrossRef]
5. Shang, Y. Estrada index of random bipartite graphs. *Symmetry* **2015**, *7*, 2195–2205. [CrossRef]
6. Mohiuddine, S.A. Statistical weighted A-summability with application to Korovkin's type approximation theorem. *J. Inequal. Appl.* **2016**, *2016*, 101. [CrossRef]
7. Karakaya, V.; Chishti, T.A. Weighted statistical convergence. *Iran. J. Sci. Technol. Trans. A* **2009**, *33*, 219–223.
8. Ansari, K.J.; Ahmad, I.; Mursaleen, M.; Hussain, I. On some statistical approximation by (p,q)-Bleimann, Butzer and Hahn operators. *Symmetry* **2018**, *10*, 731. [CrossRef]
9. Belen, C.; Mohiuddine, S.A. Generalized statistical convergence and application. *Appl. Math. Comput.* **2013**, *219*, 9821–9826.
10. Braha, N.L.; Loku, V.; Srivastava, H.M. Λ^2-Weighted statistical convergence and Korovkin and Voronovskaya type theorems. *Appl. Math. Comput.* **2015**, *266*, 675–686. [CrossRef]
11. Kadak, U.; Braha, N.L.; Srivastava, H.M. Statistical weighted $\mathcal{B}$-summability and its applications to approximation theorems. *Appl. Math. Comput.* **2017**, *302*, 80–96.
12. Özarslan, M.A.; Duman, O.; Srivastava, H.M. Statistical approximation results for Kantorovich-type operators involving some special polynomials. *Math. Comput. Model.* **2008**, *48*, 388–401. [CrossRef]
13. Srivastava, H.M.; Jena, B.B.; Paikray, S.K.; Misra, U.K. A certain class of weighted statistical convergence and associated Korovkin type approximation theorems for trigonometric functions. *Math. Methods Appl. Sci.* **2018**, *41*, 671–683. [CrossRef]
14. Jena, B.B.; Paikray, S.K.; Misra, U.K. Statistical deferred Cesàro summability and its applications to approximation theorems. *Filomat* **2018**, *32*, 2307–2319. [CrossRef]
15. Paikray, S.K.; Jena, B.B.; Misra, U.K. Statistical deferred Cesàro summability mean based on (p,q)-integers with application to approximation theorems. In *Advances in Summability and Approximation Theory*; Mohiuddine, S.A., Acar, T., Eds.; Springer: Singapore, 2019; pp. 203–222.
16. Pradhan, T.; Paikray, S.K.; Jena, B.B.; Dutta, H. Statistical deferred weighted $\mathcal{B}$-summability and its applications to associated approximation theorems. *J. Inequal. Appl.* **2018**, *2018*, 65. [CrossRef]
17. Srivastava, H.M.; Jena, B.B.; Paikray, S.K.; Misra, U.K. Generalized equi-statistical convergence of the deferred Nörlund summability and its applications to associated approximation theorems. *Rev. Real Acad. Cienc. Exactas Fís. Natur. Ser. A Mat.* **2018**, *112*, 1487–1501. [CrossRef]
18. Srivastava, H.M.; Jena, B.B.; Paikray, S.K.; Misra, U.K. Deferred weighted $\mathcal{A}$-statistical convergence based upon the (p,q)-Lagrange polynomials and its applications to approximation theorems. *J. Appl. Anal.* **2018**, *24*, 1–16. [CrossRef]
19. Srivastava, H.M.; Mursaleen, M.; Khan, A. Generalized equi-statistical convergence of positive linear operators and associated approximation theorems. *Math. Comput. Model.* **2012**, *55*, 2040–2051. [CrossRef]
20. Pringsheim, A. Zur theorie der zweifach unendlichen Zahlenfolgen. *Math. Ann.* **1900**, *53*, 289–321. [CrossRef]
21. Bardaro, C.; Mantellini, I. A Korovkin theorem in multivariate modular function spaces. *J. Funct. Spaces Appl.* **2009**, *7*, 105–120. [CrossRef]
22. Bardaro, C.; Musielak, J.; Vinti, G. *Nonlinear Integral Operators and Applications*; de Gruyter Series in Nonlinear Analysis and Applications; Walter de Gruyter Publishers: Berlin, Germany, 2003; Volume 9.
23. Musielak, J. *Orlicz Spaces and Modular Spaces*; Lecture Notes in Mathematics; Springer: Berlin, Germany, 1983; Volume 1034.
24. Moore, E.H. *An Introduction to a Form of General Analysis*; The New Haven Mathematical Colloquium; Yale University Press: New Haven, CT, USA, 1910.
25. Chittenden, E.W. On the limit functions of sequences of continuous functions converging relatively uniformly. *Trans. Am. Math. Soc.* **1919**, *20*, 179–184. [CrossRef]
26. Bohman, H. On approximation of continuous and of analytic Functions. *Arkiv Mat.* **1952**, *2*, 43–56. [CrossRef]
27. Korovkin, P.P. Convergence of linear positive operators in the spaces of continuous functions. *Doklady Akad. Nauk. SSSR* **1953**, *90*, 961–964. (In Russian)

28. Bardaro, C.; Mantellini, I. Korovkin's theorem in modular spaces. *Comment. Math.* **2007**, *47*, 239–253.
29. Orhan, S.; Demirci, K. Statistical approximation by double sequences of positive linear operators on modular spaces. *Positivity* **2015**, *19*, 23–36. [CrossRef]
30. Demirci, K.; Burçak, K. *A*-Statistical relative modular convergence of positive linear operators. *Positivity* **2016**, *21*, 847–863. [CrossRef]
31. Demirci, K.; Orhan, S. Statistical relative approximation on modular spaces. *Results Math.* **2017**, 71, 1167–1184. [CrossRef]
32. Nasiruzzaman, M.; Mukheimer, A.; Mursaleen, M. A Dunkl-type generalization of Szász-Kantorovich operators via post-quantum calculus. *Symmetry* **2019**, *11*, 232. [CrossRef]
33. Srivastava, H.M.; Özger, F.; Mohiuddine, S.A. Construction of Stancu-type Bernstein operators based on Bézier bases with shape parameter λ. *Symmetry* **2019**, *11*, 316. [CrossRef]
34. Agnew, R.P. On deferred Cesàro means. *Ann. Math.* **1932**, *33*, 413–421. [CrossRef]
35. Deshwal, S.; Ispir, N.; Agrawal, P.N. Blending type approximation by bivariate Bernstein-Kantorovich operators. *Appl. Math. Inform. Sci.* **2017**, *11*, 423–432. [CrossRef]

12

New Symmetric Differential and Integral Operators Defined in the Complex Domain

Rabha W. Ibrahim [1,*,†] **and Maslina Darus** [2,†]

1 Cloud Computing Center, University Malaya, Kuala Lumpur 50603, Malaysia
2 Center for Modelling and Data Science, Faculty of Science and Technology, Universiti Kebangsaan Malaysia, Bangi 43600, Malaysia
* Correspondence: rabhaibrahim@yahoo.com
† These authors contributed equally to this work.

Abstract: The symmetric differential operator is a generalization operating of the well-known ordinary derivative. These operators have advantages in boundary value problems, statistical studies and spectral theory. In this effort, we introduce a new symmetric differential operator (SDO) and its integral in the open unit disk. This operator is a generalization of the Sàlàgean differential operator. Our study is based on geometric function theory and its applications in the open unit disk. We formulate new classes of analytic functions using SDO depending on the symmetry properties. Moreover, we define a linear combination operator containing SDO and the Ruscheweyh derivative. We illustrate some inclusion properties and other inequalities involving SDO and its integral.

Keywords: univalent function; symmetric differential operator; unit disk; analytic function; subordination

MSC: 30C45

---

1. Introduction

Investigation of the theory of operators (differential, integral, mixed, convolution and linear) has been a capacity of apprehension for numerous scientists in all fields of mathematical sciences, such as mathematical physics, mathematical biology and mathematical computing. An additional definite field is the study of inequalities in the complex domain. Works' review shows masses of studies created by the classes of analytic functions. The relationship of geometry and analysis signifies a very central feature in geometric function theory in the open unit disk. This fast development is directly connected to the existence between analysis, construction and geometric performance [1]. In 1983, Sàlàgean introduced his famous differential operator of normalized analytic functions in the open unit disk [2]. This operator is generalized and extended to many classes of univalent functions. It plays a significant tool to develop the geometric structure of many analytic functions by suggesting different classes. Later this operator has been generalized and motivated by many researchers, for example, the Al-Oboudi differential operator [3]. Recently, a new study is presented by using the Sàlàgean operator [4]. Our research is to formulate a new symmetric differential operator and its integral by utilizing the concept of the symmetric derivative of complex variables. This concept is an operation, extending the original derivative. Note that its practical use in the the symmetry models in math modeling remains open. For example, for application in mathematical physics it is critical to employ group analysis methods. Such methods enable methods for branching solutions construction using group symmetry [5,6].

2. Preparatory

We shall need the following basic definitions throughout this paper. A function $\phi \in \Lambda$ is said to be univalent in $\mathbb{U}$ if it never takes the same value twice; that is, if $z_1 \neq z_2$ in the open unit disk $\mathbb{U} = \{z \in \mathbb{C} : |z| < 1\}$ then $\phi(z_1) \neq \phi(z_2)$ or equivalently, if $\phi(z_1) = \phi(z_2)$ then $z_1 = z_2$. Without loss of generality, we can use the notion Λ for our univalent functions taking the expansion

$$\phi(z) = z + \sum_{n=2}^{\infty} \varphi_n z^n, \quad z \in \mathbb{U}. \tag{1}$$

We let $\mathcal{S}$ denote the class of such functions $\phi \in \Lambda$ that are univalent in $\mathbb{U}$.

A function $\phi \in \mathcal{S}$ is said to be starlike with respect to origin in $\mathbb{U}$ if the linear segment joining the origin to every other point of $\phi(z : |z| = r < 1)$ lies entirely in $\phi(z : |z| = r < 1)$. In more picturesque language, the requirement is that every point of $\phi(z : |z| = r < 1)$ be visible from the origin. A function $\phi \in \mathcal{S}$ is said to be convex in $\mathbb{U}$ if the linear segment joining any two points of $\phi(z : |z| = r < 1)$ lies entirely in $\phi(z : |z| = r < 1)$. In other words, a function $\phi \in \mathcal{S}$ is said to be convex in $\mathbb{U}$ if it is starlike with respect to each and every of its points. We denote the class of functions $\phi \in \mathcal{S}$ that are starlike with respect to origin by $\mathcal{S}^*$ and convex in $\mathbb{U}$ by $\mathcal{C}$.

Neatly linked to the classes $\mathcal{S}^*$ and $\mathcal{C}$ is the class $\mathcal{P}$ of all functions ϕ analytic in $\mathbb{U}$ and having positive real part in $\mathbb{U}$ with $\phi(0) = 1$. In fact $f \in \mathcal{S}^*$ if and only if $z\phi'(z)/\phi(z) \in \mathcal{P}$ and $\phi \in \mathcal{C}$ if and only if $1 + z\phi''(z)/\phi'(z) \in \mathcal{P}$. In general, for $\epsilon \in [0,1)$ we let $\mathcal{P}(\epsilon)$ consist of functions ϕ analytic in $\mathbb{U}$ with $\phi(0) = 1$ so that $\Re(\phi(z)) > \epsilon$ ('$\Re$' represents to the real part) for all $z \in \mathbb{U}$. Note that $\mathcal{P}(\epsilon_2) \subset \mathcal{P}(\epsilon_1) \subset \mathcal{P}(0) \equiv \mathcal{P}$ for $0 < \epsilon_1 < \epsilon_2$ (e.g., see Duren [1]).

For functions ϕ and ψ in Λ we say that ϕ is subordinate to ψ, denoted by $\phi \prec \psi$, if there exists a Schwarz function ω with $\omega(0) = 0$ and $|\omega(z)| < 1$ so that $\phi(z) = \psi(\omega(z))$ for all $z \in \mathbb{U}$ (see [7]). Evidently $\phi(z) \prec \psi(z)$ is equivalent to $\phi(0) = \psi(0)$ and $\phi(\mathbb{U}) \subset \psi(\mathbb{U})$. We request the following results, which can be located in [7].

Lemma 1. *For $a \in \mathbb{C}$ and positive integer n let $\mathfrak{H}[a,n] = \{\varrho : \varrho(z) = a + a_n z^n + a_{n+1} z^{n+1} + ...\}$.*

i. If $\gamma \in \mathbb{R}$ then $\Re\Big(\varrho(z) + \gamma z\varrho'(z)\Big) > 0 \Longrightarrow \Re(\varrho(z)) > 0$. Moreover, if $\gamma > 0$ and $\varrho \in \mathfrak{H}[1,n]$, then there are constants $\lambda > 0$ and $\beta > 0$ with $\beta = \beta(\gamma, \lambda, n)$ so that

$$\varrho(z) + \gamma z\varrho'(z) \prec \left[\frac{1+z}{1-z}\right]^{\beta} \Rightarrow \varrho(z) \prec \left[\frac{1+z}{1-z}\right]^{\lambda}.$$

ii. If $\delta \in [0,1)$ and $\varrho \in \mathfrak{H}[1,n]$ then there is a constant $\lambda > 0$ with $\lambda = \lambda(\alpha, n)$ so that

$$\Re\Big(\varrho^2(z) + 2\varrho(z).z\varrho'(z)\Big) > \delta \Rightarrow \Re(\varrho(z)) > \lambda.$$

iii. If $\varrho \in \mathfrak{H}[a,n]$ with $\Re a > 0$ then $\Re\Big(\varrho(z) + z\varrho'(z) + z^2\varrho''(z)\Big) > 0$ or for $\vartheta : \mathbb{U} \to \mathbb{R}$ with $\Re\Big(\varrho(z) + \vartheta(z)\dfrac{z\varrho'(z)}{\varrho(z)}\Big) > 0$ then $\Re(\varrho(z)) > 0$.

Lemma 2. *Let h be a convex function with $h(0) = a$, and let $\mu \in \mathbb{C} \setminus \{0\}$ be a complex number with $\Re\gamma \geq 0$. If $\varrho \in \mathfrak{H}[a,n]$, and $\varrho(z) + (1/\mu)z\varrho'(z) \prec h(z), \quad z \in U$, then $\varrho(z) \prec \iota(z) \prec h(z)$, where*

$$\iota(z) = \frac{\mu}{nz^{\mu/n}} \int_0^z h(t)t^{\overline{\frac{\mu}{(n-1)}}}\,dt, \quad z \in U.$$

3. Formulas of Symmetric Operators

Let $\phi \in \Lambda$, taking the power series (1). For a function $\phi(z)$ and a constant $\alpha \in [0,1]$, we formulate the SDO as follows:

$$\begin{aligned}
\mathcal{M}_\alpha^0\phi(z) &= \phi(z)\\
\mathcal{M}_\alpha^1\phi(z) &= \alpha z\phi'(z) - (1-\alpha)z\phi'(-z)\\
&= \alpha\left(z + \sum_{n=2}^{\infty} n\varphi_n z^n\right) - (1-\alpha)\left(-z + \sum_{n=2}^{\infty} n(-1)^n\varphi_n z^n\right)\\
&= z + \sum_{n=2}^{\infty}\left[n\left(\alpha - (1-\alpha)(-1)^n\right)\right]\varphi_n z^n\\
\mathcal{M}_\alpha^2\phi(z) &= \mathcal{M}_\alpha^1[\mathcal{M}_\alpha^1\phi(z)] = z + \sum_{n=2}^{\infty}\left[n\left(\alpha - (1-\alpha)(-1)^n\right)\right]^2\varphi_n z^n\\
&\vdots\\
\mathcal{M}_\alpha^k\phi(z) &= \mathcal{M}_\alpha^1[\mathcal{M}_\alpha^{k-1}\phi(z)] = z + \sum_{n=2}^{\infty}\left[n\left(\alpha - (1-\alpha)(-1)^n\right)\right]^k\varphi_n z^n.
\end{aligned} \tag{2}$$

It is clear that when $\alpha = 1$, we have Sàlàgean differential operator [2] $\mathcal{S}^k\phi(z) = z + \sum_{n=2}^{\infty} n^k\varphi_n z^n$. We may say that SDO (2) is the symmetric Sàlàgean differential operator in the open unit disk. In the same manner of the formula of Sàlàgean integral operator, we consume that for a function $\phi \in \Lambda$, the symmetric integral operator $\mathcal{J}_\alpha^k$ satisfies

$$\mathcal{J}_\alpha^k\phi(z) = z + \sum_{n=2}^{\infty}\frac{1}{\left[n\left(\alpha - (1-\alpha)(-1)^n\right)\right]^k}\varphi_n z^n \in \Lambda.$$

Similarly, when $\alpha = 1$, we have Sàlàgean integral operator [2], Remark 5. Furthermore, we conclude the relation $\mathcal{M}_\alpha^k\left(\mathcal{J}_\alpha^k\phi(z)\right) = \mathcal{J}_\alpha^k\left(\mathcal{M}_\alpha^k\phi(z)\right) = \phi(z)$.

Next, we proceed to formulate a linear combination operator involving SDO and the Ruscheweyh derivative. For a function $\phi \in \Lambda$, the Ruscheweyh derivative achieves the formula

$$\mathcal{R}^k\phi(z) = z + \sum_{n=2}^{\infty} C_{k+n-1}^k\varphi_n z^n,$$

where the term C_{k+n-1}^k is the combination coefficients. In this note, we introduce a new operator combining R^k and $\mathcal{M}_\alpha^k$ as follows:

$$\begin{aligned}
\mathbf{C}_{\alpha,\kappa}^k\phi(z) &= (1-\kappa)\mathcal{R}^k\phi(z) + \kappa\mathcal{M}_\alpha^k\phi(z)\\
&= z + \sum_{n=2}^{\infty}\left((1-\kappa)C_{k+n-1}^k + \kappa\left[n\left(\alpha - (1-\alpha)(-1)^n\right)\right]^k\right)\varphi_n z^n.
\end{aligned} \tag{3}$$

Remark 1.

- $k = 0 \Longrightarrow \mathbf{C}_{\alpha,\kappa}^0\phi(z) = \phi(z)$;
- $\alpha = 1 \Longrightarrow \mathbf{C}_{1,\kappa}^k\phi(z) = \mathcal{L}_\kappa^k\phi(z)$; *[8] (Lupas operator)*
- $\kappa = 0 \Longrightarrow \mathbf{C}_{\alpha,0}^k\phi(z) = \mathcal{R}^k\phi(z)$;
- $\alpha = 1, \kappa = 1 \Longrightarrow \mathbf{C}_{1,1}^k\phi(z) = \mathcal{S}^k\phi(z)$;
- $\kappa = 1 \Longrightarrow \mathbf{C}_{\alpha,\kappa}^k\phi(z) = \mathcal{M}_\alpha^k\phi(z)$.

We shall deal with the following classes

$$S_k^{*\alpha}(h)=\Big\{\phi\in\Lambda:\frac{z(\mathcal{M}_\alpha^k\phi(z))'}{\mathcal{M}_\alpha^k\phi(z)}\prec h(z),\, h\in\mathcal{C}\Big\}.$$

Obviously, the subclass $S_0^*(h)=\mathcal{S}^*(h)$.

Definition 1. *If $\phi\in\Lambda$, then $\phi\in\mathbb{J}_\alpha^b(A,B,k)$ if and only if*

$$1+\frac{1}{b}\Big(\frac{2\mathcal{M}_\alpha^{k+1}\phi(z)}{\mathcal{M}_\alpha^k\phi(z)-\mathcal{M}_\alpha^k\phi(-z)}\Big)\prec\frac{1+Az}{1+Bz},$$

$$\Big(z\in\mathbb{U},\ -1\le B<A\le 1,\ k=1,2,...,\ b\in\mathbb{C}\setminus\{0\},\ \alpha\in[0,1]\Big).$$

- $\alpha=1\Longrightarrow$ [9];
- $\alpha=1,B=0\Longrightarrow$ [10];
- $\alpha=1,A=1,B=-1,b=2\Longrightarrow$ [11].

Definition 2. *Let $\epsilon\in[0,1),\alpha\in[0,1],\kappa\ge 0$, and $k\in\mathbb{N}$. A function $\phi\in\Lambda$ is said to be in the set $T_k(\alpha,\kappa,\epsilon)$ if and only if*

$$\Re\Big((\mathbf{C}_{\alpha,\kappa}^k\phi(z))'\Big)>\epsilon,\quad z\in U.$$

4. Geometric Results

In this section, we utilize the above constructions of the symmetric operators to get some geometric fulfillment.

Theorem 1. *For $\phi\in\Lambda$ if one of the following facts holds*

- *The operator $\mathcal{M}_\alpha^k\phi(z)$ in (2) is of bounded boundary rotation;*
- *ϕ achieves the subordination inequality*

$$(\mathcal{M}_\alpha^k\phi(z))'\prec\Big(\frac{1+z}{1-z}\Big)^\beta,\quad \beta>0,\ z\in\mathbb{U},\quad \alpha\in[0,\infty);$$

- *f satisfies the inequality*

$$\Re\Big((\mathcal{M}_\alpha^k\phi(z))'\frac{\mathcal{M}_\alpha^k\phi(z)}{z}\Big)>\frac{\delta}{2},\quad \delta\in[0,1),\ z\in\mathbb{U},$$

- *ϕ admits the inequality*

$$\Re\Big(z\mathcal{M}_\alpha^k\phi(z))''-\mathcal{M}_\alpha^k\phi(z))'+2\frac{\mathcal{M}_\alpha^k\phi(z))}{z}\Big)>0,$$

- *ϕ confesses the inequality*

$$\Re\Big(\frac{z\mathcal{M}_\alpha^k\phi(z))'}{\mathcal{M}_\alpha^k\phi(z))}+2\frac{\mathcal{M}_\alpha^k\phi(z)}{z}\Big)>1,$$

then $\frac{\mathcal{M}_\alpha^k\phi(z)}{z}\in\mathcal{P}(\epsilon)$ for some $\epsilon\in[0,1)$.

Proof. Define a function ϱ as follows

$$\varrho(z) = \frac{\mathcal{M}_\alpha^k \phi(z)}{z} \Rightarrow z\varrho'(z) + \varrho(z) = (\mathcal{M}_\alpha^k \phi(z))'. \tag{4}$$

By the first fact, $\mathcal{M}_\alpha^k \phi(z)$ is of bounded boundary rotation, it implies that $\Re(z\varrho'(z) + \varrho(z)) > 0$. Thus, by Lemma 1.i, we obtain $\Re(\varrho(z)) > 0$ which yields the first part of the theorem.

In view of the second fact, we have the following subordination relation

$$(\mathcal{M}_\alpha^k \phi(z))' = z\varrho'(z) + \varrho(z) \prec [\frac{1+z}{1-z}]^\beta.$$

Now, according to Lemma 1.i, there is a constant $\gamma > 0$ with $\beta = \beta(\gamma)$ such that

$$\frac{\mathcal{M}_\alpha^k \phi(z)}{z} \prec \left(\frac{1+z}{1-z}\right)^\gamma.$$

This implies that $\Re(\mathcal{M}_\alpha^k \phi(z)/z) > \epsilon$, for some $\epsilon \in [0,1)$.

Finally, consider the third fact, a simple computation yields

$$\Re\Big(\varrho^2(z) + 2\varrho(z).z\varrho'(z)\Big) = 2\Re\Big((\mathcal{M}_\alpha^k \phi(z))' \frac{\mathcal{M}_\alpha^k \phi(z)}{z}\Big) > \delta. \tag{5}$$

In virtue of Lemma 1.ii, there is a constant $\lambda > 0$ such that $\Re(\varrho(z)) > \lambda$ which implies that $\varrho(z) = \frac{\mathcal{M}_\alpha^k \phi(z)}{z} \in \mathcal{P}(\epsilon)$ for some $\epsilon \in [0,1)$. It follows from (5) that $\Re\Big(\mathcal{M}_\alpha^k \phi(z))'\Big) > 0$ and thus by Noshiro-Warschawski and Kaplan Theorems, $\mathcal{M}_\alpha^k \phi(z)$ is univalent and of bounded boundary rotation in $\mathbb{U}$.

By differentiating (4) and taking the real, we have

$$\Re\Big(\varrho(z) + z\varrho'(z) + z^2\varrho''(z)\Big) = \Re\Big(z(\mathcal{M}_\alpha^k \phi(z))'' - (\mathcal{M}_\alpha^k \phi(z))' + 2\frac{\mathcal{M}_\alpha^k \phi(z)}{z}\Big) > 0.$$

Thus, in virtue of Lemma 1.ii, we obtain $\Re(\frac{\mathcal{M}_\alpha^k \phi(z)}{z}) > 0$.

By logarithmic differentiation (4) and taking the real, we have

$$\Re\Big(\varrho(z) + \frac{z\varrho'(z)}{\varrho(z)} + z^2\varrho''(z)\Big) = \Re\Big(\frac{z(\mathcal{M}_\alpha^k \phi(z))'}{\mathcal{M}_\alpha^k \phi(z)} + 2\frac{\mathcal{M}_\alpha^k \phi(z)}{z} - 1\Big) > 0.$$

Hence, in virtue of Lemma 1.iii, with $\vartheta(z) = 1$, we conclude that $\Re(\frac{\mathcal{M}_\alpha^k \phi(z)}{z}) > 0$. This completes the proof. $\square$

Theorem 2. *Let $\phi \in S_k^{*\alpha}(h)$, where $h(z)$ is convex univalent function in $\mathbb{U}$. Then*

$$\mathcal{M}_\alpha^k \phi(z) \prec z \exp\Big(\int_0^z \frac{h(\omega(\xi)) - 1}{\xi} d\xi\Big),$$

where $\omega(z)$ is analytic in $\mathbb{U}$, with $\omega(0) = 0$ and $|\omega(z)| < 1$. Furthermore, for $|z| = \eta$, $\mathcal{M}_\alpha^k \phi(z)$ achieves the inequality

$$\exp\Big(\int_0^1 \frac{h(\omega(-\eta)) - 1}{\eta}\Big) d\eta \le \Big|\frac{\mathcal{M}_\alpha^k \phi(z)}{z}\Big| \le \exp\Big(\int_0^1 \frac{h(\omega(\eta)) - 1}{\eta}\Big) d\eta.$$

Proof. Since $\phi \in S_k^{*\alpha}(h)$, we have

$$\Big(\frac{z(\mathcal{M}_\alpha^k\phi(z))'}{\mathcal{M}_\alpha^k\phi(z)}\Big) \prec h(z), \quad z \in \mathbb{U},$$

which means that there exists a Schwarz function with $\omega(0) = 0$ and $|\omega(z)| < 1$ such that

$$\Big(\frac{z(\mathcal{M}_\alpha^k\phi(z))'}{\mathcal{M}_\alpha^k\phi(z)}\Big) = h(\omega(z)), \quad z \in \mathbb{U},$$

which implies that

$$\Big(\frac{(\mathcal{M}_\alpha^k\phi(z))'}{\mathcal{M}_\alpha^k\phi(z)}\Big) - \frac{1}{z} = \frac{h(\omega(z)) - 1}{z}.$$

Integrating both sides, we have

$$\log \mathcal{M}_\alpha^k\phi(z) - \log z = \int_0^z \frac{h(\omega(\xi)) - 1}{\xi} d\xi.$$

Consequently, this yields

$$\log \frac{\mathcal{M}_\alpha^k\phi(z)}{z} = \int_0^z \frac{h(\omega(\xi)) - 1}{\xi} d\xi. \tag{6}$$

By using the definition of subordination, we get

$$\mathcal{M}_\alpha^k\phi(z) \prec z \exp\Big(\int_0^z \frac{h(\omega(\xi)) - 1}{\xi} d\xi\Big).$$

In addition, we note that the function $h(z)$ maps the disk $0 < |z| < \eta < 1$ onto a region which is convex and symmetric with respect to the real axis, that is

$$h(-\eta|z|) \leq \Re(h(\omega(\eta z))) \leq h(\eta|z|), \quad \eta \in (0,1),$$

which yields the following inequalities:

$$h(-\eta) \leq h(-\eta|z|), \quad h(\eta|z|) \leq h(\eta)$$

and

$$\int_0^1 \frac{h(\omega(-\eta|z|)) - 1}{\eta} d\eta \leq \Re\Big(\int_0^1 \frac{h(\omega(\eta)) - 1}{\eta} d\eta\Big) \leq \int_0^1 \frac{h(\omega(\eta|z|)) - 1}{\eta} d\eta.$$

By using the above relations and Equation (6), we conclude that

$$\int_0^1 \frac{h(\omega(-\eta|z|)) - 1}{\eta} d\eta \leq \log\Big|\frac{\mathcal{M}_\alpha^k\phi(z)}{z}\Big| \leq \int_0^1 \frac{h(\omega(\eta|z|)) - 1}{\eta} d\eta.$$

This equivalence to the inequality

$$\exp\Big(\int_0^1 \frac{h(\omega(-\eta|z|)) - 1}{\eta} d\eta\Big) \leq \Big|\frac{\mathcal{M}_\alpha^k\phi(z)}{z}\Big| \leq \exp\Big(\int_0^1 \frac{h(\omega(\eta|z|)) - 1}{\eta} d\eta\Big).$$

Thus, we obtain

$$\exp\Big(\int_0^1 \frac{h(\omega(-\eta)) - 1}{\eta}\Big) d\eta \leq \Big|\frac{\mathcal{M}_\alpha^k\phi(z)}{z}\Big| \leq \exp\Big(\int_0^1 \frac{h(\omega(\eta)) - 1}{\eta}\Big) d\eta.$$

This completes the proof. □

Theorem 3. *Consider the class $\mathbb{J}^b_\alpha(A,B,k)$ in Definition 1. If $\phi \in \mathbb{J}^b_\alpha(A,B,k)$ then the odd function*

$$\mathfrak{O}(z) = \frac{1}{2}[\phi(z) - \phi(-z)], \quad z \in \mathbb{U}$$

achieves the following inequality

$$1 + \frac{1}{b}\Big(\frac{\mathcal{M}^{k+1}_\alpha \mathfrak{O}(z)}{\mathcal{M}^k_\alpha \mathfrak{O}(z)} - 1\Big) \prec \frac{1+Az}{1+Bz},$$

and

$$\Re\Big(\frac{z\mathfrak{O}(z)'}{\mathfrak{O}(z)}\Big) \geq \frac{1-r^2}{1+r^2}, \quad |z| = r < 1,$$

$$\Big(z \in \mathbb{U},\ -1 \leq B < A \leq 1,\ k = 1,2,...,\ b \in \mathbb{C}\setminus\{0\},\ \alpha \in [0,1]\Big).$$

Proof. Since $\phi \in \mathbb{J}^b_\alpha(A,B,k)$ then there is a function $P \in \mathbb{J}(A,B)$ such that

$$b(P(z)-1) = \Big(\frac{2\mathcal{M}^{k+1}_\alpha \phi(z)}{\mathcal{M}^k_\alpha \phi(z) - \mathcal{M}^k_\alpha \phi(-z)}\Big)$$

and

$$b(P(-z)-1) = \Big(\frac{-2\mathcal{M}^{k+1}_\alpha \phi(-z)}{\mathcal{M}^k_\alpha \phi(z) - \mathcal{M}^k_\alpha \phi(-z)}\Big).$$

This implies that

$$1 + \frac{1}{b}\Big(\frac{\mathcal{M}^{k+1}_\alpha \mathfrak{O}(z)}{\mathcal{M}^k_\alpha \mathfrak{O}(z)} - 1\Big) = \frac{P(z)+P(-z)}{2}.$$

Also, since

$$P(z) \prec \frac{1+Az}{1+Bz}$$

where $\frac{1+Az}{1+Bz}$ is univalent then by the definition of the subordination, we obtain

$$1 + \frac{1}{b}\Big(\frac{\mathcal{M}^{k+1}_\alpha \mathfrak{O}(z)}{\mathcal{M}^k_\alpha \mathfrak{O}(z)} - 1\Big) \prec \frac{1+Az}{1+Bz}.$$

Moreover, the function $\mathfrak{O}(z)$ is starlike in $\mathbb{U}$ which implies that

$$\frac{z\mathfrak{O}(z)'}{\mathfrak{O}(z)} \prec \frac{1-z^2}{1+z^2}$$

that is, there exists a Schwarz function $\wp \in \mathbb{U}, |\wp(z)| \leq |z| < 1, \wp(0) = 0$ such that

$$\Phi(z) := \frac{z\mathfrak{O}(z)'}{\mathfrak{O}(z)} \prec \frac{1-\wp(z)^2}{1+\wp(z)^2}$$

which yields that there is $\xi, |\xi| = r < 1$ such that

$$\wp^2(\xi) = \frac{1-\Phi(\xi)}{1+\Phi(\xi)}, \quad \xi \in \mathbb{U}.$$

A calculation gives that

$$\left|\frac{1-\Phi(\xi)}{1+\Phi(\xi)}\right| = |\wp(\xi)|^2 \leq |\xi|^2.$$

Hence, we have the following conclusion

$$\left|\Phi(\xi) - \frac{1+|\xi|^4}{1-|\xi|^4}\right|^2 \leq \frac{4|\xi|^4}{(1-|\xi|^4)^2}$$

or

$$\left|\Phi(z) - \frac{1+|\xi|^4}{1-|\xi|^4}\right| \leq \frac{2|\xi|^2}{(1-|\xi|^4)}.$$

This implies that

$$\Re(\Phi(z)) \geq \frac{1-r^2}{1+r^2}, \quad |\xi| = r < 1.$$

□

Next consequence result of Theorem 3 can be found in [9,11] respectively.

Corollary 1. *Let $\alpha = 1$ in Theorem 3. Then*

$$1 + \frac{1}{b}\Big(\frac{\mathcal{M}_1^{k+1}\mathfrak{D}(z)}{\mathcal{M}_1^k\mathfrak{D}(z)} - 1\Big) \prec \frac{1+Az}{1+Bz}.$$

Corollary 2. *Let $\alpha = 1, k = 1$ in Theorem 3. Then*

$$1 + \frac{1}{b}\Big(\frac{\mathcal{M}_1^{2}\mathfrak{D}(z)}{\mathcal{M}_1\mathfrak{D}(z)} - 1\Big) \prec \frac{1+Az}{1+Bz}.$$

Theorem 4. *The set $T_k(\alpha, \kappa, \epsilon)$ in Definition 2 is convex.*

Proof. Let $\phi_i, i = 1,2$ be two functions in the set $T_k(\alpha, \kappa, \epsilon)$ satisfying $\phi_1(z) = z + \sum_{n=2}^{\infty} a_n z^n$ and $\phi_2(z) = z + \sum_{n=2}^{\infty} b_n z^n$. It is sufficient to prove that the function

$$H(z) = c_1\phi_1(z) + c_2\phi_2(z), \quad z \in \mathbb{U}$$

is in $T_k(\alpha, \kappa, \epsilon)$, where $c_1 > 0, c_2 > 0$ and $c_1 + c_2 = 1$. By the definition of $H(z)$, a calculation implies that

$$H(z) = z + \sum_{n=2}^{\infty} (c_1 a_n + c_2 b_n) z^n$$

then under the operator $\mathbf{C}_{\alpha,\kappa}^k$, we obtain

$$\mathbf{C}_{\alpha,\kappa}^k H(z) = z + \sum_{n=2}^{\infty} (c_1 a_n + c_2 b_n)$$

$$\times[(1-\kappa)C_{k+n-1}^k + \kappa\big(n[\alpha - (1-\alpha)(-1)^n]\big)^k]z^n.$$

By taking the derivative for the last equation and following by the real, we have

$$\begin{aligned}
&\Re\Big\{(\mathbf{C}^k_{\alpha,\kappa}H(z))'\Big\} \\
&= 1 + c_1\Re\Big\{\sum_{n=2}^{\infty} n[(1-\kappa)C^k_{k+n-1} + \kappa\big(n[\alpha-(1-\alpha)(-1)^n]\big)^k a_n z^{n-1}\Big\} \\
&+ c_2\Re\Big\{\sum_{n=2}^{\infty} n[(1-\kappa)C^k_{k+n-1} + \kappa\big(n[\alpha-(1-\alpha)(-1)^n]\big)^k b_n z^{n-1}\Big\} \\
&> 1 + c_1(\epsilon-1) + c_2(\epsilon-1) \\
&= \epsilon.
\end{aligned}$$

This completes the proof. □

Next consequence result of Theorem 4 can be found in [8].

Corollary 3. *Let $\alpha = 1$ in Theorem 4. Then the set $T_k(1,\kappa,\epsilon)$ is convex.*

Theorem 5. *Let $\phi \in T_k(\alpha,\kappa,\epsilon)$, and let φ be convex. Then for a function*

$$F(z) = \frac{2+c}{z^{1+c}}\int_0^z t^c\phi(t)dt, \quad z \in U$$

the subordination

$$\Big(\mathbf{C}^k_{\alpha,\kappa}\phi(z)\Big)' \prec \varphi(z) + \frac{(z\varphi'(z))}{2+c}, \quad c > 0,$$

implies

$$\Big(\mathbf{C}^k_{\alpha,\kappa}F(z)\Big)' \prec \varphi(z),$$

and this result is sharp.

Proof. Our aim is to apply Lemma 2. By the definition of $F(z)$, we obtain

$$\Big(\mathbf{C}^k_{\alpha,\kappa}F(z)\Big)' + \frac{\Big(\mathbf{C}^k_{\alpha,\kappa}F(z)\Big)''}{2+c} = \Big(\mathbf{C}^k_{\alpha,\kappa}\phi(z)\Big)'.$$

By the assumption, we get

$$\Big(\mathbf{C}^k_{\alpha,\kappa}F(z)\Big)' + \frac{\Big(\mathbf{C}^k_{\alpha,\kappa}F(z)\Big)''}{2+c} \prec \varphi(z) + \frac{(z\varphi'(z))}{2+c}.$$

By letting

$$\varrho(z) := \Big(\mathbf{C}^k_{\alpha,\kappa}F(z)\Big)',$$

one can find

$$\varrho(z) + \frac{(z\varrho'(z))}{2+c} \prec \varphi(z) + \frac{(z\varphi'(z))}{2+c}.$$

In virtue of Lemma 2, we have

$$\Big(\mathbf{C}^k_{\alpha,\kappa}F(z)\Big)' \prec \varphi(z),$$

and φ is the best dominant. □

Theorem 6. *Let φ be convex achieving $\varphi(0) = 1$. If*

$$\left(\mathbf{C}^k_{\alpha,\kappa}\phi(z)\right)' \prec \varphi(z) + z\varphi'(z), \quad z \in U,$$

then

$$\frac{\mathbf{C}^k_{\alpha,\kappa}\phi(z)}{z} \prec \varphi(z),$$

and this result is sharp.

Proof. Our aim is to apply Lemma 1. Define the function

$$\varrho(z) := \frac{\mathbf{C}^k_{\alpha,\kappa}\phi(z)}{z} \in \mathfrak{H}[1,1] \tag{7}$$

By this assumption, yields

$$\mathbf{C}^k_{\alpha,\kappa}\phi(z) = z\varrho(z) \Longrightarrow \left(\mathbf{C}^k_{\alpha,\kappa}\phi(z)\right)' = \varrho(z) + z\varrho'(z).$$

Thus, we deduce the following subordination:

$$\varrho(z) + z\varrho'(z) \prec \varphi(z) + z\varphi'(z).$$

In view of Lemma 1, we receive

$$\frac{\mathbf{C}^k_{\alpha,\kappa}\phi(z)}{z} \prec \varphi(z),$$

and φ is the best dominant. □

Theorem 7. *If $\phi \in \Lambda$ satisfies the subordination relation*

$$(\mathbf{C}^k_{\alpha,\kappa}\phi(z))' \prec \left(\frac{1+z}{1-z}\right)^{\beta}, \quad z \in \mathbb{U},\ \beta > 0,$$

then

$$\Re\left(\frac{\mathbf{C}^k_{\alpha,\kappa}\phi(z)}{z}\right) > \epsilon$$

for some $\epsilon \in [0,1)$.

Proof. Define a function ϱ as in (7). Then, by subordination properties, we have

$$(\mathbf{C}^k_{\alpha,\kappa}\phi(z))' = z\varrho'(z) + \varrho(z) \prec [\frac{1+z}{1-z}]^{\beta}.$$

Now, in view of Lemma 1.i, there is a constant $\gamma > 0$ with $\beta = \beta(\gamma)$ such that

$$\frac{\mathbf{C}^k_{\alpha,\kappa}\phi(z)}{z} \prec \left(\frac{1+z}{1-z}\right)^{\gamma}.$$

This implies that $\Re(\mathbf{C}^k_{\alpha,\kappa}\phi(z)/z) > \epsilon$, for some $\epsilon \in [0,1)$. □

Theorem 8. *If $\phi \in \Lambda$ satisfies the inequality*

$$\Re\left((\mathbf{C}^k_{\alpha,\kappa}\phi(z))' \frac{\mathbf{C}^k_{\alpha,\kappa}\phi(z)}{z}\right) > \frac{\alpha}{2}, \quad z \in U,\ \alpha \in [0,1)$$

then $\mathbf{C}^k_{\alpha,\kappa}\phi(z) \in T_k(\alpha,\kappa,\epsilon)$ *for some* $\epsilon \in [0,1)$. *Furthermore, it is univalent and of bounded boundary rotation in* U.

We inform the readers that in virtue of Noshiro-Warschawski Theorem (Duren [1], p. 47) if a function ϕ is analytic in the simply connected complex domain $\mathbb{U}$ and $\Re\{\phi'(z)\} > 0$ in $\mathbb{U}$ then ϕ is univalent in $\mathbb{U}$ and in view of Kaplan's Theorem (Duren [1], p. 48) such functions ϕ is of bounded boundary rotation.

Proof. Define a function ϱ as in (7). A simple computation yields

$$\Re\Big(\varrho^2(z) + 2\varrho(z).z\varrho'(z)\Big) = 2\Re\Big(\mathbf{C}^k_{\alpha,\kappa}\phi(z))'\frac{\mathbf{C}^k_{\alpha,\kappa}\phi(z)}{z}\Big) > \alpha. \tag{8}$$

By virtue of Lemma 1.ii, there is a constant λ depending on α such that $\Re(\varrho(z)) > \lambda$, which implies that $\Re(\varrho(z)\ ) > \epsilon$ for some $\epsilon \in [0,1)$. It follows from (8) that $\Re\Big(\mathbf{C}^k_{\alpha,\kappa}\phi(z))'\Big) > \epsilon$ and thus by Noshiro-Warschawski and Kaplan Theorems, $\mathbf{C}^k_{\alpha,\kappa}\phi(z)$ is univalent and of bounded boundary rotation in $\mathbb{U}$. □

Example 1. *We have the following data:* $\phi(z) = z/(1-z)$, $\alpha = 0.25$. *A calculation brings*

$$\begin{aligned}\mathcal{M}^1_\alpha\phi(z) &= \alpha z\phi'(z) - (1-\alpha)z\phi'(-z)\\ &= \frac{0.25z}{(1-z)^2} + \frac{0.75z}{(1+z)^2} = \frac{z(z^2-z+1)}{(1-z)^2(1+z)^2}\\ &= z - z^2 + 3z^3 - 2z^4 + 5z^5 + o(z^6)\end{aligned} \tag{9}$$

with

$$\begin{aligned}&\Re\left((\mathcal{M}^1_\alpha\phi(z))'\frac{\mathcal{M}^1_\alpha\phi(z)}{z}\right)\\ &= \Re\left(\frac{(-z^4+2z^3-6z^2+2z-1)\left(\frac{0.25z}{(1-z)^2} + \frac{0.75z}{(1+z)^2}\right)}{z(z^2-1)^3}\right)\\ &> 0,\end{aligned} \tag{10}$$

when $z \to 1$. *Hence, in view of Theorem 1,* $\frac{\mathcal{M}^1_\alpha\phi(z)}{z} \in \mathcal{P}(\epsilon)$ *for some* $\epsilon \in [0,1)$.

5. Conclusions and Future Works

Motivated by this method, in the recent investigation we have presented new classes of univalent functions that connect to a symmetric differential operator in the open unit disk. We have obtained sufficient and necessary conditions in relation to these subclasses. Linear combinations, operator and other properties are also explored. For further research, we indicate to study the certain new classes related to other types of analytic functions such as meromorphic, harmonic and p-valent functions with respect to symmetric points associated with SDO.

Author Contributions: Conceptualization, R.W.I. and M.D.; methodology, R.W.I.; validation, R.W.I. and M.D.; formal analysis, R.W.I. and M.D.; investigation, R.W.I. and M.D.; writing-original draft preparation, R.W.I.; writing-review and editing, M.D.

Acknowledgments: The author wishes to express his profound gratitude to the anonymous referee for his/her careful reading of the manuscript and the very useful comments that have been implemented in the final version of the manuscript.

References

1. Duren, P. *Univalent Functions*; Grundlehren der mathematischen Wissenschaften; Springer: New York, NY, USA, 1983; Volume 259, ISBN 0-387-90795-5.
2. Sàlàgean, G.S. Subclasses of univalent functions, Complex Analysis Fifth Romanian-Finnish Seminar, Part 1 (Bucharest, 1981). In *Lecture Notes in Mathematics*; Springer: Berlin, Germany, 1983; Volume 1013, pp. 362–372.
3. Al-Oboudi, F.M. On univalent functions defined by a generalized Sàlàgean operator. *Int. J. Math. Math. Sci.* **2004**, *27*, 1429–1436. [CrossRef]
4. Ibrahim, R.W.; Darus, M. Subordination inequalities of a new Sàlàgean difference operator. *Int. J. Math. Comput. Sci.* **2019**, *14*, 573–582.
5. Ovsyannikov, L.V. *Group Analysis of Differential Equations*; Academic Press: New York, NY, USA, 1982.
6. Sidorov, N.; Loginov, B.; Sinitsyn, A.V.; Falaleev, M.V. *Lyapunov-Schmidt Methods in Nonlinear Analysis and Applications*; Springer: New York, NY, USA, 2013.
7. Miller, S.S.; Mocanu, P.T. *Differential Subordinations: Theory and Applications*; CRC Press: Boca Raton, FL, USA, 2000.
8. Lupas, A.A. On special differential subordinations using Salagean and Ruscheweyh operators. *Math. Inequal. Appl.* **2009**, *12*, 781–790.
9. Arif, M.; Ahmad, K.; Liu, J.L.; Sokół, J. A new class of analytic functions associated with Sălăgean operator. *J. Funct. Spaces* **2019**, 6157394. [CrossRef]
10. Sakaguchi, K. On a certain univalent mapping. *J. Math. Soc. Jpn.* **1959**, *11*, 72–75. [CrossRef]
11. Das, R.N.; Singh, P. On subclasses of schlicht mapping. *Indian J. Pure Appl. Math.* **1977**, *8*, 864–872.

13

On Periodic Solutions of Delay Differential Equations with Impulses

Mostafa Bachar

Department of Mathematics, College of Sciences, King Saud University, Riyadh 11451, Saudi Arabia; mbachar@ksu.edu.sa

Abstract: The purpose of this paper is to study the nonlinear distributed delay differential equations with impulses effects in the vectorial regulated Banach spaces $\mathcal{R}([-r,0],\mathbb{R}^n)$. The existence of the periodic solution of impulsive delay differential equations is obtained by using the Schäffer fixed point theorem in regulated space $\mathcal{R}([-r,0],\mathbb{R}^n)$.

Keywords: delay differential equations; integral operator; periodic solutions

MSC: primary 06F30; 46B20; 47E10; 34K13; 34K05

1. Introduction

In this paper, we will investigate the existence of periodic solutions for vectorial distributed delay differential equations with impulses in regulated Banach spaces. More precisely, the prototype of this delay differential equations with impulses, is of the form

$$\frac{dx(t)}{dt} = -\lambda x(t) + f(t,x_t), a.e.\ t \in [0,\omega+\tau], \lambda > 0, \omega > 0, \quad (1)$$

$$x(t_j) = x(t_j^-), \text{ and } x(t_j^+) - x(t_j) = h_j(x(t_j)), \forall j = 1,\ldots,l, \quad (2)$$

$$x_0(\theta) = \varphi(\theta),\ \theta \in [-\tau,0], \quad (3)$$

with $x_t(\theta) = x(t+\theta)$, $\theta \in [-\tau,0]$, $\tau > 0$ and where x and φ are $\mathbb{R}^n$-valued functions on $[-\tau,\omega]$, and $[-\tau,0]$, respectively. The Equation (1) is a nonlinear delay differential equation. More details about this type of equations can be found in [1]. Moreover, we assume that

(i) $h_j \in C(\mathbb{R}^n,\mathbb{R}^n), j = 1,\ldots,l,$
(ii) $\{t_1,t_2,\cdots,t_l\}$ is an increasing family of strictly positive real numbers,
(iii) there exist $\delta > 0$ and $T < \infty$, such that for any $j = 1,\ldots,l-1$, we have

$$0 < \delta \le t_{j+1} - t_j \le T < \infty.$$

We call (2) the impulses equation where, $x(t_j^-)$ (resp. $x(t_j^+)$) denotes the limit from the left (resp. from the right) of $x(t)$, as t tends to t_j. This type of differential equations without delay was initiated in 1960's by Milman and Myshkis [2,3]. This problem started to be popular mostly in Eastern Europe in the years 1960–1970, with special attention during the seventies of the last century. Later on, several investigations and important monographs appeared with more details, which show the importance of studying such systems, see for example [4–11]. In recent years, many investigations have arisen with applications to life sciences, such that the periodic treatment of some biomedical applications, where the impulses correspond to administration of a drug treatment at certain given times [12–15]. However, comparatively speaking, not much has been done in the study of impulsive functional differential equations in regulated vectorial space, taking into account the general theory of functional analysis and having an acceptable hypothesis that can be used in real life applications, see [12] for more details.

Let us first introduce for each $\tau > 0$, the regulated Banach space $\mathcal{R} = \mathcal{R}([-\tau, 0], \mathbb{R}^n)$, given by:

$$\mathcal{R} = \left\{ \varphi : [-\tau, 0] \to \mathbb{R}^n : \varphi \text{ has left and right limits at every points of } [-\tau, 0] \right\},$$

endowed with the following norm

$$\|\varphi\|_{\mathcal{R}} = \sup_{\theta \in [-\tau, 0]} \|\varphi(\theta)\|.$$

We will make the following assumptions

(I) The map $f : [0, \omega + \tau] \times \mathbb{R}^n \to \mathbb{R}^n$, $\omega > 0$, satisfies

- $\|f(t, \varphi) - f(t, \psi)\|_{\mathcal{R}} \leq K\|\varphi - \psi\|_{\mathcal{R}}, \forall t \in [0, \omega + \tau], \varphi, \psi \in \mathcal{R}$,
- $\exists M > 0, \|f(t, 0)\|_{\mathcal{R}} \leq M, \forall t \in [0, \omega + \tau]$.

(II) For each regulated map $x : [a, b] \to \mathbb{R}^n$, with $b - a > \tau$, we assume that the map $t \to f(t, x_t)$ is measurable over $[a + \tau, b]$.

(III) For each $j = 1, \ldots, l$, $h_j : \mathbb{R}^n \to \mathbb{R}^n$ is a continuous map.

We set the initial value problem as follows

Problem 1. *Let φ be an element of $\mathcal{R}$. We want to find a function x defined on $[-\tau, \omega + \tau]$ such that x satisfies (1)–(3).*

We consider the nonlinear impulsive delay differential equation in $\mathcal{R}$ as

$$\begin{cases} \dfrac{dx(t)}{dt} &= -\lambda x(t) + f(t, x_t), a.e.\ t \in [0, \omega + \tau], \lambda > 0, \omega > 0, \\ x(t_j) &= x(t_j^-), \text{ and } x(t_j^+) - x(t_j) = h_j(x(t_j)), \forall j = 1, \ldots, l, \\ x_0(\theta) &= \varphi(\theta), \theta \in [-\tau, 0] \text{ and } x(0^+) = \xi \in \mathbb{R}^n. \end{cases}$$

The aim of this paper is to extend the main results related to the existence of the ω-periodic solutions for ordinary differential equations with impulses presented by Li et al. [16] and Nieto [17]. These papers contain references which provide additional reading on this topic, i.e., differential equations with impulses by using the fixed point theory.

2. Existence and Uniqueness of Solution

Let us start first by introducing some related definitions and lemmas.

Definition 1. *A function $x : [-\tau, \omega + \tau] \to \mathbb{R}^n$ is called a solution of (1)–(3) if:*

1. *x is absolutely continuous with respect to the Lebesgue measure;*
2. *x is differentiable on the complement of a countable subset of $[0, \omega + \tau]$, and satisfies Equation (1) whenever $\frac{dx(t)}{dt}$ and the right hand side of (1) are defined on $[0, \omega + \tau]$;*
3. *x satisfies (2) at each point $t_j, t_j \geq 0, \forall j = 1, \ldots, l$, and the initial value function satisfies (3).*

Lemma 1. *Let $f : [0, \omega + \tau] \times \mathcal{R} \to \mathbb{R}^n$ be a map satisfying (I) and (II) and $t_1 \in [0, \omega + \tau]$. Then, for each $(\varphi, \xi) \in \mathcal{R} \times \mathbb{R}^n$, the problem*

$$\frac{dx(t)}{dt} = -\lambda x(t) + f(t, x_t), a.e.\ t \in [0, t_1] \tag{4}$$

$$(x_0, x(0^+)) = (\varphi, \xi) \in \mathcal{R} \times \mathbb{R}^n, \tag{5}$$

has a unique solution.

Proof. We set $S = \{y \in C([0,t_1],\mathbb{R}^n), y(0) = x(0^+) = \xi\}$. Let us define the operator T by

$$T(x)(t) \quad = \quad \xi + \int_0^t \Big(f(s,x_s) - \lambda x(s)\Big)ds, 0 \le t \le t_1. \tag{6}$$

For each $y \in S$, we consider the Nemytski operator F, defined by

$$F(y)(t) \quad = \quad f(t,z_t), \tag{7}$$

where

$$z_t(\theta) \quad = \begin{cases} y(t+\theta), & \text{if } t+\theta \ge 0, \\ \varphi(t+\theta), & \text{if } t+\theta \le 0. \end{cases} \tag{8}$$

Then, we get

$$T(y)(t) \quad = \quad \xi + \int_0^t \Big(F(y)(s) - \lambda y(s)\Big)ds. \tag{9}$$

Define, the norm of any function y in S by

$$\|y\|_S \quad = \quad \sup_{0\le t\le t_1} \Big\{\|y(t)\|e^{-\rho t}\Big\}, \tag{10}$$

where ρ is a fixed positive constant greater than $K+\lambda$. We have for each $y_1(t)$ and $y_2(t)$ in S,

$$\begin{aligned} \| T(y_1)(t) - T(y_2)(t) \| \quad &\le \quad (K+\lambda)\int_0^t \| y_1(s) - y_2(s) \| \, ds, \\ &\le \quad (K+\lambda)\int_0^t e^{\rho s - \rho s} \| y_1(s) - y_2(s) \| \, ds, \\ &\le \quad (K+\lambda) \| y_1 - y_2 \|_S \int_0^t e^{\rho s} ds, \\ &\le \quad \frac{(K+\lambda)}{\rho} \| y_1 - y_2 \|_S \, e^{\rho t}, \end{aligned}$$

and hence

$$\| T(y_1) - T(y_2) \|_S \quad \le \quad \frac{(K+\lambda)}{\rho} \| y_1 - y_2 \|_S \, .$$

Since $\frac{K+\lambda}{\rho} < 1$, then, T is a contraction on S, and the result follows immediately. □

Lemma 2. *[18] Let $f : [0,\omega+\tau] \times \mathcal{R} \to \mathbb{R}^n$ be a map satisfying (I) and (II) and h_j, for $j = 1,\cdots,l$, satisfy the condition (III). Then the problem (1)–(3) has a unique solution.*

Proof. The proof follows by using the last lemma. □

Lemma 3. *[18] Under the assumptions (I) and (II), if $x(\varphi)(t)$ is the unique solution of (4) and (5), then one has:*

$$\| x(\varphi)(t) \| \quad \le \quad e^{Kt}\Big(\| \varphi \| + \int_0^\omega \| f(s,0) \| \, ds\Big). \tag{11}$$

The next Lemma, gives a similar, key representation formula for the solutions of the delay differential equations with impulses (1)–(3) in regulated Banach space $\mathcal{R}$, see [4] for more details.

Lemma 4. *The problem (1)–(3) can be written as*

$$
\begin{aligned}
x_t &= \varphi_t^0 + H_t^0 \otimes ((\xi e^{-\lambda \max(0,\bullet)})_t - \varphi(0)) \\
&\quad + \Big(\int_0^{\max(0,\bullet)} f(s, x_s) e^{-\lambda(\bullet - s)} ds + \sum_{0 \le t_j < \bullet} e^{-\lambda(\bullet - t_j)} u_j \Big)_t,
\end{aligned}
$$

where

$$\varphi^0(\theta) = \begin{cases} \varphi(\theta), & \text{if } \theta \le 0, \\ \varphi(0), & \text{if } \theta > 0, \end{cases} \tag{12}$$

H^0 is the Heaviside function

$$H^0(\theta) = \begin{cases} 0, & \text{if } \theta \le 0, \\ 1, & \text{if } \theta > 0, \end{cases} \tag{13}$$

and the sequence

$$u_k = x(t_k^+) - x(t_k), k \ge 1$$

is determined by the following non-autonomous recurrence equation

$$u_k = h_k \Big(\xi e^{-\lambda t_k} + \int_0^{t_k} f(s, x_s) e^{-\lambda(t_k - s)} ds + \sum_{0 \le t_j < t_k} e^{-\lambda(t_k - t_j)} u_j \Big), k \ge 1,$$

starting from

$$u_1 = h_1 \Big(\xi e^{-\lambda t_1} + \int_0^{t_1} f(s, x_s) e^{-\lambda(t_1 - s)} ds \Big).$$

Proof. Let us consider $z(t) = e^{\lambda t} x(t), \forall t \in [0, \omega + \tau]$, then the problem (1)–(3) becomes

$$
\begin{aligned}
\frac{dz(t)}{dt} &= f(t, e^{-\lambda(t+\theta)} z_t) e^{\lambda t}, a.e.\ t \in [0, \omega + \tau], \lambda > 0, \omega > 0, & (14) \\
z(t_j) &= z(t_j^-), \text{ and } z(t_j^+) - z(t_j) = e^{\lambda t_j} h_j(e^{-\lambda t_j} z(t_j)), \forall j = 1, \ldots, l, & (15) \\
z_0(\theta) &= e^{\lambda \theta} \varphi(\theta) = \widetilde{\varphi}(\theta), \theta \in [-\tau, 0], \text{ and } z(0^+) = \xi \in \mathbb{R}^n. & (16)
\end{aligned}
$$

Let us consider $t \in [t_j, t_{j+1}), j = 1, \ldots, l-1$, with $t_0 = 0$, then we get

$$z(t) = z(t_j^+) + \int_{t_j}^t f(s, e^{-\lambda(s+\theta)} z_s) e^{\lambda s} ds.$$

By passing to the limit as t goes to t_j^-, and by solving the delay differential Equation (14) on the interval $[t_{j-1}, t_j)$, we have

$$z(t_j) = z(t_{j-1}^+) + \int_{t_{j-1}}^{t_j} f(s, e^{-\lambda(s+\theta)} z_s) e^{\lambda s} ds.$$

Then, by taking into account the impulses condition (15), we have

$$z(t) = z(t_{j-1}^+) + \int_{t_{j-1}}^t f(s, e^{-\lambda(s+\theta)} z_s) e^{\lambda s} ds + e^{\lambda t_j} h_j(e^{-\lambda t_j} z(t_j)),$$

for all $t \in [t_j, t_{j+1})$, for $j = 1, \cdots, l-1$. Consequently, we can rewrite the last equations in more general form for all $t > 0$

$$z(t) = \xi + \int_0^t f(s, e^{-\lambda(s+\theta)} z_s) e^{\lambda s} ds + \sum_{0 \le t_j < t} e^{\lambda t_j} u_j, t \notin \{t_k\}_{k \ge 1}, \tag{17}$$

where $z(0^+) = x(0^+) = \xi$, and

$$u_k = z(t_k^+) - z(t_k) = h_k(e^{-\lambda t_k} z(t_k)), k \ge 1. \tag{18}$$

Now, we will try to involve the $u_k' s$. To this end, we will take the limit from the left of the Formula (17) as t tends to $t_k > 0$, we obtain

$$z(t_k) = \xi + \int_0^{t_k} f(s, e^{-\lambda(s+\theta)} z_s) e^{\lambda s} ds + \sum_{0 \le t_j < t_k} e^{\lambda t_j} u_j.$$

Substituting the last expression into (18), we have

$$u_k = h_k\Big(e^{-\lambda t_k}\xi + \int_0^{t_k} f(s, e^{-\lambda(s+\theta)} z_s) e^{\lambda(s-t_k)} ds + \sum_{0 \le t_j < t_k} e^{\lambda(t_j - t_k)} u_j\Big).$$

In particular, we have $\{j : 0 \le t_j < t_1\} = \emptyset$, and therefore

$$u_1 = h_1\Big(e^{-\lambda t_1}\xi + \int_0^{t_1} f(s, e^{-\lambda(s+\theta)} z_s) e^{\lambda s} ds\Big).$$

By using, the Equation (16), we can rewrite the Equation (17) as

$$\begin{aligned} z_t(\theta) &= \xi + \int_0^{t+\theta} f(s, e^{-\lambda(s+\theta)} z_s) e^{\lambda s} ds \\ &+ \sum_{0 \le t_j < t+\theta} e^{\lambda t_j} u_j, t+\theta \notin \{t_k\}_{k \ge 0}, \text{ and } t+\theta \ge 0, \end{aligned} \tag{19}$$

and by using $x(t) = e^{-\lambda t} z(t)$, we have for $t + \theta \notin \{t_k\}_{k \ge 1}$, and $t + \theta \ge 0$

$$x_t(\varphi(\theta)) = \xi e^{-\lambda(t+\theta)} + \int_0^{t+\theta} f(s, x_s) e^{-\lambda(t+\theta-s)} ds + \sum_{0 \le t_j < t+\theta} e^{-\lambda(t+\theta-t_j)} u_j.$$

Using (12) and (13), we get

$$\begin{aligned} x_t(\varphi) &= \varphi_t^0 + H_t^0 \otimes ((\xi e^{-\lambda \max(0,\bullet)})_t - \varphi(0)) \\ &+ \Big(\int_0^{\max(0,\bullet)} f(s, x_s) e^{-\lambda(\bullet - s)} ds + \sum_{0 \le t_j < \bullet} e^{-\lambda(\bullet - t_j)} u_j\Big)_t, \end{aligned}$$

where

$$u_k = h_k\Big(\xi e^{-\lambda t_k} + \int_0^{t_k} f(s, x_s) e^{-\lambda(t_k - s)} ds + \sum_{0 \le t_j < t_k} e^{-\lambda(t_k - t_j)} u_j\Big), k \ge 1.$$

starting from

$$u_1 = h_1\Big(\xi e^{-\lambda t_1} + \int_0^{t_1} f(s, x_s) e^{-\lambda(t_1 - s)} ds\Big).$$

□

Remark 1. *Taking into account the conditions (II)–(III), we have* $u_t \in \mathcal{R}$, $\forall t \in [0, \omega + \tau]$, *and* $t \to x_t$ *is a regulated function, because the functions* $t \to \varphi_t^0$, *and* $t \to H_t^0$ *are regulated.*

In the next section, we will investigate the existence of the periodic solution(s) for the delay differential equation with impulses (1)–(3) using Schäffer's fixed point theorem [19].

3. Existence of Periodic Solutions

Let us consider the Poincaré operator, given by:

$$\begin{aligned} J : \mathcal{R} &\to \mathcal{R} \\ \varphi &\to x_\omega(\varphi), \end{aligned}$$

where $x_\omega(\varphi)$ is the solution of the delay differential equation with impulses (1)–(3). It is clear that if the Poincaré operator J admit a fixed point, then (1)–(3) has a ω-periodic solution. The following lemma is useful to prove the main theorem.

Lemma 5. *The problem (1)–(3) has a* ω*-periodic solution in* $\mathcal{R}$ *if and only if the integral equation*

$$x_t(\varphi)(\theta) = \begin{cases} e^{-\lambda\theta} \int_{t+\theta}^{t+\omega+\theta} G(t,s) f(s, x_s) ds + e^{-\lambda\theta} \sum\limits_{t+\theta \le t_j < t+\omega+\theta} G(t, t_j) u_j, & \text{if } 0 \le t+\theta \le \omega, \\ \varphi(t+\theta), & \text{if } -\tau \le t+\theta \le 0, \end{cases}$$

has a solution $\forall t \in [0, \omega + \tau]$ *and* $\omega \ge \tau$, *where*

$$G(t,s) = \frac{e^{-\lambda(t-s)}}{e^{\lambda\omega} - 1}, \tag{20}$$

and the sequence

$$u_k = x(t_k^+) - x(t_k), k \ge 1$$

is determined by the following non-autonomous recurrence equation

$$u_k = h_k\Big(\int_{t_k}^{t_k+\omega} G(t,s) f(s, x_s) ds + \sum_{t_k \le t_j < t_k+\omega} G(t, t_j) u_j \Big), k \ge 1,$$

starting from

$$u_1 = h_1\Big(\xi e^{-\lambda t_1} + \int_0^{t_1} f(s, x_s) e^{-\lambda(t_1 - s)} ds \Big).$$

Proof. Using the expression (19) for $t + \omega + \theta$, where $t \ge 0$, and $\omega \ge \tau$, we have for all $t + \theta \ge 0$

$$\begin{aligned} z_{t+\omega}(\theta) &= \xi + \int_0^{t+\omega+\theta} f(s, e^{-\lambda(s+\theta)} z_s) e^{\lambda s} ds + \sum_{0 \le t_j < t+\omega+\theta} e^{\lambda t_j} u_j, \\ &= \xi + \int_0^{t+\theta} f(s, e^{-\lambda(s+\theta)} z_s) e^{\lambda s} ds + \sum_{0 \le t_j < t+\theta} e^{\lambda t_j} u_j \\ &\quad + \int_{t+\theta}^{t+\omega+\theta} f(s, e^{-\lambda(s+\theta)} z_s) e^{\lambda s} ds + \sum_{t+\theta \le t_j < t+\omega+\theta} e^{\lambda t_j} u_j, \\ &= z_t(\theta) + \int_{t+\theta}^{t+\omega+\theta} f(s, e^{-\lambda(s+\theta)} z_s) e^{\lambda s} ds + \sum_{t+\theta \le t_j < t+\omega+\theta} e^{\lambda t_j} u_j, \end{aligned}$$

and, by using the ω-periodic condition $z_{t+\omega}(\theta) = e^{\lambda\omega} z_t(\theta)$, we get

$$z_t(\theta) = \frac{1}{e^{\lambda\omega}-1}\int_{t+\theta}^{t+\omega+\theta} f(s, e^{-\lambda(s+\theta)} z_s) e^{\lambda s} ds + \frac{1}{e^{\lambda\omega}-1} \sum_{t+\theta \le t_j < t+\omega+\theta} e^{\lambda t_j} u_j.$$

Therefore, using $z_t(\theta) = e^{\lambda(t+\theta)} x_t(\theta)$, we have

$$x_t(\theta) = e^{-\lambda\theta} \int_{t+\theta}^{t+\omega+\theta} G(t,s) f(s, x_s) ds + e^{-\lambda\theta} \sum_{t+\theta \le t_j < t+\omega+\theta} G(t, t_j) u_j,$$

where

$$G(t,s) = \frac{e^{-\lambda(t-s)}}{e^{\lambda\omega}-1}. \tag{21}$$

Then

$$u_k = h_k\Big(\int_{t_k}^{t_k+\omega} G(t,s) f(s, x_s) ds + \sum_{t_k \le t_j < t_k+\omega} G(t, t_j) u_j \Big), k \ge 1,$$

starting from

$$u_1 = h_1\Big(\xi e^{-\lambda t_1} + \int_0^{t_1} f(s, x_s) e^{-\lambda(t_1-s)} ds\Big).$$

□

Example 1. *Let us consider the scalar delay differential equation with impulses:*

$$\frac{dx(t)}{dt} = -\lambda x(t) + f(t, x(t-\tau)), a.e.\ t \in [0, 2\tau], \tag{22}$$

$$x(\tau) = x(\tau^-), \text{ and } x(\tau^+) - x(\tau) = cx(\tau), \tag{23}$$

$$x(\theta) = \varphi(\theta), \theta \in [-\tau, 0], \tag{24}$$

where $f : [0, 2\tau] \times \mathcal{R} \to \mathbb{R}^n$ is a map satisfying (II). Let us investigate the existence of the τ-periodic solution of (22)–(24) such that $x_{t+\tau}(-\tau) = x_t(-\tau), \tau \le t \le 2\tau$. The solution of the delay differential Equations (22)–(24), can be written as

$$x(t) = \begin{cases} \varphi(\theta), & \text{if } -\tau \le t \le 0, \\ \varphi(0) e^{-\lambda t} + \int_0^t e^{-\lambda(t-s)} f(s, x(s-\tau)) ds, & \text{if } 0 \le t \le \tau, \\ x(\tau^+) e^{-\lambda(t-\tau)} + \int_\tau^t e^{-\lambda(t-s)} f(s, x(s-\tau)) ds, & \text{if } \tau < t \le 2\tau. \end{cases} \tag{25}$$

Using (23), we get

$$\begin{aligned} x(\tau^+) &= x(\tau) + cx(\tau), \\ &= (c+1) x(0) e^{-\lambda\tau} + (c+1) \int_0^\tau e^{-\lambda(\tau-s)} f(s, x(s-\tau)) ds. \end{aligned}$$

Therefore, if $\tau \le t < 2\tau$, we have

$$\begin{aligned} x(t) &= \Big((c+1)\varphi(0)e^{-\lambda\tau} + (c+1)\int_0^\tau e^{-\lambda(\tau-s)} f(s,x(s-\tau))ds\Big)e^{-\lambda(t-\tau)} \\ &\quad + \int_\tau^t e^{-\lambda(t-s)} f(s,x(s-\tau))ds, \\ &= (c+1)\varphi(0)e^{-\lambda\tau}e^{-\lambda(t-\tau)} + (c+1)e^{-\lambda\tau}\int_0^{t-\tau} e^{-\lambda(t-\tau-s)} f(s,x(s-\tau))ds \\ &\quad +(c+1)e^{-\lambda\tau}\int_{t-\tau}^{\tau} e^{-\lambda(t-\tau-s)} f(s,x(s-\tau))ds + \int_\tau^t e^{-\lambda(t-s)} f(s,x(s-\tau))ds, \\ &= (c+1)e^{-\lambda\tau}\Big(x(t-\tau) + \int_{t-\tau}^{\tau} e^{-\lambda(t-\tau-s)} f(s,x(s-\tau))ds\Big) + \int_\tau^t e^{-\lambda(t-s)} f(s,x(s-\tau))ds, \end{aligned}$$

which implies

$$\begin{aligned} x_{t+\tau}(-\tau) &= (c+1)e^{-\lambda\tau}\Big(x_t(-\tau) + \int_{t-\tau}^{\tau} e^{-\lambda(t-\tau-s)} f(s,x(s-\tau))ds\Big) \\ &\quad + \int_\tau^t e^{-\lambda(t-s)} f(s,x(s-\tau))ds. \end{aligned}$$

Then, we have three cases.

(1) *If* $1-(c+1)e^{-\lambda\tau} \neq 0$, *then, we have the existence and uniqueness of a* τ*-periodic solution.*

(2) *If* $1-(c+1)e^{-\lambda\tau} = 0$, *and*

$$\int_{t-\tau}^{\tau} e^{-\lambda(t-\tau-s)} f(s,x(s-\tau))ds + \int_\tau^t e^{-\lambda(t-s)} f(s,x(s-\tau))ds = 0,$$

then, we have the existence of infinitely many τ*-periodic solutions.*

(3) *If* $1-(c+1)e^{-\lambda\tau} = 0$, *and*

$$\int_{t-\tau}^{\tau} e^{-\lambda(t-\tau-s)} f(s,x(s-\tau))ds + \int_\tau^t e^{-\lambda(t-s)} f(s,x(s-\tau))ds \neq 0,$$

then, there exists no τ*-periodic solution.*

Now, we can consider for each $t \geq -\tau$ and $\omega \geq \tau$, the Poincaré operator $J : \mathcal{R} \to \mathcal{R}$ defined by

$$J\varphi = \Big(e^{-\lambda(\bullet-t)}\int_\bullet^{\bullet+\omega} G(t,s)f(s,\varphi)ds + e^{-\lambda(\bullet-t)}\sum_{\bullet\leq t_j<\bullet+\omega} G(t,t_j)u_j\Big)_t,$$

where

$$u_k = h_k\Big(\int_{t_k}^{t_k+\omega} G(t,s)f(s,x_s)ds + \sum_{t_k\leq t_j<t_k+\omega} G(t,t_j)u_j\Big), k \geq 2,$$

and, starting from

$$u_1 = h_1\Big(\xi e^{-\lambda t_1} + \int_0^{t_1} f(s,x_s)e^{-\lambda(t_1-s)}ds\Big).$$

It is clear, that, the ω-periodic solutions in $\mathcal{R}$ of (1)–(3) are exactly the fixed points of the Poincaré operator J, i.e., $J\varphi = \varphi$.

The following theorem, is known as the Schäffer's fixed point theorem [19], which can be found for example in Deimling's book [20].

Theorem 1. *[19–22] Let X be a normed space, $\mathcal{F}$ a continuous mapping of X into X, such that the closure of $\mathcal{F}(B)$ is compact for any bounded subset B of X. Then either:*

(i) *the equation $x = \lambda \mathcal{F} x$ has a solution for $\lambda = 1$, or*
(ii) *the set of all such solutions x, for $0 < \lambda < 1$, is unbounded.*

Before, we state the main theorem of our work, we will need the following lemma.

Lemma 6. *Let $f : [0, \omega + \tau] \times \mathcal{R} \to \mathbb{R}^n$ be a map satisfying (I) and (II), where $\omega \geq \tau$, and $h_j, j = 1, \ldots, l$ are bounded and satisfy the condition (III). Then, the Poincaré operator $J : \mathcal{R} \to \mathcal{R}$ is completely continuous.*

Proof. Let $B \subset \mathcal{R}$ be a bounded set and $\varphi \in B$. Then by using the condition (I), we have

$$\|f(t,\varphi)\|_{\mathcal{R}} \leq \|f(t,0)\|_{\mathcal{R}} + \|f(t,\varphi) - f(t,0)\|_{\mathcal{R}} \leq M + K\|\varphi\|_{\mathcal{R}} < \infty.$$

Therefore, there exist two constants $\widetilde{M}$ and $\overline{M}$ such that

$$\begin{aligned} \| J\varphi(\theta) \| &= \| e^{-\lambda\theta} \int_{t+\theta}^{t+\theta+\omega} f(s,\varphi)G(t,s)ds + e^{-\lambda\theta} \sum_{t+\theta \leq t_j < t+\theta+\omega} G(t,t_j)u_j \|, \\ &\leq \| e^{-\lambda\theta} \int_{t+\theta}^{t+\theta+\omega} f(s,\varphi)G(t,s)ds \| + e^{\lambda r} \| \sum_{t+\theta \leq t_j < t+\theta+\omega} G(t,t_j)u_j \|, \\ &\leq e^{\lambda\tau}\omega\widetilde{M} + e^{\lambda r} \sum_{t+\theta \leq t_j < t+\theta+\omega} \overline{M}, \end{aligned} \tag{26}$$

where

$$\|u_k\| = \left\| h_k\Big(\int_{t_k}^{t_k+\omega} G(t,s)f(s,x_s)ds + \sum_{t_k \leq t_j < t_k+\omega} G(t,t_j)u_j \Big) \right\| < \infty, k \geq 2,$$

and starting from

$$\|u_1\| = \left\| h_1\left(\xi e^{-\lambda t_1} + \int_0^{t_1} f(s,x_s)e^{-\lambda(t_1-s)}ds \right) \right\| < \infty,$$

and, we have

$$\| J\varphi \|_{\mathcal{R}} \leq e^{\lambda\tau}\omega\widetilde{M} + e^{\lambda r}\overline{M} \sum_{t+\theta \leq t_j < t+\theta+\omega} 1,$$

which imply that $J(B)$ is uniformly bounded. For each $t \geq 0$, there exists $n \in \mathbb{N}^*$ such that $t \in [t_n, t_{n+1})$, and for any $\theta, \widetilde{\theta} \in [-r, 0]$, one can obtain for any $\varphi \in B$

$$\begin{aligned} \| J\varphi(\theta) - J\varphi(\widetilde{\theta}) \| \leq\ & \| e^{-\lambda\theta} \int_{t+\theta}^{t+\theta+\omega} f(s,\varphi)G(t,s)ds - e^{-\lambda\theta} \int_{t+\widetilde{\theta}}^{t+\widetilde{\theta}+\omega} f(s,\varphi)G(t,s)ds \| \\ & + \| e^{-\lambda\theta} \sum_{t+\theta \leq t_j < t+\theta+\omega} G(t,t_j)u_j - e^{-\lambda\theta} \sum_{t+\widetilde{\theta} \leq t_j < t+\widetilde{\theta}+\omega} G(t,t_j)u_j \|, \\ \leq\ & \frac{e^{\lambda(r+\omega)}\left(M + K\|\varphi\|\right)}{1 - e^{-\lambda\omega}} \| \int_{t+\theta}^{t+\theta+\omega} e^{-\lambda(t-s)}ds - \int_{t+\widetilde{\theta}}^{t+\widetilde{\theta}+\omega} e^{-\lambda(t-s)}ds \| \\ & + \frac{e^{\lambda(r+\omega)}}{1 - e^{-\lambda\omega}} \| \sum_{t+\theta \leq t_j < t+\theta+\omega} u_j - \sum_{t+\widetilde{\theta} \leq t_j < t+\widetilde{\theta}+\omega} u_j \| . \end{aligned}$$

Therefore, for each $t \in [t_n, t_{n+1})$, we will have as $\mid \theta - \widetilde{\theta} \mid$ goes to 0, $\| J\varphi(\theta) - J\varphi(\widetilde{\theta}) \|$ goes to 0, which imply that the Poincaré operator $J(B)$ is equicontinuous. Using Arzelà-Ascoli's theorem, we conclude that the Poincaré operator J is completely continuous. □

Now, we are ready to state the main result of our work, related to the existence of ω-periodic solution(s) of (1)–(3).

Theorem 2. *Let $f : [0, \omega + \tau] \times \mathcal{R} \to \mathbb{R}^n$ be a map satisfying (I) and (II), where $\omega \geq \tau$, and $h_j, j = 1, \ldots, l$ are bounded and satisfy the condition (III). Then, the nonlinear impulsive problem (1)–(3), has at least one ω-periodic solution in $\mathcal{R}$.*

Proof. Let us define $H(\varphi, \mu) : \mathcal{R} \times [0,1] \longrightarrow \mathcal{R}$ by

$$H(\varphi, \mu) \quad = \quad \mu J \varphi. \tag{27}$$

Then, by using (26), we have

$$\|H(\varphi, \mu)\|_{\mathcal{R}} \quad \leq \quad \mu\Big(e^{\lambda\tau}\omega\widetilde{M} + e^{\lambda r} \sum_{t+\theta \leq t_j < t+\theta+\omega} \overline{M}\Big).$$

Then, for each $\mu \in (0,1)$ the set $S = \{\varphi : \varphi = H(\varphi, \mu)\}$ is bounded. Since J is completely continuous, then by using Schäffer's fixed point theorem, the Poincaré operator J admits a fixed point. □

Next, we give the conditions of the existence and uniqueness of a ω-periodic solution of (1)–(3).

Theorem 3. *Let $f : [0, \omega + \tau] \times \mathcal{R} \to \mathbb{R}^n$ be a map satisfying (I) and (II), where $\omega \geq \tau$, and $h_j, j = 1, \ldots, l$ are bounded and satisfy the condition (III), and there exist constants $\overline{H}_j$, $j = 1, \ldots, l$, such that*

$$\| h_j(\varphi(0)) - h_j(\psi(0)) \| \quad \leq \quad \overline{H}_j \| \varphi - \psi \|_{\mathcal{R}} \, .$$

If, there exists a constant $C < 1$, such that

$$\frac{K\omega e^{\lambda r}}{1 - e^{-\lambda\omega}} + \frac{e^{\lambda r}}{1 - e^{-\lambda\omega}} \sum_{t-r+\omega \leq t_j < t+\omega} \overline{H}_j \quad \leq \quad C,$$

then, the nonlinear impulsive problem (1)–(3), has a unique ω-periodic solution in $\mathcal{R}$.

Proof. Let $\varphi, \psi \in \mathcal{R}$ be two solutions of (1)–(3), i.e., $J\varphi = \varphi$ and $J\psi = \psi$. Assume $\phi \neq \psi$. We have

$$\begin{aligned} \| \varphi(\theta) - \psi(\theta) \| &= \| J\varphi(\theta) - J\psi(\theta) \|, \\ &\leq e^{\lambda r} \int_{t+\theta}^{t+\theta+\omega} | G(t,s) | \| f(s,\varphi) - f(s,\psi) \|_{\mathcal{R}} \, ds + \\ &\quad e^{\lambda r} \sum_{t-r+\omega \leq t_j < t+\omega} | G(t,t_j) | \| h_j(\varphi(0) - h_j(\psi(0)) \|, \\ &\leq \Big(\frac{K\omega e^{\lambda r}}{1 - e^{-\lambda\omega}} + \frac{e^{\lambda r}}{1 - e^{-\lambda\omega}} \sum_{t-r+\omega \leq t_j < t+\omega} \overline{H}_j \Big) \| \varphi - \psi \|_{\mathcal{R}}, \\ &\leq C \| \varphi - \psi \|_{\mathcal{R}} \, . \end{aligned}$$

Hence

$$\begin{aligned} \| \varphi - \psi \|_{\mathcal{R}} &\leq C \| \varphi - \psi \|_{\mathcal{R}}, \\ &< \| \varphi - \psi \|_{\mathcal{R}}. \end{aligned} \tag{28}$$

This contradiction implies, the uniqueness of the ω-periodic solution of (1)–(3). □

4. Conclusions

The method described in this work presents new challenges for more investigation on more realistic models; such as the extension of the ascorbic acid model [12] and HIV model [13,14]. Taking into account the delay effect on respective compartments [23–25]. This kind of work, will need more investigation on modeling validation effort, keeping a close eye on the real life data in order to have a more realistic model. The explicit solutions presented in the technical Lemma 4 and methods of proving the existence of periodic solutions are very useful for further future investigations.

Acknowledgments: The authors would like to express their deep appreciation to the Deanship of Scientific Research at King Saud University for supporting this Research group No. (RG-1435-079). The authors profusely thank the referees for their valuable comments.

References

1. Hale, J. *Theory of Functional Differential Equations*; Springer: New York, NY, USA, 1977.
2. Milman, V.D.; Myshkis, A.D. On the stability of motion in the presence of impulses. *Sib. Math. J.* **1960**, *1*, 233–237. (In Russian)
3. Mil'man, V.D.; Myshkis, A.D. Random impulses in linear dynamical systems. In *Approximate Methods of Solution of Differential Equations*; Publishing House of the Academy of Sciences: Kiev, Ukraim, 1963; pp. 64–81. (In Russian)
4. Bachar, M.; Arino, O. Stability of a general linear delay-differential equation with impulses. *Dyn. Contin. Discret. Impuls. Syst. Ser. A Math. Anal.* **2003**, *10*, 973–990.
5. Bainov, D.D.; Barbanti, L.; Hristova, S.G. Method of quasilinearization for the periodic boundary value problem for impulsive differential-difference equations. *Commun. Appl. Anal.* **2003**, *7*, 153–170.
6. Bainov, D.D.; Hristova, S.G.; Hu, S.C.; Lakshmikantham, V. Periodic boundary value problems for systems of first order impulsive differential equations. *Differ. Integr. Equ.* **1989**, *2*, 37–43.
7. Bainov, D.D.; Simeonov, P.S. *Systems with Impulse Effect. Stability, Theory and Applications*; Ellis Horwood Limited: Hemel Hempstead, UK; John Wiley & Sons: Chichester, UK, 1989.
8. Bajo, I.; Liz, E. Periodic boundary value problem for first order differential equations with impulses at variable times. *J. Math. Anal. Appl.* **1996**, *204*, 65–73. [CrossRef]
9. Berezansky, L.; Braverman, E. Impulsive equations: Overview and open problems. *Funct. Differ. Equ.* **2008**, *15*, 39–56.
10. Pandit, S.G.; Deo, S.G. *Differential Systems Involving Impulses*; Springer: New York, NY, USA; Berlin, Germany, 1982.
11. Zhao, K. Global exponential stability of positive periodic solution of the n-species impulsive Gilpin-Ayala competition model with discrete and distributed time delays. *J. Biol. Dyn.* **2018**, *12*, 433–454. [CrossRef] [PubMed]
12. Bachar, M.; Raimann, J.G.; Kotanko, P. Impulsive mathematical modeling of ascorbic acid metabolism in healthy subjects. *J. Theor. Biol.* **2016**, *392*, 35–47. [CrossRef] [PubMed]
13. Miron, R.E.; Smith, R.J. Resistance to protease inhibitors in a model of HIV-1 infection with impulsive drug effects. *Bull. Math. Biol.* **2014**, *76*, 59–97. [CrossRef] [PubMed]
14. Miron, R.E. Impulsive Differential Equations with Applications to Infectious Diseases. Ph.D. Thesis, Department of Mathematics and Statistics, Faculty of Science, University of Ottawa, Ottawa, ON, Canada, 2014.
15. Pan, T.; Jiang, D.; Hayat, T.; Alsaedi, A. Extinction and periodic solutions for an impulsive SIR model with incidence rate stochastically perturbed. *Phys. A* **2018**, *505*, 385–397. [CrossRef]

16. Li, J.; Nieto, J.J.; Shena, J. Impulsive periodic boundary value problems of first-order differential equations. *J. Math. Anal. Appl.* **2007**, *325*, 226–236. [CrossRef]
17. Nieto, J.J. Basic theory for nonresonance impulsive periodic problems of first order. *J. Math. Anal. Appl.* **1997**, *205*, 423–433. [CrossRef]
18. Bachar, M.; Magal, P. *Existence of Periodic Solution for a Class of Delay Differential Equations with Impulses*; Topics in Functional Differential and Difference Equations (Lisbon, 1999); Fields Inst. Commun. Amer. Math. Soc.: Providence, RI, USA, 2001; Volume 29, pp. 37–49.
19. Schäffer, H. Über die Methode der a priori-Schranken. *Mathematische Annalen* **1955**, *129*, 415–416. [CrossRef]
20. Deimling, K. *Nonlinear Functional Analysis*; Springer: New York, NY, USA, 1985.
21. Evans, L.C. *Partial Differential Equations, Graduate Studies in Mathematics*; American Mathematical Society: Providence, RI, USA, 2010.
22. Smart, D.R. *Fixed Point Theorems*; Cambridge Tracts in Mathematics, No. 66; Cambridge University Press: London, UK; New York, NY, USA, 1974.
23. Morse, S.S. Factors in the emergence of infectious diseases. *Emerg. Infect. Dis.* **1995**, *1*, 7–15. [CrossRef] [PubMed]
24. Wang, G.; Yin, T.; Wang, Y. In vitro and in vivo assessment of high-dose vitamin C against murine tumors. *Exp. Ther. Med.* **2016**, *12*, 3058–3062. [CrossRef] [PubMed]
25. Yung, S.; Mayersohn, M.; Robinson, J.B. Ascorbic acid absorption in humans: A comparison among several dosage forms. *J. Pharm. Sci.* **1982**, *71*, 282–285. [CrossRef] [PubMed]

Some Improvements of the Hermite–Hadamard Integral Inequality

Slavko Simić [1,2,*] and Bandar Bin-Mohsin [3]

1 Nonlinear Analysis Research Group, Ton Duc Thang University, Ho Chi Minh City 758307, Vietnam
2 Faculty of Mathematics and Statistics, Ton Duc Thang University, Ho Chi Minh City 758307, Vietnam
3 Department of Mathematics, College of Science, King Saud University, Riyadh 11451, Saudi Arabia; balmohsen@ksu.edu.sa
* Correspondence: slavkosimic@tdtu.edu.vn

Abstract: We propose several improvements of the Hermite–Hadamard inequality in the form of linear combination of its end-points and establish best possible constants. Improvements of a second order for the class $\Phi(I)$ with applications in Analysis and Theory of Means are also given.

Keywords: Convex function; Simpson's rule; differentiable function

---

1. Introduction

A function $h : I \subset \mathbb{R} \to \mathbb{R}$ is said to be convex on a non-empty interval I if the inequality

$$h(\frac{x+y}{2}) \leq \frac{h(x)+h(y)}{2} \tag{1}$$

holds for all $x, y \in I$.

If the inequality (1) reverses, then h is said to be concave on I [1].

Let $h : I \subset \mathbb{R} \to \mathbb{R}$ be a convex function on an interval I and $a, b \in I$ with $a < b$. Then

$$h(\frac{a+b}{2}) \leq \frac{1}{b-a}\int_a^b h(t)dt \leq \frac{h(a)+h(b)}{2}. \tag{2}$$

This double inequality is well known in the literature as the Hermite–Hadamard (HH) integral inequality for convex functions. It has a plenty of applications in different parts of Mathematics; see [2,3] and references therein.

If h is a concave function on I then both inequalities in (2) hold in the reversed direction.

Our task in this paper is to improve the inequality (2) in a simple manner, i.e., to find some constants $p, q; p+q=1$ such that the relations

$$\frac{1}{b-a}\int_a^b h(t)dt \lesseqgtr p\frac{h(a)+h(b)}{2} + qh(\frac{a+b}{2}), \tag{3}$$

hold for any convex h.

It can be easily seen that the condition

$$p+q=1, \tag{4}$$

is necessary for (3) to hold for an arbitrary convex function.

Take, for example, $f(t) = Ct,\ C \in \mathbb{R}$.

Since

$$p(\frac{h(a)+h(b)}{2}) + qh(\frac{a+b}{2}) \le \max\{\frac{h(a)+h(b)}{2}, h(\frac{a+b}{2})\} = \frac{h(a)+h(b)}{2},$$

and, analogously,

$$p(\frac{h(a)+h(b)}{2}) + qh(\frac{a+b}{2}) \ge \min\{\frac{h(a)+h(b)}{2}, h(\frac{a+b}{2})\} = h(\frac{a+b}{2}),$$

it follows that the inequality of the form (3) represents a refinement of Hermite–Hadamard inequality (2) for each $p, q > 0, p + q = 1$.

Note also that the linear form $p\frac{h(a)+h(b)}{2} + qh(\frac{a+b}{2})$ is monotone increasing in p. Therefore, if the inequality

$$\frac{1}{b-a}\int_a^b h(t)dt \le p\frac{h(a)+h(b)}{2} + qh(\frac{a+b}{2}),$$

holds for some $p = p_0$, then it also holds for each $p \in [p_0, 1]$.

In the sequel we shall prove that the value $p_0 = 1/2$ is best possible for above inequality to hold for an arbitrary convex function on I.

Also, it will be shown that convexity/concavity of the second derivative is a proper condition for inequalities of the form (3) to hold (see Proposition 5 below).

This condition enables us to give refinements of second order and to increase interval of validity to $p_0 = 1/3$ as the best possible constant. In this case, coefficients $p_0 = 1/3, q_0 = 2/3$ are involved in the well-known form of Simpson's rule, which is of great importance in Numerical Analysis. Our results sharply improve Simpson's rule for this class of functions (Proposition 4).

Finally, we give some applications in Analysis and Numerical Analysis. Also, new and precise inequalities between generalized arithmetic means and power-difference means will be proved.

2. Results and Proofs

We shall begin with the basic contribution to the problem defined above.

Theorem 1. *Let $h : I \subset \mathbb{R} \to \mathbb{R}$ be a convex function on an interval I and $a, b \in I$. Then*

$$\frac{1}{b-a}\int_a^b h(t)dt \le \frac{1}{2}\frac{h(a)+h(b)}{2} + \frac{1}{2}h(\frac{a+b}{2}). \tag{5}$$

The constants $p_0 = q_0 = 1/2$ are best possible.

If h is a concave function on I then the inequality is reversed.

Proof. We shall derive the proof by Hermite–Hadamard inequality itself. Indeed, applying twice the right part of this inequality, we get

$$\frac{2}{b-a}\int_a^{\frac{a+b}{2}} h(t)dt \le \frac{1}{2}(h(a) + h(\frac{a+b}{2})),$$

and

$$\frac{2}{b-a}\int_{\frac{a+b}{2}}^b h(t)dt \le \frac{1}{2}(h(\frac{a+b}{2}) + h(b)).$$

Summing up those inequalities the result appears. Therefore, HH inequality has this self-improving property.

That the constants $p_0 = q_0 = 1/2$ are best possible becomes evident by the example $f(t) = |t|, t \in [-a, a]$.

For the second part, note that concavity of f implies convexity of $-f$ on I. Hence, applying (5) we get the result. □

For the sake of further refinements, we shall consider in the sequel functions from the class $C^{(m)}(I), m \in \mathbb{N}$ i.e., functions which are continuously differentiable up to m-th order on an interval $I \subset \mathbb{R}$.

Of utmost importance here is the class $\Phi(I)$ of functions which second derivative is convex on I. For this class we have the following

Theorem 2. *Let $\phi \in \Phi(I)$ and the inequality*

$$\frac{1}{b-a}\int_a^b \phi(t)dt \le p\frac{\phi(a)+\phi(b)}{2} + q\phi(\frac{a+b}{2}), \tag{6}$$

holds for $a, b \in I$. Then $p \ge p_0 = 1/3$.

Proof. From (6) we have

$$p \ge \frac{\frac{1}{b-a}\int_a^b \phi(t)dt - \phi(\frac{a+b}{2})}{\frac{\phi(a)+\phi(b)}{2} - \phi(\frac{a+b}{2})} =: D_\phi(a,b).$$

Since this inequality should be valid for each $a, b \in I, a < b$, let $b \to a$. We obtain that $\lim_{b\to a} D_\phi(a,b) = 1/3$ almost everywhere on I i.e, whenever $\phi''(a) \neq 0$ or $\phi''(a) = 0, \phi'''(a) \neq 0$.

Indeed, applying L'Hospital's rule 3 and 4 times to the above quotient, we get

$$\lim_{b\to a} D_\phi(a,b) = \lim_{b\to a} \frac{\phi''(b) - \frac{3}{4}\phi''(\frac{a+b}{2}) - \frac{b-a}{8}\phi'''(\frac{a+b}{2})}{\frac{3}{2}\phi''(b) - \frac{3}{4}\phi''(\frac{a+b}{2}) + (b-a)(\frac{1}{2}\phi'''(b) - \frac{1}{8}\phi'''(\frac{a+b}{2}))},$$

and

$$\lim_{b\to a} D_\phi(a,b) = \lim_{b\to a} \frac{\phi'''(b) - \frac{1}{2}\phi'''(\frac{a+b}{2}) - \frac{b-a}{16}\phi^{(4)}(\frac{a+b}{2})}{2\phi'''(b) - \frac{1}{2}\phi'''(\frac{a+b}{2}) + (b-a)(\frac{1}{2}\phi^{(4)}(b) - \frac{1}{16}\phi^{(4)}(\frac{a+b}{2}))}.$$

Therefore, the result follows. □

In the sequel we shall give sharp two-sided bounds of second order for inequalities of the type (3) involving functions from the class Φ with $p \ge 1/3$.

Main tool in all proofs will be the following relation.

Lemma 1. *For an integrable function $\phi : I \to \mathbb{R}$ and arbitrary real numbers $p, q; p + q = 1$, we have the identity*

$$p\frac{\phi(a)+\phi(b)}{2} + q\phi(\frac{a+b}{2}) - \frac{1}{b-a}\int_a^b \phi(t)dt = \frac{(b-a)^2}{16}\int_0^1 t(2p-t)(\phi''(x) + \phi''(y))dt,$$

where $x := a\frac{t}{2} + b(1-\frac{t}{2}), y := b\frac{t}{2} + a(1-\frac{t}{2})$.

Proof. It is not difficult to prove this identity by double partial integration of its right-hand side. □

For $t \in [0,1]; a,b \in I, a < b$, denote

$$\xi(a,b;t) := \phi''(a\frac{t}{2} + b(1-\frac{t}{2})) + \phi''(b\frac{t}{2} + a(1-\frac{t}{2}))$$
$$= \phi''(x) + \phi''(y).$$

Lemma 2. *If $\phi \in \Phi$ then the function $\xi(a,b;t)$ is monotone decreasing in t.*
Hence,

$$2\phi''(\frac{a+b}{2}) \leq \phi''(x) + \phi''(y) \leq \phi''(a) + \phi''(b), \tag{7}$$

for all $t \in [0,1]$.

Proof. Since $\phi''(\cdot)$ is convex, it follows that $\phi'''(\cdot)$ is increasing on I.
Also, $x \geq y$ for $t \in [0,1]$ because $x - y = (b-a)(1-t) \geq 0$.
Hence,

$$\xi'(a,b;t) = -\frac{b-a}{2}(\phi'''(x) - \phi'''(y)) \leq 0,$$

and $\xi(a,b;t)$ is decreasing in $t \in [0,1]$.

Therefore,

$$2\phi''(\frac{a+b}{2}) = \xi(a,b;1) \leq \xi(a,b;t) \leq \xi(a,b;0) = \phi''(a) + \phi''(b),$$

which is equivalent with (7).

Note that, if ϕ is concave on I, then the function $\xi(a,b;t)$ is monotone increasing and the inequality (7) is reversed. □

Remark 1. *More general assertion than (7) is contained in [4].*

Main results of this paper are given in the next two assertions.

Theorem 3. *Let $\phi \in \Phi(I)$. Then*

$$p\frac{\phi(a) + \phi(b)}{2} + q\phi(\frac{a+b}{2}) - \frac{1}{b-a}\int_a^b \phi(t)dt \leq \frac{(b-a)^2}{16} T_\phi(a,b;p),$$

where

$$T_\phi(a,b;p) = \begin{cases} \frac{4}{3}p^3(\phi''(a) + \phi''(b)) - \frac{2}{3}(1+p)(2p-1)^2\phi''(\frac{a+b}{2}) & , \frac{1}{3} \leq p \leq \frac{1}{2}; \\ (p - \frac{1}{3})(\phi''(a) + \phi''(b)) & , p \geq \frac{1}{2}. \end{cases}$$

Also, if $p \leq 0$, we have

$$p\frac{\phi(a) + \phi(b)}{2} + q\phi(\frac{a+b}{2}) - \frac{1}{b-a}\int_a^b \phi(t)dt \leq (p - \frac{1}{3})\frac{(b-a)^2}{8}\phi''(\frac{a+b}{2}).$$

Proof. If $p \geq 1/2$ we have that $2p - t \geq 0$. Therefore, applying Lemma 1 and the second part of Lemma 2, we obtain

$$p\frac{\phi(a) + \phi(b)}{2} + q\phi(\frac{a+b}{2}) - \frac{1}{b-a}\int_a^b \phi(t)dt = \frac{(b-a)^2}{16}\int_0^1 t(2p-t)(\phi''(x) + \phi''(y))dt$$
$$\leq \frac{(b-a)^2}{16}(\phi''(a) + \phi''(b))\int_0^1 t(2p-t)dt = (p - 1/3)\frac{(b-a)^2}{16}(\phi''(a) + \phi''(b)).$$

In the case $1/3 \le p < 1/2$, write

$$\int_0^1 t(2p-t)(\phi''(x)+\phi''(y))dt = \int_0^{2p} t(2p-t)(\cdot)dt - \int_{2p}^1 t(t-2p)(\cdot)dt,$$

and apply Lemma 2 to each integral separately.

It follows that

$$\int_0^1 t(2p-t)(\phi''(x)+\phi''(y))dt \le (\phi''(a)+\phi''(b))\int_0^{2p} t(2p-t)dt - 2\phi''(\frac{a+b}{2})\int_{2p}^1 t(t-2p)dt$$

$$= \frac{4p^3}{3}(\phi''(a)+\phi''(b)) - 2(\frac{1}{3}-p+\frac{4p^3}{3})\phi''(\frac{a+b}{2}),$$

which is equivalent to the stated assertion.

For $p \le 0$ we have that $2p - t \le 0$ and the proof develops in the same manner. □

Theorem 4. *If $\phi \in \Phi(I)$, then for $p \ge 1/3$ we get*

$$p\frac{\phi(a)+\phi(b)}{2} + q\phi(\frac{a+b}{2}) - \frac{1}{b-a}\int_a^b \phi(t)dt \ge (p-1/3)\frac{(b-a)^2}{8}\phi''(\frac{a+b}{2}),$$

and

$$p\frac{\phi(a)+\phi(b)}{2} + q\phi(\frac{a+b}{2}) - \frac{1}{b-a}\int_a^b \phi(t)dt \ge (p-1/3)\frac{(b-a)^2}{16}(\phi''(a)+\phi''(b)),$$

for $p \le 0$.

Proof. By Lemma 1, in terms of Lemma 2, we have

$$p\frac{\phi(a)+\phi(b)}{2} + q\phi(\frac{a+b}{2}) - \frac{1}{b-a}\int_a^b \phi(t)dt = \frac{(b-a)^2}{16}\int_0^1 t(2p-t)\xi(a,b;t)dt.$$

By partial integration, we obtain

$$\int_0^1 t(2p-t)\xi(a,b;t)dt = (pt^2 - t^3/3)\xi(a,b;t)|_0^1 - \int_0^1 t^2(p-t/3)\xi'(a,b;t)dt$$

$$\ge 2(p-1/3)\phi''(\frac{a+b}{2}),$$

since $p - t/3 \ge 0$ for $p \ge 1/3$ and, by Lemma 2, $\xi'(a,b;t) \le 0$ for $t \in [0,1]$.

If $p \le 0$ then $2p - t \le 0$ and, applying Lemmas 1 and 2, the result follows. □

Above theorems are the source of a plenty of important inequalities which sharply refine Hermite–Hadamard inequality for this class of functions.

Some of them are listed in the sequel.

Proposition 1. *Let $\phi \in \Phi(I)$. Then*

$$\frac{(b-a)^2}{24}\phi''(\frac{a+b}{2}) \le \frac{1}{b-a}\int_a^b \phi(t)dt - \phi(\frac{a+b}{2}) \le \frac{(b-a)^2}{24}\frac{\phi''(a)+\phi''(b)}{2}.$$

Proof. Put $p = 0$ in the above theorems. □

Proposition 2. *Let $\phi \in \Phi(I)$. Then*

$$\frac{(b-a)^2}{12}\phi''(\frac{a+b}{2}) \le \frac{\phi(a)+\phi(b)}{2} - \frac{1}{b-a}\int_a^b \phi(t)dt \le \frac{(b-a)^2}{12}\frac{\phi''(a)+\phi''(b)}{2}.$$

Proof. This proposition is obtained for $p = 1$. □

The next assertion represents a refinement of Theorem 1 in the case of convex functions.

Proposition 3. *Let $\phi \in \Phi(I)$. Then for each $a, b \in I, a < b$,*

$$\frac{(b-a)^2}{48}\phi''(\frac{a+b}{2}) \le \frac{1}{2}\frac{\phi(a)+\phi(b)}{2} + \frac{1}{2}\phi(\frac{a+b}{2}) - \frac{1}{b-a}\int_a^b \phi(t)dt \le \frac{(b-a)^2}{48}\frac{\phi''(a)+\phi''(b)}{2}.$$

If ϕ'' is concave on I, then

$$\frac{(b-a)^2}{48}\frac{\phi''(a)+\phi''(b)}{2} \le \frac{1}{2}\frac{\phi(a)+\phi(b)}{2} + \frac{1}{2}\phi(\frac{a+b}{2}) - \frac{1}{b-a}\int_a^b \phi(t)dt \le \frac{(b-a)^2}{48}\phi''(\frac{a+b}{2}).$$

Proof. Put $p = 1/2$ in Theorems 3 and 4.
The second part follows from a variant of Lemma 2 for concave functions. □

Note that the coefficients $p = 1/3$ and $q = 2/3$ are involved in well-known Simpson's rule which is of importance in numerical integration [5].
The next assertion sharply refines Simpson's rule for this class of functions.

Proposition 4. *For $\phi \in \Phi(I)$, we have*

$$0 \le \frac{1}{3}\frac{\phi(a)+\phi(b)}{2} + \frac{2}{3}\phi(\frac{a+b}{2}) - \frac{1}{b-a}\int_a^b \phi(t)dt$$

$$\le \frac{(b-a)^2}{162}[\frac{\phi''(a)+\phi''(b)}{2} - \phi''(\frac{a+b}{2})].$$

If ϕ'' is concave on I, then

$$0 \le \frac{1}{b-a}\int_a^b \phi(t)dt - \frac{1}{6}[\phi(a)+\phi(b)+4\phi(\frac{a+b}{2})]$$

$$\le \frac{(b-a)^2}{162}[\phi''(\frac{a+b}{2}) - \frac{\phi''(a)+\phi''(b)}{2}].$$

Proof. Applying Theorems 3 and 4 with both parts of Lemma 2 for $p = 1/3$, the proof follows. □

The next assertion gives a proper answer to the problem posed in Introduction.

Proposition 5. *If ϕ is a convex and ϕ'' is a concave function on I, then*

$$\frac{1}{3}\frac{\phi(a)+\phi(b)}{2} + \frac{2}{3}\phi(\frac{a+b}{2}) \le \frac{1}{b-a}\int_a^b \phi(t)dt \le \frac{1}{2}\frac{\phi(a)+\phi(b)}{2} + \frac{1}{2}\phi(\frac{a+b}{2}).$$

Analogously, let ϕ be concave and ϕ'' a convex function on I, then

$$\frac{1}{2}\frac{\phi(a)+\phi(b)}{2} + \frac{1}{2}\phi(\frac{a+b}{2}) \le \frac{1}{b-a}\int_a^b \phi(t)dt \le \frac{1}{3}\frac{\phi(a)+\phi(b)}{2} + \frac{2}{3}\phi(\frac{a+b}{2}).$$

Proof. Combining Proposition 4 with the results of Theorem 1, we obtain the proof. □

3. Applications in Analysis

Theorems proved above are the source of interesting inequalities from Classical Analysis. As an illustration we shall give here a couple of Cusa-type inequalities.

Theorem 5. *The inequality*

$$\frac{1}{2}\cos x + \frac{1}{2} \le \frac{\sin x}{x} \le \frac{1}{3}\cos x + \frac{2}{3},$$

holds for $|x| \le \pi/2$.

Also,

$$\frac{1}{4}\cosh x + \frac{3}{4} \le \frac{\sinh x}{x} \le \frac{1}{3}\cosh x + \frac{2}{3},$$

holds for $|x| \le (3/2)^{3/2}$.

Proof. For the first part one should apply Proposition 5 to the function $\phi(t) = \cos t$ on a symmetric interval $t \in [-x, x] \subset [-\pi/2, \pi/2]$.

For the second part, applying Proposition 4 with $\phi(t) = e^t, t \in [-x, x]$, we get

$$0 \le \frac{1}{3}\cosh x + \frac{2}{3} - \frac{\sinh x}{x} \le \frac{2x^2}{81}(\cosh x - 1).$$

Hence,

$$\frac{\sinh x}{x} \le \frac{1}{3}\cosh x + \frac{2}{3},$$

and

$$\frac{\sinh x}{x} \ge \frac{1}{3}\cosh x + \frac{2}{3} - \frac{2x^2}{81}(\cosh x - 1)$$

$$= (\frac{1}{12} - \frac{2x^2}{81})\cosh x + \frac{1}{4}\cosh x + \frac{2}{3} + \frac{2x^2}{81} \ge \frac{1}{4}\cosh x + \frac{3}{4},$$

since $\cosh x \ge 1$ and $1/12 - 2x^2/81 \ge 0$ for $|x| \le (3/2)^{3/2} \approx 1.8371$. □

We give now some numerical examples of the above inequality

$$\frac{1}{2}\cos x + \frac{1}{2} \le \frac{\sin x}{x} \le \frac{1}{3}\cos x + \frac{2}{3}, \quad |x| \le \pi/2. \tag{8}$$

Namely, using known formulae

$$\sin\frac{\pi}{2} = 1; \ \sin\frac{\pi}{4} = \frac{\sqrt{2}}{2}; \ \sin\frac{\pi}{6} = \frac{1}{2}; \ \sin\frac{\pi}{12} = \frac{\sqrt{2}}{4}(\sqrt{3} - 1) \approx 0.25882;$$

$$\sin\frac{\pi}{24} = \frac{1}{2}\sqrt{2 - \sqrt{2 + \sqrt{3}}} \approx 0.13053; \ \sin\frac{\pi}{60} = \frac{1}{16}[\sqrt{2}(\sqrt{3}+1)(\sqrt{5}-1) - 2(\sqrt{3}-1)\sqrt{5+\sqrt{5}}] \approx 0.052336,$$

and applying inequalities (8), we obtain bounds for the transcendental number π, as follows

$$x = \frac{\pi}{2} : 3 < \pi < 4; \ x = \frac{\pi}{4} : 3.1344 < \pi < 3.3137; \ x = \frac{\pi}{6} : 3.1402 < \pi < 3.2154;$$

$$x = \frac{\pi}{12} : 3.1415 < \pi < 3.1597; \ x = \frac{\pi}{24} : 3.1416 < \pi < 3.1461; \ x = \frac{\pi}{60} : 3.1416 < \pi < 3.1423.$$

Another application can be obtained by integrating both sides of (8) on the range $x \in [0, a]$, $0 < a < \pi/2$.

We get

$$\frac{1}{2}\sin a + \frac{1}{2}a \le \int_0^a \frac{\sin x}{x}dx \le \frac{1}{3}\sin a + \frac{2}{3}a,$$

that is,

$$\frac{a-\sin a}{3} \le a - \int_0^a \frac{\sin x}{x}dx \le \frac{a-\sin a}{2}.$$

By the power series expansion, we know that

$$a - \sin a = \frac{a^3}{3!} - \frac{a^5}{5!} + \frac{a^7}{7!} - \cdots.$$

Hence,

$$\frac{a^3}{18} - \frac{a^5}{360} \le a - \int_0^a \frac{\sin x}{x}dx \le \frac{a^3}{12}.$$

This estimation is effective for small values of a.
For example,

$$5.5528 \times 10^{-5} \le \frac{1}{10} - \int_0^{1/10} \frac{\sin x}{x}dx \le 8.3333 \times 10^{-5}.$$

4. Applications in Theory of Means

A *mean* $M(a,b)$ is a map $M : \mathbb{R}_+ \times \mathbb{R}_+ \to \mathbb{R}_+$, with the property

$$\min\{a,b\} \le M(a,b) \le \max\{a,b\},$$

for each $a,b \in \mathbb{R}_+$.

Some refinements of HH inequality by arbitrary means is given in [6].

An ordered set of elementary means is the following family,

$$H \le G \le L \le I \le A \le S,$$

where

$$H = H(a,b) =: 2(1/a+1/b)^{-1};\ \ G = G(a,b) =: \sqrt{ab};\ \ L = L(a,b) =: \frac{b-a}{\log b - \log a};$$

$$I = I(a,b) =: \frac{1}{e}(b^b/a^a)^{1/(b-a)};\ \ A = A(a,b) =: \frac{a+b}{2}; S = S(a,b) =: a^{\frac{a}{a+b}} b^{\frac{b}{a+b}},$$

are the harmonic, geometric, logarithmic, identric, arithmetic and Gini mean, respectively.

Generalized arithmetic mean A_α is defined by

$$A_\alpha = A_\alpha(a,b) =: \begin{cases} \left(\frac{a^\alpha+b^\alpha}{2}\right)^{1/\alpha} & ,\alpha \ne 0; \\ A_0 = G. \end{cases}$$

Power-difference mean K_α is defined by

$$K_\alpha = K_\alpha(a,b) =: \begin{cases} \frac{\alpha}{\alpha+1}\frac{a^{\alpha+1}-b^{\alpha+1}}{a^\alpha - b^\alpha} & ,\alpha \neq 0, -1; \\ K_0(a,b) = L(a,b); \\ K_{-1}(a,b) = ab/L(a,b). \end{cases}$$

It is well known that both means are monotone increasing with α and, evidently,

$$A_{-1} = H, A_1 = A, K_{-2} = H, K_{-1/2} = G, K_1 = A.$$

As an illustration of our results, we shall give firstly some sharp bounds of power-difference means in terms of the generalized arithmetic mean.

Theorem 6. *For $a, b \in \mathbb{R}^+$ and $\alpha \geq 1$, we have*

$$\frac{1}{2}(A(a,b) + A_\alpha(a,b)) \leq K_\alpha(a,b) \leq A_\alpha(a,b). \tag{9}$$

For $\alpha < 1$ the inequality (9) is reversed.

Proof. Let $g_\alpha(t) = t^{1/\alpha}, \alpha \neq 0$. Since g_α is concave for $\alpha \geq 1$, Theorem 1 combined with the HH inequality gives

$$\frac{1}{2}\Big(\frac{x+y}{2}\Big)^{1/\alpha} + \frac{1}{4}(x^{1/\alpha} + y^{1/\alpha})$$
$$\leq \frac{\alpha}{\alpha+1}\frac{x^{1+1/\alpha} - y^{1+1/\alpha}}{x-y} \leq \Big(\frac{x+y}{2}\Big)^{1/\alpha}.$$

Now, simple change of variables $x = a^\alpha, y = b^\alpha$ yields the result.
For the second part, note that g_α is convex for $\alpha < 1$ and repeat the procedure. □

The above inequality is refined by the following

Theorem 7. *We have,*

$$A_\alpha \leq K_\alpha \leq \frac{1}{3}(A + 2A_\alpha), \ \ \alpha \in (-\infty, 1/3) \cup (1/2, 1);$$

$$\frac{1}{3}(A + 2A_\alpha) \leq K_\alpha \leq A_\alpha, \ \ \alpha \in [1, \infty);$$

$$\frac{1}{3}(A + 2A_\alpha) \leq K_\alpha \leq \frac{1}{2}(A + A_\alpha), \ \ \alpha \in [1/3, 1/2].$$

Proof. Observe that g''_α is convex for $\alpha \in (-\infty, 1/3) \cup (1/2, 1)$ and concave for $\alpha \in (1/3, 1/2) \cup (1, \infty)$. Hence, applying Proposition 5 together with the HH inequality, we obtain the result. □

Remark 2. *Note that the above inequalities are so precise that in critical points for $\alpha = 1/3, 1/2, 1$ we have equality sign.*

An inequality for the reciprocals follows.

Theorem 8. *For $\beta \geq -2$ we have*

$$\frac{1}{A_{\beta+1}} \leq \frac{1}{K_\beta} \leq \frac{1}{2}\Big(\frac{1}{H} + \frac{1}{A_{\beta+1}}\Big).$$

For $\beta < -2$ the inequality is reversed.

Proof. This is a consequence of Theorem 6. Indeed, putting there $\alpha = -\beta - 1$ and using identities

$$K_\alpha = \frac{ab}{K_\beta}, A_\alpha = \frac{ab}{A_{\beta+1}}, A = \frac{ab}{H},$$

the proof appears. □

Finally, we give a new and precise double inequality for the identric mean $I(a,b)$.

Theorem 9. *For arbitrary positive a, b we have*

$$A^{4/3}S^{-1/3}\exp\left(-\frac{4}{81}\frac{(A-H)^2}{AH}\right) \le I \le A^{4/3}S^{-1/3}.$$

Proof. We need firstly an auxiliary result.

Lemma 3. *For $a, b \in \mathbb{R}^+$, we have*

$$A^{4/3}(a,b)S^{2/3}(a,b)\exp\left(-\frac{4}{81}\frac{(A(a,b)-H(a,b))^2}{A(a,b)H(a,b)}\right) \le I(a^2,b^2) \le A^{4/3}(a,b)S^{2/3}(a,b).$$

Proof. Indeed, for $\phi(t) = t\log t$ we get

$$\frac{1}{b-a}\int_a^b \phi(t)dt = \frac{1}{4}\left(\frac{b^2\log b^2 - a^2\log a^2}{b-a} - (a+b)\right) = \frac{a+b}{4}\log I(a^2,b^2).$$

Since $\phi''(t) = 1/t$, Proposition 5 yields

$$\frac{1}{6}(a\log a + b\log b) + \frac{2}{3}A\log A - \frac{(b-a)^2}{324}\left(\frac{1}{a} + \frac{1}{b} - \frac{2}{A}\right)$$
$$\le \frac{a+b}{4}\log I(a^2,b^2) \le \frac{1}{6}(a\log a + b\log b) + \frac{2}{3}A\log A,$$

and the proof follows by dividing the last expression with $(a+b)/4 = A/2$. □

Now, combining this assertion with the identity $I(a^2,b^2) = I(a,b)S(a,b)$, we obtain the desired inequality. □

Remark 3. *An equivalent form of the above result is*

$$I^{3/4}S^{1/4} \le A \le I^{3/4}S^{1/4}\exp\left(\frac{(A-H)^2}{27AH}\right),$$

which refines well-known inequality $I \le A \le S$.

Author Contributions: Theoretical part, S.S.; numerical part with numeric examples, B.B.-M. All authors have read and agreed to the published version of the manuscript.

Acknowledgments: The authors are grateful to the referees for their valuable comments.

References

1. Hardy, G.H.; Littlewood, J.E.; Polya, G. *Inequalities*; Cambridge University Press: Cambridge, UK, 1978.
2. Niculescu, C.P.; Persson, L.E. Old and new on the Hermite–Hadamard inequality. *Real Anal. Exchang.* **2003**, *29*, 663–685. [CrossRef]
3. Rostamian Delavar, M.; Dragomir, S.S.; De La Sen, M. Hermite–Hadamard's trapezoid and mid-point type inequalities on a disk. *J. Inequal. Appl.* **2019**, *2019*, 105. [CrossRef]
4. Simić, S. On a convexity property. *Krag. J. Math.* **2016**, *40*, 166–171. [CrossRef]
5. Ueberhuber, C.W. *Numerical Computation 2*; Springer: Berlin, Germany, 1997.
6. Simić, S. Further improvements of Hermite–Hadamard integral inequality. *Krag. J. Math.* **2019**, *43*, 259–265.

15

Existence of Solution for Non-Linear Functional Integral Equations of Two Variables in Banach Algebra

Hari M. Srivastava [1,2,*], Anupam Das [3], Bipan Hazarika [3,4] and S. A. Mohiuddine [5]

1. Department of Mathematics and Statistics, University of Victoria, Victoria, BC V8W 3R4, Canada
2. Department of Medical Research, China Medical University Hospital, China Medical University, Taichung 40402, Taiwan
3. Department of Mathematics, Rajiv Gandhi University, Rono Hills, Doimukh 791112, Arunachal Pradesh, India; math.anupam@gmail.com (A.D.); bh_rgu@yahoo.co.in (B.H.)
4. Department of Mathematics, Gauhati University, Guwahati 781014, Assam, India
5. Operator Theory and Applications Research Group, Department of Mathematics, Faculty of Science, King Abdulaziz University, P.O. Box 80203, Jeddah 21589, Saudi Arabia; mohiuddine@gmail.com

* Correspondence: harimsri@math.uvic.ca

Abstract: The aim of this article is to establish the existence of the solution of non-linear functional integral equations $x(l,h) = \left(U(l,h,x(l,h)) + F\left(l,h,\int_0^l\int_0^h P(l,h,r,u,x(r,u))drdu, x(l,h)\right)\right) \times G\left(l,h,\int_0^a\int_0^a Q\left(l,h,r,u,x(r,u)\right)drdu, x(l,h)\right)$ of two variables, which is of the form of two operators in the setting of Banach algebra $C\left([0,a]\times[0,a]\right), a>0$. Our methodology relies upon the measure of noncompactness related to the fixed point hypothesis. We have used the measure of noncompactness on $C\left([0,a]\times[0,a]\right)$ and a fixed point theorem, which is a generalization of Darbo's fixed point theorem for the product of operators. We additionally illustrate our outcome with the help of an interesting example.

Keywords: functional integral equations; Banach algebra; fixed point theorem; measure of noncompactness

MSC: (2010): 45G15; 47H10

1. Introduction

Many real-life problems in which we go over the investigation of various branches of mathematical physics, for example, gas kinetic theory, radiation, and neutron transportation, can be depicted and demonstrated by methods of non-linear functional integral equations (for example, we refer to [1–4]). Banaś and Lecko [5] introduced the concept of fixed points of product operators in Banach algebra. Dhage [6,7] used the concept of the fixed point theorem to find the solution of functional integral equations in Banach algebra. Banaś and Olszowy [8] used the class of measures of noncompactness to obtain the existence of solutions of nonlinear integral equations in Banach algebra. Deepmala and Pathak [9] studied the existence of the solution of nonlinear functional integral equations of a single variable in Banach algebra $C[a,b]$ of all real-valued continuous functions on the interval $[a,b]$ equipped with the maximum norm.

Kuratowski [10], in the year 1930, first introduced the idea of the measure of noncompactness (denoted by "α"). For any bounded subset A of a metric space X,

$$\alpha(A) = \inf\left\{\delta > 0 : A \subset \bigcup_{j=1}^{m} A_j, A_j \subset X, \operatorname{diam}(A_j) < \delta\ (j = 1, ..., m), m \in \mathbb{N}\right\},$$

where:

$$\operatorname{diam}(A_j) = \sup\{d(a_1, a_2) : a_1, a_2 \in A_j\}.$$

Using this idea, Darbo [11] exhibited a fixed point theorem that plays a very significant role in the finding of existence theorems. In the recent past, there have been a few fruitful endeavours to apply the idea of the measure of noncompactness in the investigation of the existence of solutions for various kinds of differential and integral equations, for example one can refer to [12–20].

In many physical problems, we come across nonlinear integral equations. The fixed point theory plays a significant role to obtain the solutions of such equations. Deepmala and Pathak, in [9], studied the following nonlinear functional integral equation, which can be considered as a particular case of many nonlinear functional integral equations that are applicable in mechanics, physics, economics, etc.,

$$\begin{aligned} x(t) = &\left(u(t, x(t)) + f\left(t, \int_0^t p(t, s, x(s))ds, x(\alpha(t))\right)\right) \\ &\times g\left(t, \int_0^a q(t, s, x(s))\,ds, x(\beta(t))\right) \quad \text{for } t \in [0, a]. \end{aligned} \tag{1}$$

The authors of [9] used the measure of noncompactness to obtain the existence of the solution of the integral Equation (1) in Banach algebra $C[0, a]$ with the help of the fixed point theorem.

Motivated by the work of [9], in this article, we study the solvability of non-linear functional integral equations of two variables, which we come across in various branches of nonlinear analysis. We consider an integral equation in the following form:

$$\begin{aligned} x(l, h) = &\left(U(l, h, x(l, h)) + F\left(l, h, \int_0^l \int_0^h P(l, h, r, u, x(r, u))drdu, x(l, h)\right)\right) \\ &\times G\left(l, h, \int_0^a \int_0^a Q(l, h, r, u, x(r, u))\,drdu, x(l, h)\right) \quad \text{for } l, h \in [0, a]. \end{aligned} \tag{2}$$

The right-hand side of the above integral equation that we are considering is the product of two functional operators involving integral operators and applying a fixed point theorem, which is a generalization of Darbo's fixed point theorem for the product of operators to check the existence of the solution of the integral equation in Banach algebra. It can be seen that Equation (2) is a generalization of Equation (1) in two variables. Here, we used a fixed point theorem associated with Darbo's condition of the measure of noncompactness in Banach algebra of continuous functions in $[0, a] \times [0, a]$ to establish the solvability of Equation (2). Furthermore, we used the modified homotopy perturbation analytic method to find the solution of Equation (2).

2. Preliminaries

Let $\mathbb{R}$ denote the set of real numbers, and write $\mathbb{R}_+ = [0, \infty)$. Suppose $\bar{E}$ is a real Banach space with the norm $\|\ .\ \|$, and let $X(\neq \phi) \subseteq \bar{E}$. The closure and convex closure of X will be denoted by $\bar{X}$ and conv$X$, respectively. The convex closure of a set X of points in the Euclidean plane or in a Euclidean space over the reals is the smallest convex set that contains X. A closed ball in $\bar{E}$ centred at a and with radius b is denoted by $B(a, b)$. In addition, we use the symbol $\mathcal{M}_{\bar{E}}$ to denote the family of all non-empty and bounded subsets of $\bar{E}$ and use $\mathcal{N}_{\bar{E}}$ to denote its subfamily consisting of all relatively compact sets.

Definition 1. *Let X be a linear space over $\mathbb{R}$. A norm on X is a function from X to $\mathbb{R}_+$, commonly denoted $||.||$ such that:*

(N1) $||x|| \geq 0$ *and* $||x|| = 0 \iff x = 0$;
(N2) $||\alpha x|| = |\alpha|||x||$;
(N3) $||x+y|| \leq ||x|| + ||y||$ *for all* $x, y \in X$ *and* $\alpha \in \mathbb{R}$.

The pair $(X, ||.||)$ is called a normed space. A complete normed space is called a Banach space.

Definition 2. *An algebra A is a vector space A over a field K such that for each ordered pair of elements $x, y \in A$, a unique product $xy \in A$ is defined with the properties:*

(A1) $(xy)z = x(yz)$,
(A2) $x(y+z) = xy + xz$,
(A3) $(x+y)z = xz + yz$,
(A4) $\alpha(xy) = (\alpha x)y = x(\alpha y)$ *for all* $x, y, z \in A$ *and scalars* α.

A normed algebra A is normed space, which is an algebra such that for all $x, y \in A$:

$$\| xy \| \leq \| x \| \| y \|$$

and if A has an identity e, then $\| e \| = 1$.

A Banach algebra is a normed algebra that is complete, considered as a normed space.
The notion of the measure of noncompactness due to Banaś and Goebel [21] is as follows:

Definition 3. *A function $\mu : \mathcal{M}_{\bar{E}} \to [0, \infty)$ is said to be a measure of noncompactness in $\bar{E}$ if:*

(i) *for all* $X \in \mathcal{M}_{\bar{E}}$, *we have that* $\mu(X) = 0$ *implies that X is precompact.*
(ii) *the family* $\ker \mu = \{X \in \mathcal{M}_{\bar{E}} : \mu(X) = 0\}$ *is non-empty, and* $\ker \mu \subset \mathcal{N}_{\bar{E}}$.
(iii) $X \subseteq Z \implies \mu(X) \leq \mu(Z)$.
(iv) $\mu(\bar{X}) = \mu(X)$.
(v) $\mu(convX) = \mu(X)$ *where convX is the convex closure of set X.*
(vi) $\mu(\lambda X + (1-\lambda) Z) \leq \lambda\mu(X) + (1-\lambda)\mu(Z)$ *for* $\lambda \in [0,1]$.
(vii) *if* $X_n \in \mathcal{M}_{\bar{E}}, X_n = \bar{X}_n$, $X_{n+1} \subset X_n$ *for* $n = 1, 2, 3, \ldots$ *and* $\lim_{n\to\infty} \mu(X_n) = 0$, *then* $\bigcap_{n=1}^{\infty} X_n \neq \phi$.

The family $\ker \mu$ is called the *kernel of measure* μ. Note that the intersection set X_∞ from the above condition (vii) is a member of the family $\ker \mu$. Since $\mu(X_\infty) \leq \mu(X_n)$ for any n, we deduce $\mu(X_\infty) = 0$. Consequently, $X_\infty \in \ker\mu$.

For given subsets X, Y in a Banach algebra E, the product XY defined by:

$$XY = \{xy : x \in X, y \in Y\}.$$

In [8], Banaś and Olszowy defined the measure of noncompactness μ on the Banach algebra E, which *satisfies condition* (m) if for arbitrary sets $X, Y \subset \mathcal{M}_E$ such that:

$$\mu(XY) \leq ||X||\mu(Y) + ||y||\mu(X).$$

Deepmala and Pathak [9] used this concept of measure of noncompactness and obtained the existence of the solution of Equation (1).

Definition 4 ([21]). *Let E be a Banach space. Consider a non-empty subset X of E and a continuous operator $T : X \to E$ transforming the bounded subset of X to the bounded ones. We say that T satisfies the Darbo condition with a constant k with respect to measure μ provided $\mu(TY) \leq k\mu(Y)$ for each $Y \in \mathcal{M}_E$ such that $Y \subset X$. If $k < 1$, then T is called a contraction with respect to μ.*

Remark 1. *The Darbo condition has many applications, particularly in fixed point theorems, which can be applied to check the existence of the solution of different types of integral, differential, and integro differential equations. The Darbo condition can be potentially applied to extend the linear space in the work of Shang [22]. The assumptions (1)–(4) of the next section have been utilized in the study of consensus problems (see [23,24]).*

We recall the following important theorems:

Theorem 1 ([11]). *Assume that Z is a non-empty, closed, bounded, and convex subset of a Banach space $\bar{E}$. Let $S : Z \to Z$ be a continuous mapping. Suppose that there is a constant $k \in [0, 1)$ such that:*

$$\mu(SM) \leq k\mu(M), \ M \subseteq Z.$$

Then, S has a fixed point.

Theorem 2 ([8]). *Suppose that X is a non-empty, bounded, convex, and closed subset of a Banach algebra E, and the operators P and T transform continuously the set X into E such that $P(X)$ and $T(X)$ are bounded. Furthermore, suppose that the operator $S = P.T$ transforms X into itself. If P and T satisfy on the set X the Darbo condition with respect to the measure of noncompactness μ with the constants k_1 and k_2, respectively, then S satisfies on X the Darbo condition with constant $\| P(X) \| k_2 + \| T(X) \| k_1$. Particularly, if:*

$$\| P(X) \| k_2 + \| T(X) \| k_1 < 1,$$

then S is a contraction with respect to the measure of noncompactness μ and has at least one fixed point in X.

We consider the space $E = C([0, a] \times [0, a])$, which consists of the set of real-valued continuous functions on $[0, a] \times [0, a]$. It is obvious that E is the vector space over the field of scalars $\mathbb{R}$ with the following operations:

$$(x + y)(t, s) = x(t, s) + y(t, s)$$

and:

$$(\alpha x)(t, s) = \alpha x(t, s),$$

where $x, y \in E$, $\alpha \in \mathbb{R}$ and $t, s \in [0, a]$. Since $x, y \in E$, i.e., both x, y are real-valued continuous functions on $[0, a] \times [0, a]$ and the product of two real-valued continuous functions is also a real-valued continuous function, therefore $xy \in E$, where:

$$(xy)(t, s) = x(t, s)y(t, s), \ t, s \in [0, a].$$

Let $z \in E$. For all $t, s \in [0, a]$,

$$\begin{aligned} ((xy)\, z)\, (t, s) &= (xy)\, (t, s)z(t, s) \\ &= x(t, s)y(t, s)z(t, s) \\ &= x(t, s)\, (yz)\, (t, s) \\ &= (x\, (yz))\, (t, s). \end{aligned}$$

Since t, s are arbitrary, therefore $(xy)z = x(yz)$.

Similarly, it can be shown that:

$$x(y + z) = xy + xz,$$

$$(x + y)z = xz + yz$$

and:

$$\alpha\, (xy) = (\alpha x)\, y = x\, (\alpha y)\, .$$

Therefore E is an algebra.

The space E is also a normed space with the norm:

$$\| x \| = \sup\{|x(l,h)| : l,h \in [0,a],\ a > 0\},\ x \in E.$$

For all $x, y \in E$ and $l, h \in [0,a]$,

$$|(xy)(l,h)| = |x(l,h)y(l,h)| = |x(l,h)|\,|y(l,h)|$$

and so:

$$\sup_{l,h\in[0,a]} |(xy)(l,h)| \leq \sup_{l,h\in[0,a]} |x(l,h)| \sup_{l,h\in[0,a]} |y(l,h)|,$$

i.e.,

$$\| xy \| \leq \| x \| \| y \| .$$

Thus, E is a normed algebra.

Let $(x_n(t,s))_{n=1}^{\infty}$ be a Cauchy sequence in E where $x_n(t,s) \in \mathbb{R} \times \mathbb{R}$ for all $n \in \mathbb{N}$ and $t, s \in [0,a]$. Then:

$$\| x_n - x_m \| \to 0 \quad (n, m \to \infty).$$

Therefore, for all $t, s \in [0,a]$, we get:

$$|x_n(t,s) - x_m(t,s)| \to 0 \quad (n, m \to \infty).$$

For fixed $t, s \in [0,a]$, the sequence $(x_n(t,s))$ is a Cauchy sequence of real numbers, so it is a convergent sequence and converging to $x_0(t,s) \in E$ (say) as the limit of the continuous function is also continuous. Therefore, for all $t, s \in [0,a]$:

$$|x_n(t,s) - x_0(t,s)| \to 0 \quad (n, m \to \infty)$$

which yields:

$$\sup_{t,s\in[0,a]} |x_n(t,s) - x_0(t,s)| \to 0 \quad (n, m \to \infty)$$

Thus:

$$\| x_n - x_0 \| \to 0 \quad (n \to \infty)$$

which proves that E is complete normed space. Hence, we conclude that the space E has the Banach algebra structure.

Let X be a fixed non-empty and bounded subset of $E = C([0,a] \times [0,a])$, and for $x \in X$ and $\epsilon > 0$, the modulus of the continuity function (denoted by $\omega(x,\epsilon)$) is given by the formula:

$$\omega(x,\epsilon) = \sup\{|x(l,h) - x(v,w)| : l,h,v,w \in [0,a], |l-v| \leq \epsilon, |h-w| \leq \epsilon\}.$$

Further, we define:

$$\omega(X,\epsilon) = \sup\{\omega(x,\epsilon) : x \in X\},\ \ \omega_0(X) = \lim_{\epsilon\to 0} \omega(x,\epsilon).$$

Similar to [5], it can be shown that the function $\omega_0(X)$ is a regular measure of non-compactness in the space $C([0,a] \times [0,a])$. Apart from this, it is easy to check that the measure $\omega_0(X)$ satisfies condition (m).

3. Main Result

In this section, we study the existence of solutions of the integral Equation (2). We consider the following assumptions:

(1) The functions $U : [0, a] \times [0, a] \times \mathbb{R} \to \mathbb{R}$, $F : [0, a] \times [0, a] \times \mathbb{R} \times \mathbb{R} \to \mathbb{R}$, and $G : [0, a] \times [0, a] \times \mathbb{R} \times \mathbb{R} \to \mathbb{R}$ are continuous, and there exist nonnegative constants L, M such that:

$$|U(l,h,0)| \leq L, \quad |F(l,h,M_1,0)| \leq M \text{ and } |F(l,h,M_2,0)| \leq M,$$

where $M_1, M_2 \in \mathbb{R}$.

(2) Let $A_i : [0, a] \times [0, a] \to \mathbb{R}_+$ $(i = 1, 2, 3, 4, 5)$ be continuous functions such that:

$$|U(l,h,x_1) - U(l,h,x_2)| \leq A_1(l,h)\,|x_1 - x_2|,$$

$$|F(l,h,y,x_1) - F(l,h,y,x_2)| \leq A_2(l,h)\,|x_1 - x_2|,$$

$$|G(l,h,y,x_1) - G(l,h,y,x_2)| \leq A_3(l,h)\,|x_1 - x_2|,$$

$$|F(l,h,y_1,x) - F(l,h,y_2,x)| \leq A_4(l,h)\,|y_1 - y_2|$$

and:

$$|G(l,h,y_1,x) - G(l,h,y_2,x)| \leq A_5(l,h)\,|y_1 - y_2|,$$

where $l, h \in [0, a]$ and $x, x_1, x_2, y, y_1, y_2 \in \mathbb{R}$. Furthermore, let:

$$K = \max\{A_i(l,h) : i = 1,2,3,4,5; l, h \in [0,a]\},$$

where $K \geq 0$.

(3) The functions P, Q are continuous functions from $[0, a] \times [0, a] \times [0, a] \times [0, a] \times \mathbb{R}$ to $\mathbb{R}$.

(4) Furthermore, $4\alpha\beta < 1$ for $\alpha = 2k$, $\beta = L + M$.

Theorem 3. *Under the hypotheses (1)–(4), Equation (2) has at least one solution in* $E = C(I \times I)$, *where* $I = [0, a]$.

Proof. Let us consider the operators $\hat{F}$ and $\hat{G}$ defined on E by:

$$(\hat{F}x)(l,h) = U(l,h,x(l,h)) + F\left(l, h, \int_0^l \int_0^h P(l,h,r,u,x(r,u))drdu, x(l,h)\right)$$

and:

$$(\hat{G}x)(l,h) = G\left(l, h, \int_0^a \int_0^a Q\,(l,h,r,u,x(r,u))\,drdu, x(l,h)\right), \text{ where } l,h \in [0,a].$$

From Assumptions (1)–(3), we get that $\hat{F}$ and $\hat{G}$ map $C(I \times I)$ into itself. Furthermore, let us define another operator $\hat{T}$ on $C(I \times I)$ as follows:

$$\hat{T}x = (\hat{F}x)(\hat{G}x).$$

It is obvious that $\hat{T}$ maps $C(I \times I)$ into itself.

Let:

$$I_1(x) = \int_0^l \int_0^h P(l,h,r,u,x(r,u))drdu$$

and:

$$I_2(x) = \int_0^a \int_0^a Q\,(l,h,r,u,x(r,u))\,drdu.$$

Let $x \in C(I \times I)$ be fixed and $l, h \in I$. We get:

$$\begin{aligned}
\left|(\hat{T}x)(l,h)\right| &= \left|(\hat{F}x)(l,h)\right| . \left|(\hat{G}x)(l,h)\right| \\
&= |U(l,h,x(l,h)) + F(l,h,I_1(x),x(l,h))| \times |G(l,h,I_2(x),x(l,h))| \\
&\leq (|U(l,h,x(l,h)) - U(l,h,0)| + |U(l,h,0)| + |F(l,h,I_1(x),x(l,h)) \\
&\quad -F(l,h,I_1(x),0)| + |F(l,h,I_1(x),0)|) \times (|G(l,h,I_2(x),x(l,h)) \\
&\quad -G(l,h,I_2(x),0)| + |G(l,h,I_2(x),0)|) \\
&\leq (A_1(l,h)\,|x(l,h)| + L + A_2(l,h)\,|x(l,h)| + M) \times (A_3(l,h)\,|x(l,h)| + M) \\
&\leq (2K \parallel x \parallel + L + M)\,(K \parallel x \parallel + M) \\
&\leq (2K \parallel x \parallel + L + M)^2 .
\end{aligned}$$

Let $\alpha = 2k$, $\beta = L + M$. Then, we have:

$$\parallel \hat{F}x \parallel \leq \alpha \parallel x \parallel + \beta,$$

$$\parallel \hat{G}x \parallel \leq \alpha \parallel x \parallel + \beta$$

and:

$$\parallel \hat{T}x \parallel \leq (\alpha \parallel x \parallel + \beta)^2 \tag{3}$$

for $x \in C(I \times I)$.

From (3), we have that the operator $\hat{T}$ maps $B_d \subset C(I \times I)$ into B_d, where:

$$B_d = \{x(l,h) \in I : \parallel x(l,h) \parallel \leq d\}$$

for $d_2 \leq d \leq d_1$, where:

$$d_1 = \frac{1 - 2\alpha\beta - \sqrt{1 - 4\alpha\beta}}{2\alpha^2}$$

and:

$$d_2 = \frac{1 - 2\alpha\beta + \sqrt{1 - 4\alpha\beta}}{2\alpha^2}.$$

Furthermore, we have:

$$\parallel \hat{F}B_d \parallel \leq \alpha d + \beta \tag{4}$$

and:

$$\parallel \hat{G}B_d \parallel \leq \alpha d + \beta. \tag{5}$$

Let $\epsilon > 0$ be fixed and $x(l,h), y(l,h) \in B_d$ such that:

$$\parallel x - y \parallel \leq \epsilon, \;\; (l, h \in I).$$

Then, we have:

$$\begin{aligned}
\left|(\hat{F}x)(l,h)-(\hat{F}y)(l,h)\right| &= \Big|U\left(l,h,x(l,h)\right)+F\left(l,h,I_1(x),x(l,h)\right) \\
&\quad -U\left(l,h,y(l,h)\right)-F\left(l,h,I_1(y),y(l,h)\right)\Big| \\
&\le \left|U\left(l,h,x(l,h)\right)-U\left(l,h,y(l,h)\right)\right| \\
&\quad +\left|F\left(l,h,I_1(x),x(l,h)\right)-F\left(l,h,I_1(x),y(l,h)\right)\right| \\
&\quad +\left|F\left(l,h,I_1(x),y(l,h)\right)-F\left(l,h,I_1(y),y(l,h)\right)\right| \\
&\le A_1(l,h)\left|x(l,h)-y(l,h)\right|+A_2(l,h)\left|x(l,h)-y(l,h)\right| \\
&\quad +A_4(l,h)\left|I_1(x)-I_1(y)\right| \\
&\le 2K \parallel x-y \parallel \\
&\quad +K\int_0^l\int_0^h \left|P(l,h,r,u,x(r,u))-P(l,h,r,u,y(r,u))\right| drdu \\
&\le 2K \parallel x-y \parallel +Ka^2\omega(P,\epsilon),
\end{aligned}$$

where:

$$\omega(P,\epsilon)=\sup\left\{\begin{array}{c}\left|P(l,h,r,u,x(r,u))-P(l,h,r,u,y(r,u))\right| : l,h,r,u\in I, \\ x,y\in[-d,d], \parallel x-y \parallel<\epsilon\end{array}\right\}.$$

Since P is continuous, so it is uniformly continuous on the compact set $I\times I\times I\times I\times[-d,d]$; therefore:

$$\omega(P,\epsilon)\to 0 \text{ as } \epsilon\to 0.$$

Thus, $\hat{F}$ is continuous on B_d. Similarly, one can prove that $\hat{G}$ is continuous on B_d. Thus, we can conclude that $\hat{T}$ is continuous on B_d.

Let us consider a non-empty subset X of B_d and $x \in X$. Then, for a fixed $\epsilon > 0$ and $l_1, l_2, h_1, h_2 \in I$ such that $l_1 \leq l_2$, $h_1 \leq h_2$, $l_1 - l_2 \leq \epsilon$, $h_1 - h_2 \leq \epsilon$, one obtains:

$$
\begin{aligned}
&\left|(\hat{F}x)(l_2,h_2) - (\hat{F}x)(l_1,h_1)\right| \\
&= \Bigg| U\left(l_2,h_2,x(l_2,h_2)\right) + F\left(l_2,h_2,\int_0^{l_2}\int_0^{h_2} P(l_2,h_2,r,u,x(r,u))drdu, x(l_2,h_2)\right) \\
&\quad - U\left(l_1,h_1,x(l_1,h_1)\right) - F\left(l_1,h_1,\int_0^{l_1}\int_0^{h_1} P(l_1,h_1,r,u,x(r,u))drdu, x(l_1,h_1)\right)\Bigg| \\
&\leq \left|U\left(l_2,h_2,x(l_2,h_2)\right) - U\left(l_2,h_2,x(l_1,h_1)\right)\right| + \left|U\left(l_2,h_2,x(l_1,h_1)\right) - U\left(l_1,h_1,x(l_1,h_1)\right)\right| \\
&\quad + \Bigg|F\left(l_2,h_2,\int_0^{l_2}\int_0^{h_2} P(l_2,h_2,r,u,x(r,u))drdu, x(l_2,h_2)\right) \\
&\quad - F\left(l_2,h_2,\int_0^{l_1}\int_0^{h_1} P(l_1,h_1,r,u,x(r,u))drdu, x(l_2,h_2)\right)\Bigg| \\
&\quad + \Bigg|F\left(l_2,h_2,\int_0^{l_1}\int_0^{h_1} P(l_1,h_1,r,u,x(r,u))drdu, x(l_2,h_2)\right) \\
&\quad - F\left(l_1,h_1,\int_0^{l_1}\int_0^{h_1} P(l_1,h_1,r,u,x(r,u))drdu, x(l_2,h_2)\right)\Bigg| \\
&\quad + \Bigg|F\left(l_1,h_1,\int_0^{l_1}\int_0^{h_1} P(l_1,h_1,r,u,x(r,u))drdu, x(l_2,h_2)\right) \\
&\quad - F\left(l_1,h_1,\int_0^{l_1}\int_0^{h_1} P(l_1,h_1,r,u,x(r,u))drdu, x(l_1,h_1)\right)\Bigg| \\
&\leq A_1(l,h)\left|x(l_2,h_2) - x(l_1,h_1)\right| + \left|U\left(l_2,h_2,x(l_1,h_1)\right) - U\left(l_1,h_1,x(l_1,h_1)\right)\right| \\
&\quad + A_4(l,h)\left|\int_0^{l_2}\int_0^{h_2} P(l_2,h_2,r,u,x(r,u))drdu - \int_0^{l_1}\int_0^{h_1} P(l_1,h_1,r,u,x(r,u))drdu\right| \\
&\quad + \Bigg|F\left(l_2,h_2,\int_0^{l_1}\int_0^{h_1} P(l_1,h_1,r,u,x(r,u))drdu, x(l_2,h_2)\right) \\
&\quad - F\left(l_1,h_1,\int_0^{l_1}\int_0^{h_1} P(l_1,h_1,r,u,x(r,u))drdu, x(l_2,h_2)\right)\Bigg| \\
&\quad + A_2(l,h)\left|x(l_2,h_2) - x(l_1,h_1)\right|
\end{aligned}
$$

which yields:

$$
\begin{aligned}
&\left|(\hat{F}x)(l_2,h_2) - (\hat{F}x)(l_1,h_1)\right| \leq 2K\left|x(l_2,h_2) - x(l_1,h_1)\right| \\
&+ \left|U\left(l_2,h_2,x(l_1,h_1)\right) - U\left(l_1,h_1,x(l_1,h_1)\right)\right| \\
&+ K\Bigg|\int_0^{l_2}\int_0^{h_2} P(l_2,h_2,r,u,x(r,u))drdu
\end{aligned}
$$

$$
\begin{aligned}
&- \int_0^{l_1} \int_0^{h_1} P(l_1, h_1, r, u, x(r,u)) drdu \Bigg| \\
&+ \Bigg| F\left(l_2, h_2, \int_0^{l_1} \int_0^{h_1} P(l_1, h_1, r, u, x(r,u)) drdu, x(l_2, h_2) \right) \\
&- F\left(l_1, h_1, \int_0^{l_1} \int_0^{h_1} P(l_1, h_1, r, u, x(r,u)) drdu, x(l_2, h_2) \right) \Bigg| .
\end{aligned}
$$

Let:

$$
\omega(U, \epsilon) = \sup \left\{ \begin{array}{c} |U(l_2, h_2, x(l_2, h_2)) - U(l_1, h_1, x(l_1, h_1))| : l_1, l_2, h_1, h_2 \in I, \\ |l_2 - l_1| \le \epsilon, |h_2 - h_1| \le \epsilon, x \in [-d, d] \end{array} \right\},
$$

$$
\omega(P, \epsilon) = \sup \left\{ \begin{array}{c} |P(l_2, h_2, r, u, x(r,u)) - P(l_1, h_1, r, u, x(r,u))| : l_1, l_2, h_1, h_2, r, u \in I, \\ |l_2 - l_1| \le \epsilon, |h_2 - h_1| \le \epsilon, x \in [-d, d] \end{array} \right\},
$$

$$
\bar{k} = \sup \left\{ |P(l, h, r, u, x(r,u))| : l, h, r, u \in I, x \in [-d, d] \right\}
$$

and:

$$
\omega(F, \epsilon) = \sup \left\{ \begin{array}{c} |F(l_2, h_2, z, x(l_2, h_2)) - F(l_1, h_1, z, x(l_1, h_1))| : l_1, l_2, h_1, h_2 \in I, \\ |l_2 - l_1| \le \epsilon, |h_2 - h_1| \le \epsilon, x \in [-d, d], z \in [-\bar{k}a^2, \bar{k}a^2] \end{array} \right\}.
$$

Furthermore:

$$
\begin{aligned}
&\left| \int_0^{l_2} \int_0^{h_2} P(l_2, h_2, r, u, x(r,u)) drdu - \int_0^{l_1} \int_0^{h_1} P(l_1, h_1, r, u, x(r,u)) drdu \right| \\
&\le \left| \int_0^{l_2} \int_0^{h_2} \left(P(l_2, h_2, r, u, x(r,u)) - P(l_1, h_1, r, u, x(r,u)) \right) drdu \right| \\
&+ \left| \int_{l_1}^{l_2} \int_{h_1}^{h_2} P(l_1, h_1, r, u, x(r,u)) drdu \right| \\
&\le a^2 \omega(P, \epsilon) + \bar{k}\epsilon^2.
\end{aligned}
$$

Therefore:

$$
\begin{aligned}
|(\hat{F}x)(l_2, h_2) - (\hat{F}x)(l_1, h_1)| \le\ & 2K |x(l_2, h_2) - x(l_1, h_1)| + \omega(U, \epsilon) \\
& + K\left(a^2 \omega(P, \epsilon) + \bar{k}\epsilon^2 \right) + \omega(F, \epsilon).
\end{aligned}
$$

This gives:

$$
\omega(\hat{F}x, \epsilon) \le 2K\omega(x, \epsilon) + \omega(U, \epsilon) + K\left[a^2 \omega(P, \epsilon) + \bar{k}a^2 \right] + \omega(F, \epsilon).
$$

Since U and F are continuous on $I \times I \times \mathbb{R}$ and $I \times I \times \mathbb{R} \times \mathbb{R}$, respectively, therefore we get:

$$
\omega(U, \epsilon) \to 0,\ \omega(P, \epsilon) \to 0 \text{ and } \omega(F, \epsilon) \to 0 \text{ as } \epsilon \to 0.
$$

Thus:

$$
\omega_0(\hat{F}X) \le 2K\omega_0(X). \tag{6}
$$

Similarly, we can show that:

$$
\omega_0(\hat{G}X) \le 2K\omega_0(X). \tag{7}
$$

From (4)–(7) and Theorem 2 (for the details of this theorem, we refer to [8]), we get that $\hat{T}$ satisfies the Darbo condition on B_d with respect to measure ω_0 with constant:

$$\begin{aligned} 2K(\alpha d+\beta)+2K(\alpha d+\beta) &= 4K(\alpha d+\beta) \\ &= 4K(\alpha d_1+\beta) \\ &= 4K\left[\alpha\left(\frac{1-2\alpha\beta-\sqrt{1-4\alpha\beta}}{2\alpha^2}\right)+\beta\right] \\ &= 2K\left(\frac{1-\sqrt{1-4\alpha\beta}}{\alpha}\right) \\ &< 1. \end{aligned}$$

This implies that $\hat{T}$ is a contraction operator on B_d with respect to ω_0. Thus, by Theorem 2, we have that $\hat{T}$ has at least one fixed point in B_d. Hence, Equation (2) has at least one solution in $B_d \subset C([0,a]\times[0,a])$. This completes the proof. □

4. An Illustrative Example

We construct the following example to illustrate the obtained result in the previous section.

Example 1. *Consider the following integral equation:*

$$x(l,h)=\left(\frac{1}{6}\cos\left(\frac{l+h}{2}\right)+\frac{1}{9}\int_0^l\int_0^h\frac{rue^{-lh}}{3+x^2(r,u)}drdu\right)\left(\frac{1}{8}\int_0^1\int_0^1\frac{lh}{6+|x(r,u)|}drdu\right) \tag{8}$$

for $l,h\in[0,1]=I$. Here, we have:

$$U(l,h,x(l,h))=\frac{1}{6}\cos\left(\frac{l+h}{2}\right),$$

$$F(l,h,y,x(l,h))=\frac{y}{9},$$

$$G(l,h,y,x(l,h))=\frac{y}{8},$$

$$P(l,h,r,u,x(r,u))=\frac{rue^{-lh}}{3+x^2(r,u)},$$

$$Q(l,h,r,u,x(r,u))=\frac{lh}{6+|x(r,u)|}$$

and $a=1$; $x,y\in\mathbb{R}$.

It is obvious that all the functions U,F,G,P, and Q are continuous. We have:

$$|U(l,h,x_1)-U(l,h,x_2(l,h))|=0.\,|x_1(l,h)-x_2(l,h)|,$$

$$|F(l,h,y,x_1(l,h))-F(l,h,y,x_2(l,h))|=0.\,|x_1(l,h)-x_2(l,h)|,$$

$$|G(l,h,y,x_1(l,h))-G(l,h,y,x_2(l,h))|=0.\,|x_1(l,h)-x_2(l,h)|,$$

$$|F(l,h,y_1,x(l,h))-F(l,h,y_2,x(l,h))|=\frac{1}{9}|y_1-y_2|,$$

and:

$$|G(l,h,y_1,x(l,h))-G(l,h,y_2,x(l,h))|=\frac{1}{8}|y_1-y_2|.$$

It follows that:

$$A_1(l,h) = A_2(l,h) = A_3(l,h) = 0,\ A_4(l,h) = \frac{1}{9} \text{ and } A_5(l,h) = \frac{1}{8}.$$

Consequently, we get $K = \frac{1}{8}$.

Furthermore,

$$|U(l,h,0)| \le \frac{1}{6},$$

$$|F(l,h,y_1,0)| \le \frac{1}{27e^2} \approx \frac{1}{198.29}$$

and:

$$|G(l,h,y_2,0)| \le \frac{1}{48}.$$

Thus:

$$M = \frac{1}{48},\ L = \frac{1}{6}$$

and:

$$4\alpha\beta = \frac{9}{48} < 1.$$

Hence, all the assumption from (1)–(4) are satisfied. Thus, by applying Theorem 3, we conclude that Equation (8) has at least one solution in the Banach algebra $C([0,1] \times [0,1])$.

5. An Iterative Algorithm Created by a Coupled Semi-Analytic Method to Find the Solution of the Integral Equation

To find an approximation of solution for Equation (8), we make an iterative algorithm by a coupled method created by modified homotopy perturbation and the Adomian decomposition method in the case of two-dimensional functions. Applications of the modified homotopy perturbation method to solve nonlinear integral equations, nonlinear singular integral equations, and nonlinear differential equations can be seen in [25–27], respectively. The Adomian decomposition method to solve physical problems was used in [28] and also to solve integro-differential equations system in [29]. However, in this article, we introduce a modified homotopy perturbation method in terms of a function with two variables, and for simplification of nonlinear terms, we use the Adomian decomposition method in the suitable form; therefore, we make an effective algorithm by the above process. Equation (8) can be shown in a general form of the two-dimensional nonlinear problem:

$$A\left(x(l,h)\right) - f(l,h) = 0$$

with $(l,h) \in I \times I$, where A is a general nonlinear operator and f is a known analytic function. Similar to [26,27], we divide the general operator A into two nonlinear operators as M_1 and M_2. Of course, M_1 or M_2 can be linear operators in the special case that also f is converted to f_1 and f_2 functions; in other words, we have:

$$M_1\left(x(l,h)\right) - f_1(l,h) + M_2\left(x(l,h)\right) - f_2(l,h) = 0.$$

A modified homotopy perturbation for the above problem can be introduced as follows:

$$H\left(u(l,h),p\right) = M_1\left(u(l,h)\right) - f_1(l,h) + p\left[M_2\left(u(l,h)\right) - f_2(l,h)\right] = 0,\ p \in [0,1], \tag{9}$$

where p is an embedding parameter and u is an approximation of x. According to the variations of $p = 0$ to $p = 1$, it can be observed that $M_1\left(u(l,h)\right) = f_1(l,h)$ to $A\left(u(l,h)\right) = f(l,h)$. This implies that for $p = 1$ in (9), we get the solution of (5).

We consider the above solution as the series:

$$x(l,h) \sim u(l,h) = \sum_{k=0}^{\infty} p^k u_k(l,h) \tag{10}$$

and:

$$x(l,h) = \lim_{p\to 1} u(l,h). \tag{11}$$

To solve Equation (8), M_1, M_2 and f can be defined as follows:

$$M_1(x(l,h)) = x(l,h), \tag{12}$$

$$M_2(x(l,h)) = -\left(\frac{1}{6}\cos\left(\frac{l+h}{2}\right) + \frac{1}{9}\int_0^l\int_0^h \frac{rue^{-lh}}{3+x^2(r,u)}drdu\right)\left(\frac{1}{8}\int_0^1\int_0^1 \frac{lh}{6+|x(r,u)|}drdu\right), \tag{13}$$

and:

$$f(l,h) = f_1(l,h) + f_2(l,h). \tag{14}$$

Since in (8) $f(l,h) = 0$, therefore $f_1(l,h) = f_2(l,h) = 0$. From (9)–(14), we have:

$$\sum_{k=0}^{\infty} p^k u_k(l,h) - p\left(\frac{1}{6}\cos\left(\frac{l+h}{2}\right) + \frac{1}{9}\int_0^l\int_0^h \frac{rue^{-lh}}{3+\left(\sum\limits_{k=0}^{\infty} p^k u_k(r,u)\right)^2}drdu\right)\left(\frac{1}{8}\int_0^1\int_0^1 \frac{lh}{6+\left|\sum\limits_{k=0}^{\infty} p^k u_k(r,u)\right|}drdu\right) = 0.$$

Now, we use Adomain polynomials for simplicity for the nonlinear terms:

$$\frac{1}{6}\cos\left(\frac{l+h}{2}\right) + \frac{1}{9}\int_0^l\int_0^h \frac{rue^{-lh}}{3+\left(\sum\limits_{k=0}^{\infty} p^k u_k(r,u)\right)^2}drdu = \sum_{k=0}^{\infty} p^k A_k(l,h)$$

and:

$$\frac{1}{8}\int_0^1\int_0^1 \frac{lh}{6+\left|\sum\limits_{k=0}^{\infty} p^k u_k(r,u)\right|}drdu = \sum_{k=0}^{\infty} p^k \hat{A}_k(l,h),$$

where the Adomain polynomials are given by:

$$A_k(l,h) = \frac{1}{k!}\left[\frac{d^k}{dp^k}\left\{\frac{1}{6}\cos\left(\frac{l+h}{2}\right) + \frac{1}{9}\int_0^l\int_0^h \frac{rue^{-lh}}{3+\left(\sum\limits_{k=0}^{\infty} p^k u_k(r,u)\right)^2}drdu\right\}\right]_{p=0}$$

and:

$$\hat{A}_k(l,h) = \frac{1}{k!}\left[\frac{d^k}{dp^k}\left\{\frac{1}{8}\int_0^1\int_0^1 \frac{lh}{6+\left|\sum\limits_{k=0}^{\infty} p^k u_k(r,u)\right|}drdu\right\}\right]_{p=0}$$

Therefore, we have:

$$\sum_{k=0}^{\infty} p^k u_k(l,h) - p\left\{\sum_{k=0}^{\infty} p^k A_k(l,h)\right\}\left\{\sum_{k=0}^{\infty} p^k \hat{A}_k(l,h)\right\} = 0. \tag{15}$$

By rearranging the terms in powers in p of (15) and using modified homotopy perturbation (9), the coefficients of p powers must be equal to zero, so we obtain an iterative algorithm (Algorithm 1) to solve for the numerical solution of Equation (8).

Algorithm 1. Algorithm of calculating $u_k(l,h)$

$u_0(l,h) = 0,$
$u_1(l,h) = A_0(l,h)\hat{A}_0(l,h),$
$u_k(l,h) = \sum_{i=0}^{k-1} A_i(l,h)\hat{A}_{k-1-i}(l,h),\ k = 2,3,\cdots.$

Calculating the sequence $\{u_0(l,h), u_1(l,h), ...\}$, we can obtain a closed form of the solution for (8) using the above algorithm.

We compute the Adomain polynomial for $k = 0$,

$$\begin{aligned} &A_0(l,h) \\ &= \frac{1}{6}\cos\left(\frac{l+h}{2}\right) + \frac{1}{9}\int_0^l\int_0^h \frac{rue^{-lh}}{3+u_0^2(r,u)}drdu \\ &= \frac{1}{6}\cos\left(\frac{l+h}{2}\right) + \frac{l^2h^2e^{-lh}}{108} \end{aligned}$$

and:

$$\hat{A}_0(l,h) = \frac{1}{8}\int_0^1\int_0^1 \frac{lh}{6+|u_0(r,u)|}drdu = \frac{lh}{48}.$$

Therefore, we obtain by the algorithm:

$$u_1(l,h) = \left\{\frac{1}{6}\cos\left(\frac{l+h}{2}\right) + \frac{l^2h^2e^{-lh}}{108}\right\}\frac{lh}{48}.$$

We use (10) to approximate $x(l,h)$ by a few term of $u_k(l,h)$ as follows:

$$x_1(l,h) = u_0(l,h) + u_1(l,h) = \left\{\frac{1}{6}\cos\left(\frac{l+h}{2}\right) + \frac{l^2h^2e^{-lh}}{108}\right\}\frac{lh}{48}.$$

6. Conclusions

In our present investigation, we have established the existence of the solution of a functional integral equation of two variables, which is of the form of the product of two operators in the Banach algebra $C([0,a]\times[0,a])$, $a > 0$ and illustrated our results with the help of an example. We also constructed an iteration algorithm to get the solution of Equation (8). Further, one can solve Equation (8) using different numerical, as well as analytical methods in the setting of Banach sequence spaces and Banach algebra. Moreover, due our existence theorem for Equation (8) of two variables, we therefore conclude that our existence result is more general than the one obtained earlier by Deepmala and Pathak [9].

Author Contributions: All authors contributed equally in this work.

References

1. Chandrasekhar, S. *Radiative Transfer*; Oxford University Press: London, UK, 1950.
2. Hu, S.; Khavanin, M.; Zhuang, W. Integral equations arising in the kinetic theory of grass. *Appl. Anal.* **1989**, *34*, 261–266. [CrossRef]
3. Kelly, C.T. Approximation of solution of some quadratic integral equations in transport theory. *J. Integral Equ.* **1982**, *4*, 221–237.
4. Zabrejko, P.P.; Koshelev, A.I.; Krasnosel'skii, M.A.; Mikhlin, S.G.; Rakovshchik, L.S.; Stetsenko, V.J. *Integral Equations*; Nauka: Moscow, Russia, 1968.
5. Banaś, J.; Lecko, M. Fixed points of the product of operators in Banach algebra. *Panamer Math. J.* **2002**, *12*, 101–109.
6. Dhage, B.C. A flxed point theorem in Banach algebras with applications to functional integral equations. *Kyungpook Math. J.* **2004**, *44*, 145–155.
7. Dhage, B.C. On a flxed point theorem in Banach algebras with applications. *Appl. Math. Lett.* **2005**, *18*, 273–280. [CrossRef]
8. Banaś, J.; Olszowy, L. On a class of measure of noncompactness in Banach algebras and their application to nonlinear integral equations. *J. Anal. Appl.* **2009**, *28*, 1–24. [CrossRef]
9. Pathak, H.K. A study on some problems on existence of solutions for nonlinear functional-integral equations. *Acta Math. Sci.* **2013**, *33*, 1305–1313.
10. Kuratowski, K. Sur les espaces complets. *Fund. Math.* **1930**, *15*, 301–309. [CrossRef]
11. Darbo, G. Punti uniti in trasformazioni a codominio non compatto. *Rend. Sem. Mater. Univ. Padova* **1955**, *24*, 84–92.
12. Alotaibi, A.; Mursaleen, M.; Mohiuddine, S.A. Application of measure of noncompactness to infinite system of linear equations in sequence spaces. *Bull. Iran. Math. Soc.* **2015**, *41*, 519–527.
13. Aghajani, A.; Allahyari, R.; Mursaleen, M. A generalization of Darbo's theorem with application to the solvability of systems of integral equations. *J. Comput. Appl. Math.* **2014**, *260*, 68–77. [CrossRef]
14. Arab, R.; Allahyari, R.; Haghighi, A.S. Existence of solutions of infinite systems of integral equations in two variables via measure of noncompactness. *Appl. Math. Comput.* **2014**, *246*, 283–291. [CrossRef]
15. Darwish, M.A. On quadratic integral equation of fractional orders. *J. Math. Anal. Appl.* **2005**, *311*, 112–119. [CrossRef]
16. Mishra, L.N.; Sen, M.; Mohapatra, R.N. On existence theorems for some generalized nonlinear functional-integral equations with applications. *Filomat* **2017**, *31*, 2081–2091.
17. Mohiuddine, S.A.; Srivastava, H.M.; Alotaibi, A. Application of measures of noncompactness to the infinite system of second-order differential equations in l_p spaces. *Adv. Differ. Equ.* **2016**, *2016*, 317. [CrossRef]
18. Mursaleen, M.; Mohiuddine, S.A. Applications of measures of noncompactness to the infinite system of differential equations in l_p spaces. *Nonlinear Anal.* **2012**, *75*, 2111–2115. [CrossRef]
19. Mursaleen, M.; Rizvi, S.M.H. Solvability of infinite systems of second order differential equations in c_0 and ℓ_1 by Meir-Keeler condensing operators. *Proc. Am. Math. Soc.* **2016**, *144*, 4279–4289. [CrossRef]
20. Srivastava, H.M.; Das, A.; Hazarika, B.; Mohiuddine, S.A. Existence of solutions of infinite systems of differential equations of general order with boundary conditions in the spaces c_0 and l_1 via the measure of noncompactness. *Math. Meth. Appl. Sci.* **2018**, *41*, 3558–3569. [CrossRef]
21. Banaś, J.; Goebel, K. *Measure of Noncompactness in Banach Spaces, Lecture Notes in Pure and Applied Mathematics*; Marcel Dekker: New York, NY, USA, 1980; Volume 60.
22. Shang, Y. Subspace confinement for switched linear systems. *Forum Math.* **2017**, *29*, 693–699. [CrossRef]
23. Shang, Y. Fixed-time group consensus for multi-agent systems with non-linear dynamics and uncertainties. *IET Control Theory Appl.* **2018**, *12*, 395–404. [CrossRef]
24. Shang, Y. Resilient consensus of switched multi-agent systems. *Syst. Control Lett.* **2018**, *122*, 12–18. [CrossRef]
25. Glayeri, A.; Rabbani, M. New Technique in Semi-Analytic Method for Solving Non-Linear Differential Equations. *Math. Sci.* **2011**, *5*, 395–404.
26. Rabbani, M. Modified homotopy method to solve non-linear integral equations. *Int. J. Nonlinear Anal. Appl.* **2015**, *6*, 133–136.

27. Rabbani, M. New Homotopy Perturbation Method to Solve Non-Linear Problems. *J. Math. Comput. Sci.* **2013**, *7*, 272–275. [CrossRef]
28. Adomian, G. *Solving Frontier Problem of Physics: The Decomposition Method*; Kluwer Academic Press: Dordrecht, The Netherlands, 1994.
29. Rabbani, M.; Arab, R.; Hazarika, B. Solvability of nonlinear quadratic integral equation by using simulation type condensing operator and measure of noncompactness. *Appl. Math. Comput.* **2019**, *349*, 102–117. [CrossRef]

Class of Analytic Functions Defined by q-Integral Operator in a Symmetric Region

Lei Shi [1], Mohsan Raza [2,*], Kashif Javed [2], Saqib Hussain [3] and Muhammad Arif [4]

1 School of Mathematics and Statistics, Anyang Normal University, Anyang 455002, China
2 Department of Mathematics, Government College University, Faisalabad 38000, Pakistan
3 Department of Mathematics, COMSATS University Islamabad, Abbottabad Campus 22010, Pakistan
4 Department of Mathematics, Abdul Wali Khan University Mardan, 23200 Mardan, Pakistan
* Correspondence: mohsan976@yahoo.com

Abstract: The aim of the present paper is to introduce a new class of analytic functions by using a q-integral operator in the conic region. It is worth mentioning that these regions are symmetric along the real axis. We find the coefficient estimates, the Fekete–Szegö inequality, the sufficiency criteria, the distortion result, and the Hankel determinant problem for functions in this class. Furthermore, we study the inverse coefficient estimates for functions in this class.

Keywords: analytic functions; q-integral operator; conic region

1. Introduction

Let $\mathcal{A}$ denote the class of functions f of the form:

$$f(z) = z + \sum_{m=2}^{\infty} a_m z^m, \quad z \in \mathbb{D}. \tag{1}$$

which are analytic in $\mathbb{D} = \{z \in \mathbb{C} : |z| < 1\}$ and $\mathcal{S}$ denotes a subclass of $\mathcal{A}$, which contains univalent functions in $\mathbb{D}$. Let f be a univalent function in $\mathbb{D}$. Then, its inverse function f^{-1} exists in some disc $|w| < r \leq 1/4$, of the form:

$$f^{-1}(w) = w + B_2 w^2 + B_3 w^3 + \cdots . \tag{2}$$

For any analytic functions f of the form (1) and g of the form:

$$g(z) = z + \sum_{m=2}^{\infty} b_m z^m, \qquad z \in \mathbb{D}, \tag{3}$$

the convolution (Hadamard product) is given as:

$$(f * g)(z) = z + \sum_{m=2}^{\infty} a_m b_m z^m, \ (z \in \mathbb{D}).$$

Let f and g be analytic functions in $\mathbb{D}$. Then, f is said to be subordinate to g, written as $f(z) \prec g(z)$, if there exists a function w analytic in $\mathbb{D}$ with $w(0) = 0$ and $|w(z)| < 1$ such that $f(z) = g(w(z))$. Moreover, if g is univalent in $\mathbb{D}$, then the following equivalent relation holds:

$$f(z) \prec g(z) \Longleftrightarrow f(0) = g(0) \quad \text{and} \quad f(\mathbb{D}) \subset g(\mathbb{D}).$$

The classes of k-uniformly starlike and k-uniformly convex functions were introduced by Kanas and Wiśniowska [1,2]. A function $f \in \mathcal{S}$ is in $k-\mathcal{ST}$, if and only if:

$$\Re\frac{zf'(z)}{f(z)} > k\left|\frac{zf'(z)}{f(z)}-1\right|,$$

where $k \in [0,\infty)$ and $z \in \mathbb{D}$. Similarly, for $k \in [0,\infty)$, a function $f \in \mathcal{S}$ is in $k-\mathcal{UCV}$, if and only if:

$$\Re\left(1+\frac{zf''(z)}{f'(z)}\right) > k\left|\frac{zf''(z)}{f'(z)}\right|.$$

In particular, the classes $0-\mathcal{ST}=\mathcal{ST}$ and $0-\mathcal{UCV}=\mathcal{UCV}$ are the familiar classes of uniformly-starlike and uniformly-convex functions, respectively. These classes have been studied extensively. For some details, see [1–5].

Recently, a vivid interest has been shown by many researchers in quantum calculus due to its wide-spread applications in many branches of sciences especially in mathematics and physics. Among the contributors to the study, Jackson was the first to provide the basic notions and established results for the theory of q-calculus [6,7]. The idea of the q-derivative was first time used by Ismail et al. [8], and they introduced the q-extension of the class of starlike functions. A remarkable usage of the q-calculus in the context of geometric function theory was basically furnished, and the basic (or q-) hypergeometric functions were first used in geometric function theory in a book chapter by Srivastava (see, for details, p. 347 of [9]). The idea of q-starlikeness was further extended to certain subclasses of q-starlike functions. Recently, the q-analogue of the Ruscheweyh operator was introduced in [4], and it was studied in [10]. Many researchers contributed to the development of the theory by introducing certain classes with the help of q-calculus. For some details about these contributions, see [11–25]. We contribute to the subject by studying the q-integral operator in the conic region.

Now, we write some notions and basic concepts of q-calculus, which will be useful in our discussions. Throughout our discussion, we suppose that $q \in (0,1)$, $\mathbb{N}=\{1,2,3,\cdots\}$, and $\mathbb{N}=\mathbb{N}_0\setminus\{0\}$, unless otherwise mentioned.

Definition 1. *Let $q \in (0,1)$. Then, the q-number $[t]_q$ is defined as:*

$$[t]_q=\begin{cases}\frac{1-q^t}{1-q}, & t \in \mathbb{C},\\ \sum\limits_{j=0}^{m-1} q^j = 1+q+q^2+\cdots+q^{m-1}, & t=m \in \mathbb{N}.\end{cases}$$

Definition 2. *Let $q \in (0,1)$. Then, the q-factorial $[m]_q!$ is defined as:*

$$[m]_q!=\begin{cases}1, & m=0,\\ \prod\limits_{j=1}^{m}[j]_q, & m\in\mathbb{N}.\end{cases}$$

Definition 3. *Let $q \in (0,1)$. Then, the q-Pochhammer symbol $[t]_{m,q}$, $(z \in \mathbb{C},\ m \in \mathbb{N}_0)$ is defined as:*

$$[t]_{m,q}=\frac{(q^t;q)_m}{(1-q)^m}=\begin{cases}1, & m=0,\\ [t]_q[t+1]_q[t+2]_q\cdots[t+m-1]_q & m\in\mathbb{N}.\end{cases}$$

Furthermore, the gamma function in the q-analogue is defined by the following relation:

$$\Gamma_q(1)=1 \text{ and } \Gamma_q(t+1)=[t]_q\Gamma_q(t).$$

Definition 4. *Let $q \in (0,1)$. Then, the q-derivative D_q of a function f is defined as:*

$$D_q f(z) = \begin{cases} \frac{f(z)-f(qz)}{z(1-q)}, & z \neq 0, \\ f'(0) & z = 0 \end{cases} \tag{4}$$

provided that $f'(0)$ exists.

We observe that:

$$\lim_{q\to 1^-} D_q f(z) = \lim_{q\to 1^-} \frac{f(z) - f(qz)}{z(1-q)} = f'(z).$$

From Definition 4 and (1), it is clear that:

$$D_q f(z) = 1 + \sum_{m=2}^{\infty} [m]_q a_m z^{m-1}.$$

Now, take the function:

$$F_{q,\mu+1}(z) = z + \sum_{m=2}^{\infty} \Lambda_m z^m, \tag{5}$$

where $\mu > -1$, $\Lambda_m = \frac{[\mu+1]_{m-1,q}}{[m-1]_q!}$ and $z \in \mathbb{D}$. Now, consider a function $F^{(-1)}_{q,\mu+1}$ by:

$$F^{(-1)}_{q,\mu+1}(z) * F_{q,\mu+1}(z) = zD_q f(z),$$

then the q-Noor integral operator is define by:

$$\mathrm{I}^{\mu}_q f(z) = F^{(-1)}_{q,\mu+1}(z) * f(z) = z + \sum_{m=2}^{\infty} \Phi_{m-1} a_m z^m, \ (\mu > -1, \ z \in \mathbb{D}), \tag{6}$$

where:

$$\Phi_{m-1} = \frac{[m]_q!}{[\mu+1]_{m-1,q}}. \tag{7}$$

It is clear that $\mathrm{I}^0_q f(z) = zD_q f(z)$ and $\mathrm{I}^1_q f(z) = f(z)$. From (6), we obtain:

$$[\mu+1,q]\mathrm{I}^{\mu}_q f(z) = [\mu,q]\mathrm{I}^{\mu+1}_q f(z) + q^{\mu} z D_q\left(\mathrm{I}^{\mu+1}_q f(z)\right). \tag{8}$$

The q-Noor integral operator was recently defined by Arif et al. [26]. By taking $q \to 1^-$, the operator defined in (6) coincides with the Noor integral operator defined in [27,28]. For some details about the q-analogues of various differential operators, see [29–33]. The main aim of the current paper is to study the q-Noor integral operator by defining a class of analytic functions. Now, we introduce it as follows:

Definition 5. *A function f belongs to the class $\mathcal{K} - \mathcal{UST}^{\mu}_q(\gamma)$, $\gamma \in \mathbb{C} - \{0\}$, if:*

$$\Re\left\{\frac{1}{\gamma}\left(\frac{zD_q \mathrm{I}^{\mu}_q f(z)}{\mathrm{I}^{\mu}_q f(z)} - 1\right) + 1\right\} > k\left|\frac{1}{\gamma}\left(\frac{zD_q \mathrm{I}^{\mu}_q f(z)}{\mathrm{I}^{\mu}_q f(z)} - 1\right)\right|, \ \mu > -1, \ k \in [0,\infty), \ z \in \mathbb{D}. \tag{9}$$

Geometric Interpretation

Let $f \in \mathcal{K} - \mathcal{UST}^{\mu}_q(\gamma)$. Then, $\frac{zD_q \mathrm{I}^{\mu}_q f(z)}{\mathrm{I}^{\mu}_q f(z)}$ assumes all the values in the domain $\Delta_{k,\gamma} = h_{k,r}(\mathbb{D})$ such that:

$$\Delta_{k,\gamma} = \gamma\Delta_k + (1-\gamma),$$

where:

$$\Delta_k = \left\{ u + iv : u > k\sqrt{(u-1)^2 + v^2} \right\},$$

or equivalently,

$$\frac{zD_q I_q^\mu f(z)}{I_q^\mu f(z)} \prec h_{k,\gamma}(z). \tag{10}$$

The boundary $\partial\Delta_{k,\gamma}$ of the above region is the imaginary axis when $k = 0$. It is a hyperbola in the case of $k \in (0,1)$. When $k \in [0,1)$, we have:

$$h_{k,\gamma}(z) = 1 + \frac{2\gamma}{1-k^2}\left\{ \left(\frac{2}{\pi} arccos k \right) arctanh\sqrt{z} \right\}, \quad z \in \mathbb{D}.$$

In the case of $k = 1$, $\partial\Delta_{k,\gamma}$ is a parabola, and in this case:

$$h_{1,\gamma}(z) = 1 + \frac{2\gamma}{\pi}\left(log\frac{1+\sqrt{z}}{1-\sqrt{z}} \right)^2, \quad z \in \mathbb{D}.$$

When $k > 1$, $\partial\Delta_{k,\gamma}$ is an ellipse and:

$$h_{k,\gamma}(z) = 1 + \frac{2\gamma}{k^2-1} sin\left(\frac{\pi}{2\mathcal{F}(s)} \int_0^{v(z)/\sqrt{s}} \left(1-y^2\right)^{-1/2} \left(1-(sy)^2\right)^{-1/2} dy \right) + \frac{2\gamma}{1-k^2},$$

where $v(z) = \frac{z-\sqrt{s}}{1-\sqrt{sz}}$, $0 < s < 1$, $z \in \mathbb{D}$, and z is selected so that $k = \cosh\left(\frac{\pi\mathcal{F}'(s)}{4\mathcal{F}(s)}\right)$, where $\mathcal{F}$ is the first kind of Legendre's complete elliptic integral and $\mathcal{F}'$ is the complementary integral of $\mathcal{F}$; see [1,2]. Kanas and Wiśniowska [1,2] showed that the function $h_{k,\gamma}(\mathbb{D})$ is convex and univalent. All the curves discussed above have a vertex at $(k+\gamma)/(k+1)$. Now, it is clear that the domain $\Delta_{k,\gamma}$ is the right half plane for $k = 0$, hyperbolic for $k \in (0,1)$, parabolic when $k = 1$, and elliptic when $k > 1$. It is worth mentioning that the domain $\Delta_{k,\gamma}$ is symmetric with respect to the real axis. The function $h_{k,\gamma}(\mathbb{D}) = \Delta_{k,\gamma}$ is the extremal function in many problems for the classes of uniformly-starlike and uniformly-convex functions. For more about the conic domain; see [3,34].

Let $\mathcal{P}$ denote the class of functions h of the form:

$$h(z) = 1 + \sum_{m=1}^{\infty} c_m z^m, \quad z \in \mathbb{D}, \tag{11}$$

which are analytic with a positive real part in $\mathbb{D}$. If $k \in [0,\infty)$, $\gamma \in \mathbb{C} - \{0\}$, then the class $\mathcal{P}(h_{k,\gamma})$ can be defined as:

$$\mathcal{P}(h_{k,\gamma}) = \left\{ h \in \mathcal{P} : h(\mathbb{D}) \subset \Delta_{k,\gamma} \right\}.$$

Lemma 1 ([35]). *Let $k \in [0,\infty)$ and $h_{k,\gamma}$ be introduced above. If:*

$$h_{k,\gamma}(z) = 1 + \sum_{m=1}^{\infty} Q_m z^m, \tag{12}$$

then:

$$Q_1 = \begin{cases} \frac{2\gamma A^2}{1-k^2}, & 0 \le k < 1, \\ \frac{8\gamma}{\pi^2}, & k = 1, \\ \frac{\pi^2\gamma}{4\sqrt{s}(k^2-1)R^2(s)(1+s)}, & k > 1, \end{cases} \tag{13}$$

and:

$$Q_2 = \begin{cases} \frac{A^2+2}{3} Q_1 & 0 \leq k < 1, \\ \frac{2}{3} Q_1 & k = 1, \\ \frac{4R^2(s)(s^2+6s+1)-\pi^2}{24\sqrt{s}R^2(s)(1+s)} Q_1 & k > 1, \end{cases} \tag{14}$$

where:

$$A = \frac{2\cos^{-1} k}{\pi},$$

and $0 < s < 1$, *which is selected so that* $k = cosh\left(\frac{\pi\mathcal{F}'(s)}{\mathcal{F}(s)}\right)$.

Let:

$$f_{k,\gamma}(z) = z + \sum_{m=2}^{\infty} A_m z^m$$

be the extremal function in class $\mathcal{K} - \mathcal{UST}_q^{\mu}(\gamma)$ and $h_{k,\gamma}$ be of the form (12). Then, these functions can be related by the relation:

$$\frac{zD_q \mathrm{I}_q^{\mu} f_{k,\gamma}(z)}{\mathrm{I}_q^{\mu} f_{k,\gamma}(z)} = h_{k,\gamma}(z). \tag{15}$$

From (15), we have:

$$zD_q \mathrm{I}_q^{\mu} f_{k,\gamma}(z) = p_{k,\gamma}(z) \mathrm{I}_q^{\mu} f_{k,\gamma}(z).$$

Furthermore:

$$z + \sum_{m=2}^{\infty} [m]_q \Phi_{m-1} A_m z^m = \left(\sum_{m=0}^{\infty} Q_m z^m\right)\left(z + \sum_{m=2}^{\infty} \Phi_{m-1} A_m z^m\right).$$

Equating the coefficients of z^m in the above relation, we obtain:

$$[m]_q \Phi_{m-1} A_m = \Phi_{m-1} A_m + \sum_{j=1}^{m-1} \Phi_{j-1} A_j Q_{m-j}$$

and:

$$A_m = \frac{1}{q[m-1]_q \Phi_{m-1}} \sum_{j=1}^{m-1} \Phi_{j-1} A_j Q_{m-j}. \tag{16}$$

This implies that:

$$A_2 = \frac{Q_1}{q\Phi_1}, \tag{17}$$

$$A_3 = \frac{Q_1^2 + qQ_2}{q^2(1+q)\Phi_2}, \tag{18}$$

$$A_4 = \frac{1}{(1+q+q^2)q\Phi_3}\left\{Q_3 + \frac{Q_1 Q_2}{q} + \frac{Q_1^3 + qQ_1 Q_2}{q^2(1+q)}\right\}. \tag{19}$$

Lemma 2 ([36]). *If* $h \in \mathcal{P}$ *satisfies* (11), *then:*

$$|c_2 - vc_1^2| \leq 2\ \max\{1; |2v-1|\} \quad (v \in \mathbb{C}).$$

Lemma 3 ([37]). *If* $h \in \mathcal{P}$ *satisfies* (11), *then:*

$$|c_n - c_{n-m} c_m| < 2, n > m,\ n = 1, 2, 3, \cdots.$$

Lemma 4 ([38]). *If $h \in \mathcal{P}$ satisfies* (11)*, then:*

$$|c_3 - 2c_1c_2 + c_1^3| \leq 2.$$

2. Main Results

Theorem 1. *If $f \in \mathcal{K} - \mathcal{UST}_q^{\mu}(\gamma)$, then:*

$$|a_2| \leq A_2, \quad |a_3| \leq A_3, \tag{20}$$

and:

$$|a_4| \leq \frac{Q_1}{4q\,[3]_q\,\Phi_3}\{|F| + |(E-2F)| + |(F-E+4)|\}, \tag{21}$$

where:

$$E = 4 - \frac{4Q_2}{Q_1} - \frac{2Q_1}{q} - \frac{2Q_1}{q[2]_q}, \tag{22}$$

with:

$$F = 1 + \frac{Q_3}{Q_1} - \frac{2Q_2}{Q_1} + \frac{1+[2]_q}{q[2]_q}(Q_2 - Q_1) + \frac{Q_1^2}{q[2]_q}. \tag{23}$$

Proof. Suppose that:

$$\frac{zD_q\mathrm{I}_q^{\mu} f(z)}{\mathrm{I}_q^{\mu} f(z)} = p(z), \tag{24}$$

where p is analytic in $\mathbb{D}$. Then, from (24), we have:

$$zD_q\mathrm{I}_q^{\mu} f(z) = p(z)\mathrm{I}_q^{\mu} f(z).$$

Consider:

$$p(z) = 1 + \sum_{m=1}^{\infty} p_m z^m \tag{25}$$

and $\mathrm{I}_q^{\mu} f(z)$ is given in the relation (6). Then:

$$z + \sum_{m=2}^{\infty} [m]_q \Phi_{m-1} a_m z^m = \left(\sum_{m=0}^{\infty} p_m z^m\right)\left(z + \sum_{m=2}^{\infty} \Phi_{m-1} a_m z^m\right).$$

It follows from the above relation that:

$$[m]_q \Phi_{m-1} a_m = \Phi_{m-1} a_m + \sum_{j=1}^{m-1} \Phi_{j-1} a_j p_{m-j}$$

and:

$$a_m = \frac{1}{q[m-1]_q \Phi_{m-1}} \sum_{j=1}^{m-1} \Phi_{j-1} a_j p_{m-j}. \tag{26}$$

Furthermore, consider the function:

$$h(z) = (1 + w(z))(1 - w(z))^{-1} = 1 + c_1 z + c_2 z^2 + \cdots. \tag{27}$$

Then, h is analytic in $\mathbb{D}$ with $Re(h(z)) > 0$. By using (12) and (27), we have:

$$\begin{aligned} p(z) &= p_{k,\gamma}\left(\frac{-1+h(z)}{1+h(z)}\right) = 1+\frac{1}{2}c_1Q_1z+\left(\frac{1}{2}c_2Q_1+\frac{1}{4}c_1^2(Q_2-Q_1)\right)z^2 \\ &+\left\{\frac{1}{8}(Q_1-2Q_2+Q_3)c_1^3+\frac{1}{2}(Q_2-Q_1)c_2c_1+\frac{1}{2}Q_1c_3\right\}z^3+\cdots. \end{aligned} \tag{28}$$

Now, from (26) and (28), we obtain:

$$a_2 = \frac{p_1}{q\Phi_1} = \frac{c_1Q_1}{2q\Phi_1}. \tag{29}$$

Now, using the fact that $|c_m| \leq 2$, we get:

$$|a_2| = \left|\frac{p_1}{q\Phi_1}\right| = \left|\frac{c_1Q_1}{2q\Phi_1}\right| \leq \frac{|Q_1|}{q\Phi_1} = \frac{Q_1}{q\Phi_1} = A_2.$$

Similarly:

$$a_3 = \frac{1}{q[2]_q\Phi_2}\{p_2+p_1a_2\Phi_1\} = \frac{qp_2+p_1^2}{(1+q)\,q^2\Phi_2}. \tag{30}$$

In view of the relation $|p_1|^2+|p_2| \leq Q_1^2+Q_2$ (see [5]) and (17), we obtain:

$$\begin{aligned} |a_3| &= \frac{|qp_2+p_1^2|}{(1+q)\,q^2\Phi_2} \leq \frac{q\left(|p_2|+|p_1^2|\right)+(1-q)\left|p_1^2\right|}{(1+q)\,q^2\Phi_2} \\ &\leq \frac{q\left(|Q_2|+|Q_1^2|\right)+(1-q)\left|Q_1^2\right|}{(1+q)\,q^2\Phi_2} \\ &\leq \frac{q\,|Q_2|+|Q_1^2|}{(1+q)\,q^2\Phi_2} = A_3, \end{aligned}$$

which implies the required result. Now, equating the coefficients of z^3, we have:

$$a_4 = \frac{Q_1}{8\,[3]_q\,q\Phi_3}(4c_3-Ec_1c_2+Fc_1^3), \tag{31}$$

where E and F are given by (22) and (23), respectively. This implies that:

$$\begin{aligned} |a_4| &= \frac{Q_1}{8q\,[3]_q\,\Phi_3}|F(c_3-2c_1c_2+c_1^3)+(E-2F)(c_3-c_1c_2)+(F-E+4)c_3| \\ &\leq \frac{Q_1}{8q\,[3]_q\,\Phi_3}|F(c_3-2c_1c_2+c_1^3)|+|(E-2F)(c_3-c_1c_2)|+|(F-E+4)c_3| \\ &\leq \frac{Q_1}{2q\,[3]_q\,\Phi_3}|F|+|(E-2F)|+|(F-E+4)|, \end{aligned}$$

where we have used Lemmas 3 and 4. □

Theorem 2. *Let $0 \leq k < \infty$, $q \in (0,1)$, and $\gamma \in \mathbb{C}-\{0\}$. If $f \in \mathcal{K}-\mathcal{UST}_q^{\mu}(\gamma)$ of the form (1), then:*

$$|a_m| \leq \frac{Q_1\,(Q_1+q)\left(Q_1+q\,[2]_q\right)\ldots\left(Q_1+q\,[m-2]_q\right)}{q^{m-1}\Phi_{m-1}\prod\left(1+q+\ldots+q^{k-1}\right)}, \quad m \geq 2.$$

Proof. The result is clearly true for $m = 2$. That is:

$$|a_2| \leq \frac{Q_1}{q} = A_2.$$

Let $m \geq 2$, and suppose that the relation is true for $j \leq m-1$, then we obtain:

$$\begin{aligned}
|a_m| &= \frac{1}{q\,[m-1]_q\,\Phi_{m-1}}\left|p_{m-1} + \sum_{j=2}^{m-1} \Phi_{j-1} a_j p_{m-j}\right| \\
&\leq \frac{1}{q\,[m-1]_q\,\Phi_{m-1}}\left\{Q_1 + \sum_{j=2}^{m-1} \Phi_{j-1}|a_j| Q_1\right\} \\
&\leq \frac{1}{q\,[m-1]_q\,\Phi_{m-1}} Q_1 \left\{1 + \sum_{j=2}^{m-1} \Phi_{j-1}|a_j|\right\} \\
&\leq \frac{1}{q\,[m-1]_q\,\Phi_{m-1}} Q_1 \left\{1 + \sum_{j=2}^{m-1} \Phi_{j-1} \frac{Q_1 (Q_1 + q)\left(Q_1 + q\,[2]_q\right)\cdots\left(Q_1 + q\,[j-2]_q\right)}{q^{j-1}\Phi_{j-1}\prod(1+q+\ldots+q^{k-1})}\right\},
\end{aligned}$$

where we applied the induction hypothesis to $|a_j|$ and the Rogosinski result $|p_m| \leq Q_1$ (see [39]). This implies that:

$$|a_m| \leq \frac{1}{q\,[m-1]_q\,\Phi_{m-1}} Q_1 \left\{1 + \sum_{j=2}^{m-1} \frac{Q_1 (Q_1 + q)\left(Q_1 + q\,[2]_q\right)\cdots\left(Q_1 + q\,[j-2]_q\right)}{q^{j-1}\prod(1+q+\ldots+q^{k-1})}\right\}.$$

Applying the principal of mathematical induction, we find:

$$\begin{aligned}
& 1 + \sum_{j=2}^{m-1} \frac{Q_1 (Q_1 + q)\left(Q_1 + q\,[2]_q\right)\cdots\left(Q_1 + q\,[j-2]_q\right)}{q^{j-1}\prod(1+q+\ldots+q^{k-1})} \\
= \; & \frac{Q_1 (Q_1 + q)\left(Q_1 + q\,[2]_q\right)\cdots\left(Q_1 + q\,[m-2]_q\right)}{q^{m-2}\prod(1+q+\ldots+q^{k-2})}.
\end{aligned}$$

Hence, the desired result. □

Theorem 3. *If $f \in \mathcal{A}$ is given in (1) and the inequality:*

$$\sum_{m=2}^{\infty} \left\{q[m-1]_q(k+1) + |\gamma|\right\} \Phi_{m-1}|a_m| \leq |\gamma| \tag{32}$$

holds true for some $0 \leq k < \infty$, $q \in (0,1)$ and $\gamma \in \mathbb{C} - \{0\}$, then $f \in \mathcal{K} - \mathcal{UST}_q^{\mu}(\gamma)$.

Proof. Using (9), we have:

$$k\left|\frac{1}{\gamma}\left(\frac{zD_q \mathrm{I}_q^{\mu} f(z)}{\mathrm{I}_q^{\mu} f(z)} - 1\right)\right| - \Re\left\{\frac{1}{\gamma}\left(\frac{zD_q \mathrm{I}_q^{\mu} f(z)}{\mathrm{I}_q^{\mu} f(z)} - 1\right)\right\} < 1.$$

This implies that:

$$k\left|\frac{1}{\gamma}\left(\frac{zD_q\mathrm{I}_q^\mu f(z)}{\mathrm{I}_q^\mu f(z)}-1\right)\right|-\Re\left\{\frac{1}{\gamma}\left(\frac{zD_q\mathrm{I}_q^\mu f(z)}{\mathrm{I}_q^\mu f(z)}-1\right)\right\}$$
$$\leq \frac{k}{|\gamma|}\left|\frac{zD_q\mathrm{I}_q^\mu f(z)}{\mathrm{I}_q^\mu f(z)}-1\right|+\frac{1}{|\gamma|}\left|\frac{zD_q\mathrm{I}_q^\mu f(z)}{\mathrm{I}_q^\mu f(z)}-1\right|$$
$$\leq \frac{(k+1)}{|\gamma|}\left|\frac{zD_q\mathrm{I}_q^\mu f(z)}{\mathrm{I}_q^\mu f(z)}-1\right|.$$

We see that:

$$\left|\frac{zD_q\mathrm{I}_q^\mu f(z)}{\mathrm{I}_q^\mu f(z)}-1\right|=\left|\frac{z+\sum_{m=2}^{\infty}[m]_q\,\Phi_{m-1}a_mz^m-z-\sum_{m=2}^{\infty}\Phi_{m-1}a_mz^m}{z+\sum_{m=2}^{\infty}\Phi_{m-1}a_mz^m}\right|$$
$$=\left|\frac{\sum_{m=2}^{\infty}q[m-1]_q\Phi_{m-1}a_mz^m}{z+\sum_{m=2}^{\infty}\Phi_{m-1}a_mz^m}\right|$$
$$\leq\frac{\sum_{m=2}^{\infty}q[m-1]_q\Phi_{m-1}|a_m|}{1-\sum_{m=2}^{\infty}\Phi_{m-1}|a_m|}.$$

From the above, we have:

$$k\left|\frac{1}{\gamma}\left(\frac{zD_q\mathrm{I}_q^\mu f(z)}{\mathrm{I}_q^\mu f(z)}-1\right)\right|-\Re\left\{\frac{1}{\gamma}\left(\frac{zD_q\mathrm{I}_q^\mu f(z)}{\mathrm{I}_q^\mu f(z)}-1\right)\right\}$$
$$\leq\frac{(k+1)}{|\gamma|}\frac{\sum_{m=2}^{\infty}q[m-1,q]\Phi_{m-1}|a_m|}{1-\sum_{m=2}^{\infty}\Phi_{m-1}|a_m|}$$
$$\leq 1.$$

This completes the proof. □

Theorem 4. *If $f\in\mathcal{K}-\mathcal{UST}_q^\mu(\gamma)$, then $f(\mathbb{D})$ contains an open disk of radius:*

$$\frac{q(1+q)}{q|Q_1|[\mu+1]_q+2q(1+q)},$$

where Q_1 is defined by (11).

Proof. Let $w_0\in\mathbb{C}$ and $w_0\neq 0$ with $f(z)\neq w_0$ in $\mathbb{D}$. Then:

$$f_1(z)=w_0f(z)(w_0-f(z))^{-1}=z+\left(\frac{1}{w_0}+a_2\right)z^2+....$$

Since $f_1\in\mathcal{S}$,

$$\left|\frac{1}{w_0}+a_2\right|\leq 2.$$

Now, by applying Theorem 1, we obtain:

$$\left|\frac{1}{w_0}\right|\leq 2+\frac{2|Q_1|[\mu+1]_q}{q(1+q)}.$$

Hence:

$$|w_0|\geq\frac{q(1+q)}{|Q_1|q[\mu+1]_q+2q(1+q)}.$$

□

Theorem 5. *If $f \in \mathcal{K}-\mathcal{UST}_q^{\mu}(\gamma)$, then:*

$$\mathrm{I}_q^{\mu} f(z) \prec z exp \int_0^z \frac{h_{k,\gamma}(w(\xi))-1}{\xi} d\xi, \tag{33}$$

where w is analytic in $\mathbb{D}$ with $w(0)=0$ and $|w(z)|<1$. Moreover, for $|z|=\rho$, we have:

$$\left(exp \int_0^1 \frac{h_{k,\gamma}(-\rho)-1}{\rho} d\rho \right) \le \left| \frac{\mathrm{I}_q^{\mu} f(z)}{z} \right| \le \left(exp \int_0^1 \frac{h_{k,\gamma}(\rho)-1}{\rho} d\rho \right),$$

where $h_{k,\gamma}$ is given in (10).

Proof. From (10), we obtain:

$$\frac{D_q \mathrm{I}_q^{\mu} f(z)}{\mathrm{I}_q^{\mu} f(z)} = \frac{h_{k,\gamma}(w(z))-1}{z} + \frac{1}{z},$$

for a function w, which is analytic in $\mathbb{D}$ with $w(0)=0$ and $|w(z)|<1$. Integrating the above relation with respect to z, we have:

$$\mathrm{I}_q^{\mu} f(z) \prec z exp \int_0^z \frac{h_{k,\gamma}(w(\xi))-1}{\xi} d\xi. \tag{34}$$

Since the function $h_{k,\gamma}$ is univalent and maps the disk $|z|<\rho(0<\rho\le 1)$ onto a convex and symmetric region with respect to the real axis,

$$\frac{k+\gamma}{\gamma+1} < h_{k,\gamma}(-\rho|z|) \le \Re\{h_{k,\gamma}(w(\rho z))\} \le h_{k,\gamma}(\rho|z|). \tag{35}$$

Using the above inequality, we have:

$$\int_o^1 \frac{h_{k,\gamma}(-\rho|z|)-1}{\rho} d\rho \le \Re \int_o^1 \frac{h_{k,\gamma} w(\rho z)-1}{\rho} d\rho \le \int_o^1 \frac{h_{k,\gamma}(\rho|z|)-1}{\rho} d\rho, \quad z \in \mathbb{D}.$$

Consequently, the subordination (24) implies that:

$$\int_0^1 \frac{h_{k,\gamma}(-\rho|z|)-1}{\rho} d\rho \le log \left| \frac{\mathrm{I}_q^{\mu} f(z)}{z} \right| \le \int_0^1 \frac{h_{k,\gamma}(\rho|z|)-1}{\rho} d\rho.$$

Furthermore, the relations $h_{k,\gamma}(-\rho) \le h_{k,\gamma}(-\rho|z|)$, $h_{k,\gamma}(\rho|z| \le h_{k,\gamma}(\rho)$ leads to:

$$\left(exp \int_0^1 \frac{h_{k,\gamma}(-\rho|z|)-1}{\rho} d\rho \right) \le \left| \frac{\mathrm{I}_q^{\mu} f(z)}{z} \right| \le \left(exp \int_0^1 \frac{h_{k,\gamma}(\rho|z|)-1}{\rho} d\rho \right).$$

This completes the proof. □

Theorem 6. *Let $k \in [0,\infty)$ and $f \in \mathcal{K}-\mathcal{UST}_q^{\mu}(\gamma)$ of the form (1). Then:*

$$|a_3 - \sigma a_2^2| \le \frac{|Q_1|}{2q\,[2]_q\,\Phi_2} max\{1; |2v-1|\}, \quad \sigma \in \mathbb{C},$$

where:

$$v = \frac{1}{2}\left(1 - \frac{Q_2}{Q_1} - \frac{Q_1}{q} + \frac{\sigma Q_1 \Phi_2 (1+q)}{q\Phi_1^2}\right). \tag{36}$$

The values of Q_1 and Q_2 are given by (13) and (14), respectively, and that of Φ_2 is given in (7).

Proof. If $f \in \mathcal{K} - \mathcal{UST}_q^{\mu}(\gamma)$, then using (29) and (30), we have:

$$\begin{aligned} a_2 &= \frac{Q_1 c_1}{2q\Phi_1}, \\ a_3 &= \frac{1}{4q[2]_q\Phi_2}\left\{2c_2Q_1 + c_1^2(Q_2 - Q_1) + \frac{Q_1^2c_1^2}{q}\right\}, \end{aligned}$$

which together imply that:

$$\begin{aligned} \left|a_3 - \sigma a_2^2\right| &= \frac{1}{4q[2]_q\Phi_2}\left|\left\{\left(2c_2Q_1 + c_1^2(Q_2 - Q_1)\right) + \frac{Q_1^2c_1^2}{q}\right\} - \frac{\sigma Q_1^2 c_1^2}{4q^2\Phi_1^2}\right| \\ &= \frac{Q_1}{4q[2]_q\Phi_2}\left|c_2 - vc_1^2\right|, \end{aligned}$$

where v is defined by (36). Applying Lemma 2, we have the desired result. □

Theorem 7. *If $f \in \mathcal{K} - \mathcal{UST}_q^{\mu}(\gamma)$ is given in (1), then:*

$$|a_2a_3 - a_4| \leq \frac{|Q_1|}{4q\,[3]_q\,\Phi_3}\left\{|A| + |(B-2A)| + |A - B + 4|\right\},$$

where:

$$B = E + \frac{2Q_1\Phi_3\,[3]_q}{q\,[2]_q\,\Phi_1\Phi_2}, \quad A = F + \frac{Q_1\Phi_3\,[3]_q}{q\,[2]_q\,\Phi_1\Phi_2}\left(Q_2 - Q_1 + \frac{Q_1^2}{q}\right),$$

with E and F given in (22) and (23), respectively.

Proof. By using (29)–(31), it is easy to see that:

$$\begin{aligned} |a_2a_3 - a_4| &= \frac{|-Q_1|}{8q\,[3]_q\,\Phi_3}\left|4c_3 - Bc_1c_2 + Ac_1^3\right| \\ &= \frac{|Q_1|}{8q\,[3]_q\,\Phi_3}\left|(A - B + 4)c_3 + (B - 2A)(c_3 - c_1c_2) + A(c_3 - 2c_1c_2 + c_1^3)\right| \\ &\leq \frac{|Q_1|}{4q\,[3]_q\,\Phi_3}\left\{|A| + |(B-2A)| + |A - B + 4|\right\}, \end{aligned}$$

where we used Lemmas 3 and 4. This completes the proof. □

Theorem 8. *If $k \in [0, \infty)$ and letting $f \in \mathcal{K} - \mathcal{UST}_q^{\mu}(\gamma)$ and having the inverse coefficients of the form (2), then the following results hold:*

$$|B_2| \leq \frac{|Q_1|}{q\Phi_1},$$

$$|B_3| \leq \frac{|Q_1|}{q\,[2]_q\,\Phi_2} max\left\{1; \left|\frac{Q_1H}{q} + \frac{Q_2}{Q_1}\right|\right\},$$

and:

$$H = \frac{2[2]_q\Phi_2}{\Phi_1^2} - 1. \tag{37}$$

Proof. Since $f(f^{-1}(\omega)) = \omega$; therefore, using (2), we have:

$$B_2 = -a_2, \quad B_3 = 2a_2^2 - a_3.$$

Putting the value of a_2 and a_3 in the above relation, it follows easily that:

$$B_2 = -a_2 = -\frac{c_1 Q_1}{2q\Phi_1}. \tag{38}$$

Using the coefficient bound $|c_1| \leq 2$, we can write:

$$|B_2| = \left|\frac{-c_1 Q_1}{2q\Phi_1}\right| \leq \frac{|Q_1|}{q\Phi_1}. \tag{39}$$

Now with the help of Lemma 2, we obtain:

$$\begin{aligned} B_3 &= 2a_2^2 - a_3 \\ &= -\frac{Q_1}{2q\,[2]_q\,\Phi_2}\left\{c_2 - \frac{c_1^2}{2}\left(1 - \frac{Q_2}{Q_1} - \frac{Q_1}{q}\right) - \frac{c_1^2 Q_1}{q\Phi_1^2}[2]_q\Phi_2\right\} \\ &= -\frac{Q_1}{2q\,[2]_q\,\Phi_2}\left\{c_2 - \frac{c_1^2}{2}\left(1 - \frac{Q_2}{Q_1} - \frac{Q_1}{q}\left(\frac{2[2]_q\Phi_2}{\Phi_1^2} - 1\right)\right)\right\} \\ &= -\frac{Q_1}{2q\,[2]_q\,\Phi_2}\left\{c_2 - \frac{c_1^2}{2}\left(1 - \frac{Q_2}{Q_1} - \frac{Q_1 H}{q}\right)\right\}. \end{aligned} \tag{40}$$

Taking the absolute value of the above relation, we have:

$$\begin{aligned} |B_3| &\leq \frac{|Q_1|}{q\,[2]_q\,\Phi_2}\left|c_2 - \frac{c_1^2}{2}\left(1 - \frac{Q_2}{Q_1} - \frac{Q_1 H}{q}\right)\right| \\ &\leq \frac{|Q_1|}{q\,[2]_q\,\Phi_2} max\left\{1; \left|\frac{Q_1 H}{q} + \frac{Q_2}{Q_1}\right|\right\}. \end{aligned}$$

□

Theorem 9. *If $f \in \mathcal{K} - \mathcal{UST}_q^{\mu}(\gamma)$ with inverse coefficients given by (2), then for a complex number λ, we have:*

$$|B_3 - \lambda B_2^2| \leq \frac{|Q_1|}{q\,[2]_q\,\Phi_2} max\left\{1; \left|\left(\frac{(2-\lambda)\,[2]_q\,\Phi_2 Q_1}{q\Phi_1^2} - 1\right)\frac{Q_1}{q} + \frac{Q_2}{Q_1}\right|\right\}.$$

Proof. From (38) and (40), we have:

$$\begin{aligned} B_3 - \lambda B_2^2 &= \frac{c_1^2 Q_1^2}{2q^2\Phi_1^2} - \frac{Q_1}{2q\,[2]_q\,\Phi_2}\left(c_2 - \frac{c_1^2}{2}\left(1 - \frac{Q_2}{Q_1} - \frac{Q_1}{q}\right)\right) - \frac{\lambda c_1^2 Q_1^2}{4q^2\Phi_1^2} \\ &= \frac{c_1^2 Q_1^2}{4q^2\Phi_1^2}(2-\lambda) - \frac{Q_1}{2q\,[2]_q\,\Phi_2}\left(c_2 - \frac{c_1^2}{2}\left(1 - \frac{Q_2}{Q_1} - \frac{Q_1}{q}\right)\right) \\ &= -\frac{Q_1}{2q\,[2]_q\,\Phi_2}\left\{c_2 - \frac{c_1^2}{2}\left(1 - \frac{Q_2}{Q_1} - \frac{Q_1}{q}\left(\frac{(2-\lambda)\,[2]_q\,\Phi_2 Q_1}{q\Phi_1^2} - 1\right)\right)\right\}. \end{aligned}$$

Now, by applying Lemma 2, the absolute value of the above equation becomes:

$$|B_3 - \lambda B_2^2| \leq \frac{|Q_1|}{2q\,[2]_q\,\Phi_2}\left|c_2 - \frac{c_1^2}{2}\left(1 - \frac{Q_2}{Q_1} - \frac{Q_1}{q}\left(\frac{(2-\lambda)\,[2]_q\,\Phi_2 Q_1}{q\Phi_1^2} - 1\right)\right)\right|$$
$$\leq \frac{|Q_1|}{q\,[2]_q\,\Phi_2} max\left\{1; \left|\left(\frac{(2-\lambda)\,[2]_q\,\Phi_2 Q_1}{q\Phi_1^2} - 1\right)\frac{Q_1}{q} + \frac{Q_2}{Q_1}\right|\right\}.$$

This completes the proof. □

3. Future Work

The idea presented in this paper can easily be implemented to introduce some more subfamilies of analytic and univalent functions connected with different image domains.

4. Conclusions

In this article, we defined a new class of analytic functions by using the q-Noor integral operator. We investigated some interesting properties, which are useful to study the geometry of the image domain. We found the coefficient estimates, the Fekete–Szegö inequality, the sufficiency criteria, the distortion result, and the Hankel determinant problem for this class.

Author Contributions: Conceptualization, M.R. and M.A.; methodology, M.R. and K.J.; software, L.S. and S.H.; validation, M.R., K.J. and M.A.; formal analysis, M.R., L.S. and K.J.; investigation, M.R. and K.J.; writing–original draft preparation, K.J.; and M.R. writing–review and editing, S.H., M.A. and L.S.; visualization, S.H.; supervision, M.R.; funding acquisition, L.S.

References

1. Kanas, S.; Wiśniowska, A. Conic regions and k-uniform convexity. *J. Comput. Appl. Math.* **1999**, *105*, 327–336. [CrossRef]
2. Kanas, S.; Wiśniowska, A. Conic regions and k-starlike functions. *Revue Roumaine Mathématique Pures Appliquées* **2000**, *45*, 647–657.
3. Kanas, S. Alternative characterization of the class $k-$UCV and related classes of univalent functions. *Serdica Math. J.* **1999**, *25*, 341–350.
4. Kanas, S.; Răducanu, D. Some class of analytic functions related to conic domains. *Math. Slovaca* **2014**, *64*, 1183–1196. [CrossRef]
5. Kanas, S.; Wiśniowska, A. Conic regions and k-uniform convexity II. *Zeszyty Naukowe Politechniki Rzeszowskiej Matematyka* **1998**, *170*, 65–78.
6. Jackson, F.H. On q-definite integrals. *Q. J. Pure Appl. Math.* **1910**, *41*, 193–203.
7. Jackson, F.H. On q-functions and a certain difference operator. *Trans. R. Soc. Edinb.* **1909**, *46*, 253–281. [CrossRef]
8. Ismail, M.E.H.; Merkes, E.; Styer, D. A generalization of starlike functions. *Complex Var. Theory Appl.* **1990**, *14*, 77–84. [CrossRef]
9. Srivastava, H.M. Univalent functions, fractional calculus, and associated generalized hypergeometric functions. In *Univalent Functions, Fractional Calculus, and Their Applications*; Srivastava, H.M., Owa, S., Eds.; Halsted Press: Chichester, UK, 1989; pp. 329–354.
10. Aldweby, H.; Darus, M. Some subordination results on q-analogue of Ruscheweyh differential operator. *Abstr. Appl. Anal.* **2014**, *2014*, 1–6. [CrossRef]
11. Arif, M.; Ahmad, B. New subfamily of meromorphic starlike functions in circular domain involving q-differential operator. *Math. Slovaca* **2018**, *68*, 1049–1056. [CrossRef]

12. Arif, M.; Srivastava, H,M.; Umar, S. Some applications of a q-analogue of the Ruscheweyh type operator for multivalent functions. *RACSAM* **2019**, *113*, 1211–1221. [CrossRef]
13. Mahmood, S.; Ahmad, Q.Z.; Srivastava, H.M.; Khan, N.; Khan, B.; Tahir, M. A certain subclass of meromorphically q-starlike functions associated with the Janowski functions. *J. Inequal. Appl.* **2019**, *2019*, 88. [CrossRef]
14. Mahmood, S.; Jabeen, M.; Malik, S.N.; Srivastava, H.M.; Manzoor, R.; Riaz, S.M.J. Some coefficient inequalities of q-starlike functions associated with conic domain defined by q-derivative. *J. Funct. Spaces* **2018**, *2018*, 8492072. [CrossRef]
15. Mahmood, S.; Srivastava, H.M.; Khan, N.; Ahmad, Q.Z.; Khan, B.; Ali, I. Upper bound of the third Hankel determinant for a subclass of q-starlike functions. *Symmetry* **2019**, *11*, 347. [CrossRef]
16. Srivastava, H.M.; Ahmad, Q.Z.; Khan, N.; Khan, N.; Khan, B. Hankel and Toeplitz determinants for a subclass of q-starlike functions associated with a general conic domain. *Mathematics* **2019**, *7*, 181. [CrossRef]
17. Srivastava, H.M.; Altınkaya, S.; Yalcin, S. Hankel determinant for a subclass of bi-univalent functions defined by using a symmetric q-derivative operator. *Filomat* **2018**, *32*, 503–516. [CrossRef]
18. Srivastava, H.M.; Bansal, D. Close-to-convexity of a certain family of q-Mittag-Leffler functions. *J. Nonlinear Var. Anal.* **2017**, *1*, 61–69.
19. Srivastava, H.M.; Khan, B.; Khan, N.; Ahmad, Q.Z. Coefficient inequalities for q-starlike functions associated with the Janowski functions. *Hokkaido Math. J.* **2019**, *48*, 407–425. [CrossRef]
20. Srivastava, H.M.; Mostafa, A.O.; Aouf, M.K.; Zayed, H.M. Basic and fractional q-calculus and associated Fekete–Szegö problem for p-valently q-starlike functions and p-valently q-convex functions of complex order. *Miskolc Math. Notes* **2019**, *20*, 489–509. [CrossRef]
21. Srivastava, H.M.; Tahir, M.; Khan, B.; Ahmad, Q.Z.; Khan, N. Some general classes of q-starlike functions associated with the Janowski functions. *Symmetry* **2019**, *11*, 292. [CrossRef]
22. Ahmad, k.; Arif, M.; Liu, J.-L. Convolution properties for a family of analytic functions involving q-analogue of Ruscheweyh differential operator. *Turk. J. Math.* **2019**, *43*, 1712–1720. [CrossRef]
23. Huda, A.; Darus, M. Integral operator defined by q-analogue of Liu-Srivastava operator. *Stud. Univ. Babes-Bolyai Math.* **2013**, *58*, 529–537.
24. Kanas, S.; Srivastava, H.M. Linear operators associated with k-uniformly convex functions. *Integral Transforms Spec. Funct.* **2000**, *9*, 121–132. [CrossRef]
25. Mahmood, S.; Raza, N.; AbuJarad, E.S.A.; Srivastava, G.; Srivastava, H.M.; Malik, S.N. Geometric properties of certain classes of analytic functions associated with a q-Integral operator. *Symmetry* **2019**, *11*, 719. [CrossRef]
26. Arif,M.; Haq, M.U.; Liu, J.-L. A subfamily of univalent functions associated with q-analogue of Noor integral operator. *J. Func. Spaces* **2018**, *2018*, 3818915. [CrossRef]
27. Noor, K. I. On new classes of integral operators. *J. Nat. Geom.* **1999**, *16*, 71–80.
28. Noor, K.I.; Noor, M.A. On integral operators. *J. Math. Anal. Appl.* **1999**, *238*, 341–352. [CrossRef]
29. Aldawish, I.; Darus, M. Starlikness of q-differential operator involving quantum calculus. *Korean J. Math.* **2014**, *22*, 699–709. [CrossRef]
30. Aldweby, H.; Darus, M. A subclass of harmonic univalent functions associated with q-analogue of Dziok-Srivastava operator. *ISRN Math. Anal.* **2013**, *2013*, 382312. [CrossRef]
31. Mohammed, A.; Darus, M. A generalized operator involving the q-hypergeometric function. *Matematički Vesnik* **2013**, *65*, 454–465.
32. Noor, K.I.; Riaz, S. Generalized q-starlike functions. *Stud. Sci. Math. Hung.* **2017**, *54*, 509–522. [CrossRef]
33. Noor, K.I.; Shahid, H. On dual sets and neighborhood of new subclasses of analytic functions involving q-derivative. *Iran. J. Sci. Technol. Trans. A Sci.* **2018**, *42*, 1579–1585. [CrossRef]
34. Srivastava, H.M.; Khan, M. R.; Arif, M. Some subclasses of close-to-convex mappings associated with conic regions. *Appl. Math. Comput.* **2016**, *285*, 94–102. [CrossRef]
35. Sim, Y.J.; Kwon, O.S.; Cho, N.E.; Srivastava, H.M. Some classes of analytic functions associated with conic regions. *Taiwan J. Math.* **2012**, *16*, 387–408. [CrossRef]
36. Ma, W.; Minda, D. A unified treatment of some special classes of univalent functions. In *Proceedings of the Conferene on Complex Analysis*; Li, Z., Ren, F., Yang, L., Zhang, S., Eds.; International Press Inc.: Tianjin, China, 1992; pp. 157–169.

37. Livingston, A.E. The coefficients of multivalent close-to-convex functions. *Proc. Am. Math. Soc.* **1969**, *21*, 545–552. [CrossRef]
38. Libera, R.J.; Zlotkiewicz, E.J. Early coefficients of the inverse of a regular convex function. *Proc. Am. Math. Soc.* **1982**, *85*, 225–230. [CrossRef]
39. Rogosinski, W. On the coefficients of subordinate functions. *Proc. Lond. Math. Soc.* **1943**, *48*, 48–82. [CrossRef]

Permissions

All chapters in this book were first published in MDPI; hereby published with permission under the Creative Commons Attribution License or equivalent. Every chapter published in this book has been scrutinized by our experts. Their significance has been extensively debated. The topics covered herein carry significant findings which will fuel the growth of the discipline. They may even be implemented as practical applications or may be referred to as a beginning point for another development.

The contributors of this book come from diverse backgrounds, making this book a truly international effort. This book will bring forth new frontiers with its revolutionizing research information and detailed analysis of the nascent developments around the world.

We would like to thank all the contributing authors for lending their expertise to make the book truly unique. They have played a crucial role in the development of this book. Without their invaluable contributions this book wouldn't have been possible. They have made vital efforts to compile up to date information on the varied aspects of this subject to make this book a valuable addition to the collection of many professionals and students.

This book was conceptualized with the vision of imparting up-to-date information and advanced data in this field. To ensure the same, a matchless editorial board was set up. Every individual on the board went through rigorous rounds of assessment to prove their worth. After which they invested a large part of their time researching and compiling the most relevant data for our readers.

The editorial board has been involved in producing this book since its inception. They have spent rigorous hours researching and exploring the diverse topics which have resulted in the successful publishing of this book. They have passed on their knowledge of decades through this book. To expedite this challenging task, the publisher supported the team at every step. A small team of assistant editors was also appointed to further simplify the editing procedure and attain best results for the readers.

Apart from the editorial board, the designing team has also invested a significant amount of their time in understanding the subject and creating the most relevant covers. They scrutinized every image to scout for the most suitable representation of the subject and create an appropriate cover for the book.

The publishing team has been an ardent support to the editorial, designing and production team. Their endless efforts to recruit the best for this project, has resulted in the accomplishment of this book. They are a veteran in the field of academics and their pool of knowledge is as vast as their experience in printing. Their expertise and guidance has proved useful at every step. Their uncompromising quality standards have made this book an exceptional effort. Their encouragement from time to time has been an inspiration for everyone.

The publisher and the editorial board hope that this book will prove to be a valuable piece of knowledge for researchers, students, practitioners and scholars across the globe.

List of Contributors

Rania Saadeh, Ahmad Qazza and Aliaa Burqan
Department of Mathematics, Zarqa University, Zarqa 13132, Jordan

Amalia Luque and Jesús Gómez-Bellido
Ingeniería del Diseño, Escuela Politécnica Superior, Universidad de Sevilla, 41004 Sevilla, Spain

Alejandro Carrasco
Tecnología Electrónica, Escuela Ingeniería Informática, Universidad de Sevilla, 41004 Sevilla, Spain

Julio Barbancho
Tecnología Electrónica, Escuela Politécnica Superior, Universidad de Sevilla, 41004 Sevilla, Spain

Shahid Mahmood
Department of Mechanical Engineering, Sarhad University of Science & I. T Landi Akhun Ahmad, Hayatabad Link. Ring Road, Peshawar 25000, Pakistan

Gautam Srivastava
Department of Mathematics and Computer Science, Brandon University, 270 18th Street, Brandon, MB R7A 6A9, Canada
Research Center for Interneural Computing, China Medical University, Taichung 40402, Taiwan, Republic of China

Hari Mohan Srivastava
Department of Mathematics and Statistics, University of Victoria, Victoria, BC V8W 3R4, Canada
Department of Medical Research, China Medical University Hospital, China Medical University, Taichung 40402, Taiwan

Eman S.A. Abujarad
Department of Mathematics, Aligarh Muslim University, Aligarh 202002, India

Fazal Ghani
Department of Mathematics, Abdul Wali Khan University Mardan, Mardan 23200, Pakistan

Ndolane Sene
Laboratoire Lmdan, Département de Mathématiques de la Décision, Université Cheikh Anta Diop de Dakar, Faculté des Sciences Economiques et Gestion, BP 5683 Dakar Fann, Senegal

Tianyu Huang, Xijuan Guo and Yue Zhang
Colleage of Information Science and Engineering, Yanshan University, Qinhuangdao 066004, China

Zheng Chang
Department of Mathematical Information Technology, University of Jyväskylä, FIN-40014 Jyväskylä, Finland

Ron Kerman
Department of Mathematics, Brock University, St. Catharines, ON L2S 3A1, Canada

Nusrat Raza
Mathematics Section, Women's College, Aligarh Muslim University, Aligarh 202001, Uttar Pradesh, India

Eman S. A. Abujarad
Department of Mathematics, Aligarh Muslim University, Aligarh 202001, Uttar Pradesh, India

H. M. Srivastava
Department of Mathematics and Statistics, University of Victoria, Victoria, BC V8W 3R4, Canada
Department of Medical Research, China Medical University Hospital, China Medical University, Taichung 40402, Taiwan, Republic of China

Sarfraz Nawaz Malik
Department of Mathematics, COMSATS University Islamabad, Wah Campus 47040, Pakistan

Hari M. Srivastava
Department of Mathematics and Statistics, University of Victoria, Victoria, BC V8W 3R4, Canada
Department of Medical Research, China Medical University Hospital, China Medical University, Taichung 40402, Taiwan

Faruk Özger
Department of Engineering Sciences, ˙Izmir Katip Çelebi University, ˙Izmir 35620, Turkey

Lei Shi
School of Mathematics and Statistics, Anyang Normal University, Anyan 455002, Henan, China

Shehzad Hussain and Hassan Khan
Department of Mathematics, Abdul Wali Khan University Mardan, Mardan 23200, Pakistan

Nak Eun Cho
Department of Applied Mathematics, Pukyong National University, Busan 48513, Korea

Mohamed Kamal Aouf
Department of Mathematics, Faculty of Science, Mansoura University, 35516 Mansoura, Egypt

Rekha Srivastava
Department of Mathematics and Statistics, University of Victoria, Victoria, BC V8W 3R4, Canada

Bidu Bhusan Jena and Susanta Kumar Paikray
Department of Mathematics, Veer Surendra Sai University of Technology, Burla, Odisha 768018, India

Umakanta Misra
Department of Mathematics, National Institute of Science and Technology, Palur Hills, Golanthara, Odisha 761008, India

Rabha W. Ibrahim
Cloud Computing Center, University Malaya, Kuala Lumpur 50603, Malaysia

Maslina Darus
Center for Modelling and Data Science, Faculty of Science and Technology, Universiti Kebangsaan Malaysia, Bangi 43600, Malaysia

Mostafa Bachar
Department of Mathematics, College of Sciences, King Saud University, Riyadh 11451, Saudi Arabia

Slavko Simić
Nonlinear Analysis Research Group, Ton Duc Thang University, Ho Chi Minh City 758307, Vietnam
Faculty of Mathematics and Statistics, Ton Duc Thang University, Ho Chi Minh City 758307, Vietnam

Bandar Bin-Mohsin
Department of Mathematics, College of Science, King Saud University, Riyadh 11451, Saudi Arabia

Hari M. Srivastava
Department of Mathematics and Statistics, University of Victoria, Victoria, BC V8W 3R4, Canada
Department of Medical Research, China Medical University Hospital, China Medical University, Taichung 40402, Taiwan

Anupam Das
Department of Mathematics, Rajiv Gandhi University, Rono Hills, Doimukh 791112, Arunachal Pradesh, India

Bipan Hazarika
Department of Mathematics, Rajiv Gandhi University, Rono Hills, Doimukh 791112, Arunachal Pradesh, India
Department of Mathematics, Gauhati University, Guwahati 781014, Assam, India

S. A. Mohiuddine
Operator Theory and Applications Research Group, Department of Mathematics, Faculty of Science, King Abdulaziz University, Jeddah 21589, Saudi Arabia

Lei Shi
School of Mathematics and Statistics, Anyang Normal University, Anyang 455002, China

Mohsan Raza and Kashif Javed
Department of Mathematics, Government College University, Faisalabad 38000, Pakistan

Saqib Hussain
Department of Mathematics, COMSATS University Islamabad, Abbottabad Campus 22010, Pakistan

Muhammad Arif
Department of Mathematics, Abdul Wali Khan University Mardan, 23200 Mardan, Pakistan

Index

R

S

T

U

V

Printed in the USA
CPSIA information can be obtained
at www.ICGtesting.com
JSHW051408091023
49903JS00006B/327